Southern Lights

The Scottish Contribution to New Zealand's Lighthouses

Guinevere Nalder

Whittles Publishing

Dedication

Dedicated with love to my brother, Conrad Howard Nalder, who set me on the path to civil engineering by teaching me how to calculate areas with the aid of a block of chocolate.

Published by
Whittles Publishing Ltd.,
Dunbeath,
Caithness, KW6 6EG,
Scotland, UK
www.whittlespublishing.com

ISBN 978-184995-156-2

Printed by
Typeset by Raspberry Creative Type, Edinburgh

CONTENTS

Acknowledgements

I wish to thank the staff of the National Library of Scotland, both in the main library at George IV Bridge and at the map collection in Causewayside, for their professional support, and to acknowledge the help provided by the NLS in allowing maps and drawings to be reproduced in this book. The National Lighthouse Museum at Fraserburgh also made me most welcome, providing information and allowing access to its library.

In New Zealand, the staff at the New Zealand Archives have been most helpful, not just those at the main archive in Wellington, but also the staff at the branch archives in Dunedin, Christchurch, and the cheery staff at Auckland. I am indebted to New Zealand Archives, the National Library of New Zealand, and the Alexander Turnbull Library in Wellington for many of the illustrations in this book.

I am also grateful for the help provided by the staffs of other museums and libraries. Marleen Boyd of the Voyager Museum unearthed information on the Auckland lighthouses and the minutes of the Auckland Harbour Board. Similarly the Hawkes Bay Museum provided photocopies from their records of the Napier Bluff lighthouse, even though the museum archive was closed to update the premises. The staff of the Hokitika Museum provided me with information on the Hokitika lighthouse and construction drawings. Naomi Boult of the Wellington Museum City and Sea invited me to visit the museum store where I was able to examine some of the oil burners used in New Zealand's lighthouses and provided me with illustrations. The staff on the South Canterbury Museum in Timaru provided information on and illustrations of both the Timaru lighthouses. The staff of the Alexander Heritage and Research Library, Wanganui, also helped me with bound copies of the *New Zealand Gazette.*

Craig Terris of Ports of Nelson Ltd provided me with access to PNL's files on the Nelson lighthouse and took photographs especially for this study, even going to the extent of cleaning the lighthouse lantern windows to ensure a sharp photograph of the 1862 Chance Brothers lens.

Other illustrations were provided by Monica and Christopher North, who gave up their time to join me on a jaunt to the windswept Manukau South Head. Also, Jim Foye, Lighthouse Engineer, and Maritime NZ for colour illustrations.

Not all the help I received was within the walls of libraries or museums. I could not have done much of the research for this book without the uncomplaining help of friends who house watched, cat fed, lawn mowed, and provided airport transport. Thanks here to

Jasmine and the late Captain Jack Priest
The Hall family including Ryan, lawnmower extraordinaire

Pat and Bruce Chalmers, and
Rita Lum

and also to my landlords Ingrid and Hec Ramsay, and Sue and Graham Booker who generously allowed me access to their internet connections even while running their businesses.

Grateful thanks to my editor Caroline Petherick, who was understanding about my Oxford commas.

Introduction

Establishing and maintaining New Zealand's lighthouses was an exercise in technology transfer from Scotland to New Zealand over a period of almost 90 years. After an initial flirtation with English technology (eventually upgraded with Scottish) all of New Zealand's lighthouses were built using Scottish materials either partially or fully.

Lanterns were all based on Alan Stevenson's design, and manufactured in Edinburgh. The oil lamps supplied were Argand derivatives and were generally, but not always, Scottish made. In contrast, the lenses were made either by Chance Brothers of Smethwick (near Birmingham) or French glassmakers including Henri Lepaute, Sautier et Cie, and, most often, Barbier & Fenestre and its successor firms. The machines for revolving lights were also manufactured in Edinburgh. As late as 1942 an order was sent to Scotland for replacement bearings for the lighthouse at Castlepoint.

The overriding Scottish contribution was the professional service provided by the firm founded by Robert Stevenson. By the time the first lighthouses were built in New Zealand, Robert had died and his eldest son Alan had retired. Consequently it was to David and Thomas Stevenson (Robert's remaining two sons), practising as D & T Stevenson, that the first orders were sent. The New Zealand Marine Department continued to rely on the successor firms: from 1890 to 1936, David's two sons, David Alan and Charles, trading as D & C Stevenson; and then from 1936 to 1952, David Alan's son, D Alan, in turn, with Charles still hard at work, practising as A & C Stevenson. It was to A & C Stevenson that the last New Zealand order was sent.

For several decades the Stevenson firms acted as consulting engineers, advising, designing, arranging the supply of lighthouse materials, and updating the New Zealand Marine Department on the latest technological advances, many the result of the Stevensons' own researches. A less obvious but nevertheless valuable additional role played by the Stevenson firms was that of recruiting engineers and other experts to build and operate the New Zealand lighthouses.

However over the first half of the 20th century, with the move away from oil lamps, and towards electrification and automation specifically, Scottish technology was no longer of major importance to the New Zealand lighthouse service. Even so, incandescent lamps were apparently still ordered through D & C Stevenson although manufactured by non-Scottish firms. But acetone-acetylene automatic lights and equipment for electrification were obtained by the Marine Department directly from the manufacturers in other countries.

The main sources of information for this book were the Stevenson Archive in the National Library of Scotland and the New Zealand Government Archive. Unfortunately both sources could provide only partial information. The relevant part of the Stevenson archive only goes up to 1901. Hence there are no records of the purchase of incandescent burners, and the only information for the newer lighthouses comes from the New Zealand Archives. However, the New Zealand Archives suffered a major fire in 1952 and a large number of records were lost; consequently more information is available on the earlier lighthouses than the later.

Although this is disappointing from an historical viewpoint, sufficient material is available to demonstrate the importance of the Scottish contribution to New Zealand's lighthouses. Read on!

Author's note

During the period covered by this study, the spelling of New Zealand place names was not standardised. Consequently there were often variations, particularly with Maori names. The author has followed the usual convention when quoting from a source. The spelling used by the source is retained even if it differs from the current accepted usage. Quotes are either indented or placed in inverted commas.

1 Provincial Governments and the Marine Boards

In the period 1859 to 1941 the New Zealand governments built 38 major lighthouses; 30 of them were constructed during the period 1865–1897. Despite New Zealand's limited industrial base, these usually represented the latest technology of the time, primarily because of the direct connection with D & T Stevenson (the noted Scottish civil engineering firm founded by Robert Stevenson) and its successor firms along with access to the firms which manufactured lanterns and apparatus for the Commissioners for the Northern Lights, now the Northern Lighthouse Board who were (and still are) responsible for the lighthouses of Scotland (often called the Scottish Lights) and the Isle of Man

Lighthouse construction and operation in New Zealand was the responsibility of regional governments at first, and then three different central government organisations in succession. New Zealand's constitutional history was an important factor in the emergence of Scottish lighthouse technology as pre-eminent there. Hence the Scottish contribution is best appreciated when this background is understood.

New Zealand as colony and dominion

In February 1840 the first Māori chiefs signed the Treaty of Waitangi. This Treaty placed New Zealand under British sovereignty while guaranteeing Māori rights to lands, fisheries etc. For the next 13 years New Zealand's status was that of a crown colony ruled by governors loosely under the authority of the older and less respectable colony of New South Wales. The city of Auckland was founded in February 1841 by Lieutenant Hobson, the first governor, to provide an administrative centre for the new colony. It remained the capital until 1856, when the honour was transferred to Wellington.

The period after 1840 saw the arrival of European settlers, chiefly from the United Kingdom. Immigration was systematically organised, and individual settlements often developed their own flavour. Settlers landing midway down the east coast of the South island of New Zealand showed their origins by naming their province Canterbury and its main settlement Christchurch, and christening a rather small stream the Avon. Further south, Scottish settlers retained the Māori name for the area – now spelt Otago – but assuaged homesickness by naming their settlement after Scotland's capital but in the Gaelic; now Dunedin.

By 1852 the increasing immigrant population was settled predominantly around Auckland and the five original planned settlements, with intermittent coastal shipping the only form of communication between them. These demographics were no longer conducive to the personal rule by a governor. On 30 June of that year the British Parliament passed an 'Act to Grant a Representative Constitution to the Colony of New Zealand'. This came into force on 17 January 1853, and provided a semi-federal system of government that recognised the realities of the colony's limited communications. The country was divided into six provinces based on the existing settlement patterns. Each province was a miniature Westminster system with a superintendent and a provincial council, both directly elected by men aged at least 21 years and owning land worth at least £50 per year. The first six provinces were Auckland, New Plymouth (renamed Taranaki in 1856), Wellington, Nelson, Canterbury, and Otago. Subsequently some of these were subdivided: Hawkes Bay was separated from Wellington in 1858, Marlborough from Nelson in 1859, Southland from Otago in 1861, and Westland from Canterbury in 1873.

Above the provincial legislatures was a general assembly which was based initially in Auckland and after 1856 in Wellington. This consisted of an elected House of Representatives with 24 to 42 members, and a Crown-nominated Legislative Council of not fewer than 10 members. These were usually referred to as the General Government. The term Central Government only came into use later. The executive of the General Government consisted of the governor (hence 'His Excellency's Government'), assisted by the colonial secretary (later to become premier and prime minister), the colonial treasurer, and the attorney general.

The provincial governments and the General Government maintained representatives in the United Kingdom to act for them. Hence D & T Stevenson received instructions initially from the agents for Otago Province, and then from the Agent General who represented the General Government and the New Zealand Government after the abolition of the provincial system. After 1907 the Agent General became the high commissioner. As orders were usually sent by the Marine Department to the Agent General, the Marine Department reports frequently spoke of lighthouses being ordered 'from England', thus obscuring the importance of Scottish technology to the development of New Zealand's lighthouses.

One of the first acts of the general assembly was to appoint a Select Committee on Beacons and Lighthouses to consider the

> expediency of establishing and maintaining any and what beacons and lighthouses on the Coast of New Zealand, and as to best means.[1]

In the committee's report it was suggested that

> Lighthouses which are coast lights and harbour lights should be established and maintained by the Provincial Governments under the supervision of the General Government,
>
> Lighthouses which are harbour lights only to be the responsibility of the Provincial Government entirely, and
>
> Sealights should be responsibility of the General Government.

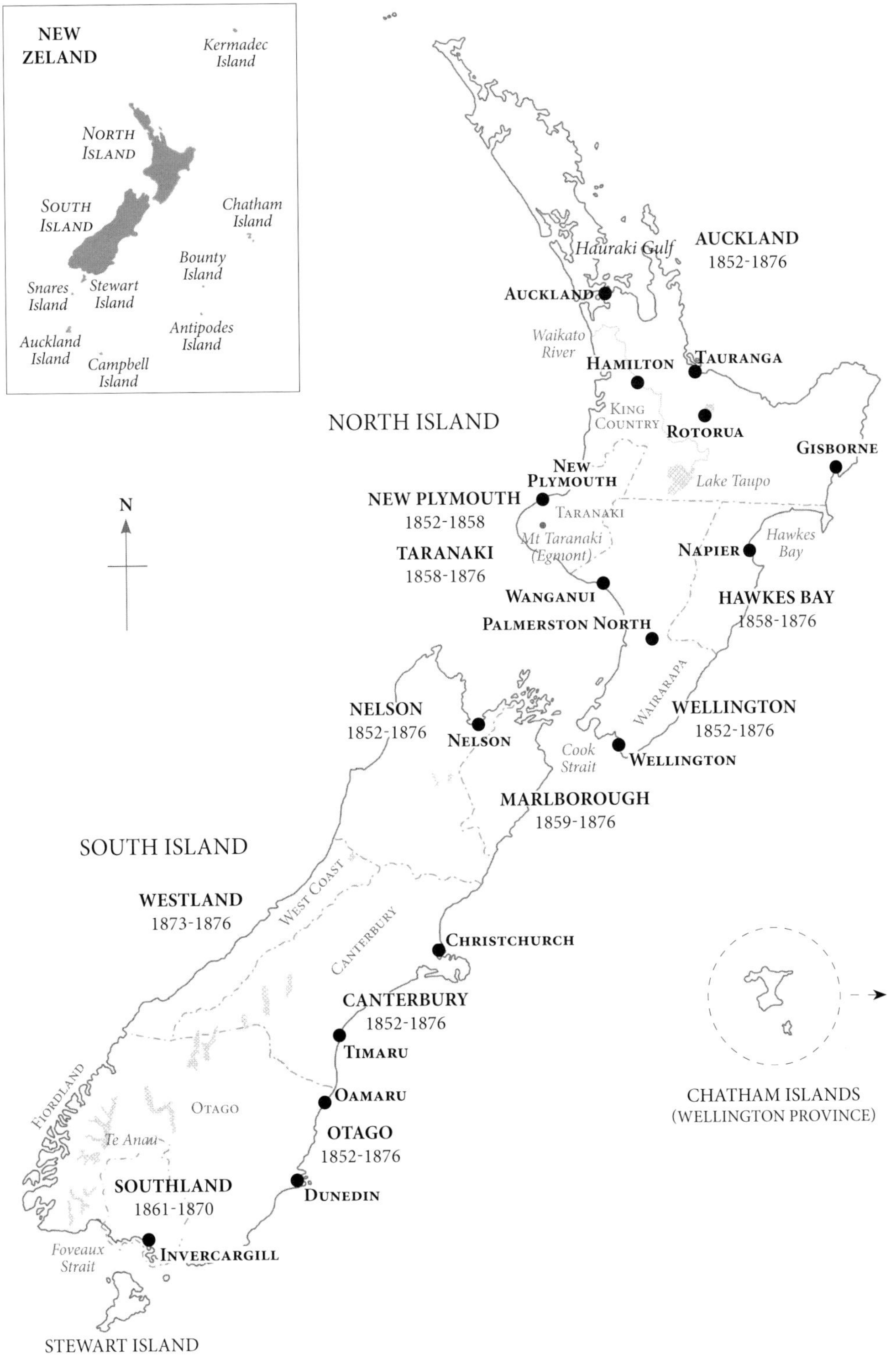

Figure 1.1: New Zealand's provincial boundaries redrawn after The Penguin History of New Zealand, *Penguin Books, 2003 © King M.*

By 1875 communications had improved and stresses had built up between the General Government and the provincial legislatures. On 12 October 1875 the Abolition of Provinces Act was assented to, and it became fully operative on 1 Nov 1876.

The two significant constitutional changes after this were independence in 1907, when the Dominion of New Zealand came into being, and the abolition of the Legislative Council in 1950.

Provincial governments and lighthouses

It was during the period of provincial governments (often called in New Zealand 'Settler Rule') that the first lighthouses were constructed. The actual responsibility was not always clear. The Imperial Government endeavoured to exercise some control. A despatch from Downing Street dated 8 January 1857 stated that where lighthouse funds were obtained

> by levying tolls under the Merchant Shipping Amendment Act 1855 or by means of a grant from the Imperial Parliament, applications will of course be made through the Board of Trade.[2]

Where the home government had not been asked for financial assistance or advice for the erection or altering of lighthouses they endeavoured to impress on colonial authorities that

> It is most desirable that a full intimation of such intention, and of the plans by which it may be proposes to carry it into effect should be transmitted to the Board of Trade at the earliest possible period,

and emphasised that

> The benefits to be derived from an arrangement, under which the Colonial authorities may obtain the advantage of the experience of the nautical advisers of the Board of Trade, and of the three Lighthouse Boards of the United Kingdom, are so fully adverted to in previous correspondence on the subject, that it is unnecessary to advert to them further.

The provinces erecting lighthouses showed a marked disinclination to draw on the 'experience of the nautical advisers' through the Board of Trade. Wellington and Nelson provinces sent orders directly to England for towers, lanterns, and light apparatus. The Otago Provincial Government in Dunedin demonstrated that they regarded the best 'nautical advisers' to be those to the Northern Lighthouse Board, and consulted directly with David and Thomas Stevenson, thus beginning a relationship with the Stevenson firm which was to last from 1862 to 1944.

A further complication arose in April 1857 when the General Government disallowed the Act to Authorise the Raising of a Further Sum of £25,000 passed by the Wellington Provincial Government. The Act was disallowed on the grounds that the loan was to pay for a lighthouse among other public works and '19th Section of Constitution Act' (i.e. the

Act of 1862) 'prohibits any Provincial Legislature from making any law for erection and maintenance of lighthouses'.[3] However as the General Government had provided plans for the proposed lighthouse (Pencarrow) to the Provincial Government, and as the lighthouse had been ordered, the General Government could only lamely indicate that 'it will support the Province in building the lighthouse'.

Reactions to this varied. The Canterbury Provincial Council, anxious to build the lighthouse at Godley Head, presented an address to the superintendent, stating that

> The Council is aware that under the existing state of the law, it is debarred from passing any Bill for that purpose. The Council therefore respectfully pray that your Honor will direct the attention of His Excellency's Government to the matter. And urge upon the Colonial Secretary the necessity of legislation upon the subject.[4]

In contrast, the provinces of Nelson, Otago, and Southland took advantage of the semi-autonomy of the provincial system, simply ignored the 19th section and went ahead with procuring, and in the case of Nelson erecting, the desired lighthouses.

The Otago Provincial Government was particularly well placed to benefit from this semi-autonomy. Dunedin was the provincial seat furthest from Wellington, and it was easier to go by steamer from Dunedin to Sydney in New South Wales than it was to go to Wellington, let alone Auckland.

D & T Stevenson were already the province's consulting engineers, designing and advising on infrastructure and recruiting the skilled professionals needed for the settlement. When the Provincial Government decided to erect lighthouses, it was a logical step for it to turn to D & T Stevenson directly without referring to either Wellington or London.

The marine boards

In November 1862 an attempt was made to bring lighthouse construction and management under the nominal control of a single body with the passing of the 'Marine Board Act, 1862'.[5,6] This allowed for the establishment of local marine boards under the supervision of a Chief Marine Board. Captain Charles Sharp (Wellington) was appointed president and in an echo of Trinity House 'Master Warden'. The Hon. Reader Gibson Wood (Auckland) was a member by virtue of his position as commissioner of customs and colonial secretary, and Captain Thomas Robertson (Otago) acted as secretary.

At the first meeting of the Chief Marine Board in January 1863

> numerous applications were received by them from various parts of the Colony, showing the great want of Lighthouses in certain localities.[7]

However, in the same report of October 1864, Captain Sharp wrote that the Act

> was in many respects, found to be too cumbrous, and unsuited to the circumstances of the Colony; it being also of little avail until local Marine

> Boards were brought into operation, their concurrent action being required for the performance of the most important duties for which this new department had been considered necessary.[8]

Some local marine boards had been established at the time of Sharp's report. On 7 April 1863 the Auckland Provincial Legislature passed the Local Marine Board Act which established local marine boards; one at each of the province's eight main ports.[9] Should all of the provinces have followed this example the number of local marine boards would have soon made supervision difficult, especially given the colony's communications.

Erecting and maintaining lighthouses was only part of the Board's responsibilities. Wood, in his letter to the secretary, Chief Marine Board of 11 June 1863, showed how 'the whole question of administration under the Marine Board Act is a very difficult one' by quoting the example of appointing harbour pilots:

> By that Act the appointment of Pilots and the management of Pilot Establishments vests in the Chief Marine Board. But the Board cannot appoint any person to act as Pilot for any Port until he has first been examined by Examiners to be appointed by the Local Marine Board of the Port to which he is to be attached; and as by some singular fatality no Local Marine Board is yet known to be in existence in any of the Provinces, the Chief Marine Board cannot appoint a Pilot.[10]

Apparently the Auckland Province's local marine boards had either not come into operation or the Chief Marine Board had not been informed of the Auckland Act.

The Chief Marine Board also had cash flow problems:

> The Chief Marine Board is empowered within the limits prescribed by the Schedule to fix Lighthouse dues; and the Attorney-General has given his opinion that dues cannot be levied until the lights are burning.

This problem was eventually overcome when the Canterbury Province voted a grant of £5,000 towards the establishment of a lighthouse near the entrance to the port of Lyttelton, and the General Government obtained a credit for the Board on the Bank of New Zealand for a sum of £10,000 for the construction of other lighthouses.[11]

The difficulties inherent in the Chief Marine Board were addressed by the passing of the Marine Board Act 1863 on 14 December 1863.[12] Section 56 repealed

> The Marine Boards Act 1862 and all Provincial ordinances passed under the authority of the said Act … and all Local Boards constituted under any such Provincial Ordinances, and set up a single national organisation, The Marine Board of New Zealand, to be headed by a master warden 'and not less than three nor more than five other persons to be Wardens'.

In Section 19 of the Act, the Marine Board was given responsibility for 'several Light-Houses Lightships Sea-marks and Beacons erected or to be erected on the Coast of this Colony'. The Act's draftsman showed a limited understanding of the reality of lighthouse administration in the United Kingdom when it was hopefully stated in the later part of Section 19 that the Marine Board

> shall have all the powers and jurisdiction in respect of such light-houses and light-ships which the Guild or Fraternity commonly called the Corporation of the Trinity House of Deptford, Strond has and exercises or may hereafter have and exercise in the United Kingdom [sic].

A common thread throughout the history of lighthouse procurement, construction, maintenance, and administration in New Zealand was the disinclination to follow the example of the Corporation of the Trinity House etc., although some cooperation did occur.

Section 20 gave the Board authority to erect new light-houses and light-ships, and Section 21 made it 'lawful for the Marine Board to fix Lighthouse Dues'.

Although it was not specified in the Act, the Marine Board was placed under the jurisdiction of the postmaster general, no doubt reflecting the importance of safe shipping to mail delivery. The Marine Board provided its own reports to the House of Representatives, but these were transmitted by the postmaster-general.

Once established the Marine Board took over construction of five lighthouses which had been ordered by the Provinces but not yet built. (See figure 1.2)

Wellington Provincial Government had already constructed the lighthouse at Pencarrow Head, and Nelson Provincial Government the one on Boulder Bank.

Although construction was in hand there was still the question of land ownership. Hence on 23 March 1865 Sharp, as president of the Marine Board, wrote to the General Government reminding them that although the land had been acquired for the Tiri Tiri Matangi light, arrangements still needed to be made for the other four lighthouses in the above list, as well as Cape Saunders, Pencarrow Head, and Nelson Boulder Bank.[13] This was eventually done.

Presciently he also suggested that land should be reserved for further lighthouses. (See figure 1.2)

Lighthouses were eventually placed at all but three of these above sites.

As the members of the Marine Board were all master mariners, they retained the services of James Balfour, marine engineer for the Otago Province, as a consulting engineer. Balfour had trained with D & T Stevenson and was the brother of Thomas Stevenson's wife, Margaret. Balfour's services were detailed in his report attached to the main Marine Board report for 1864.[14]

The Marine Department under J M Balfour

The Marine Board was to have a limited lifespan. The Marine Board Act Amendment Act 1865 (passed 26 September 1856) abolished the Marine Board and repealed certain sections of the 1863 Act. The situation was clarified on 8 October 1866 when the General

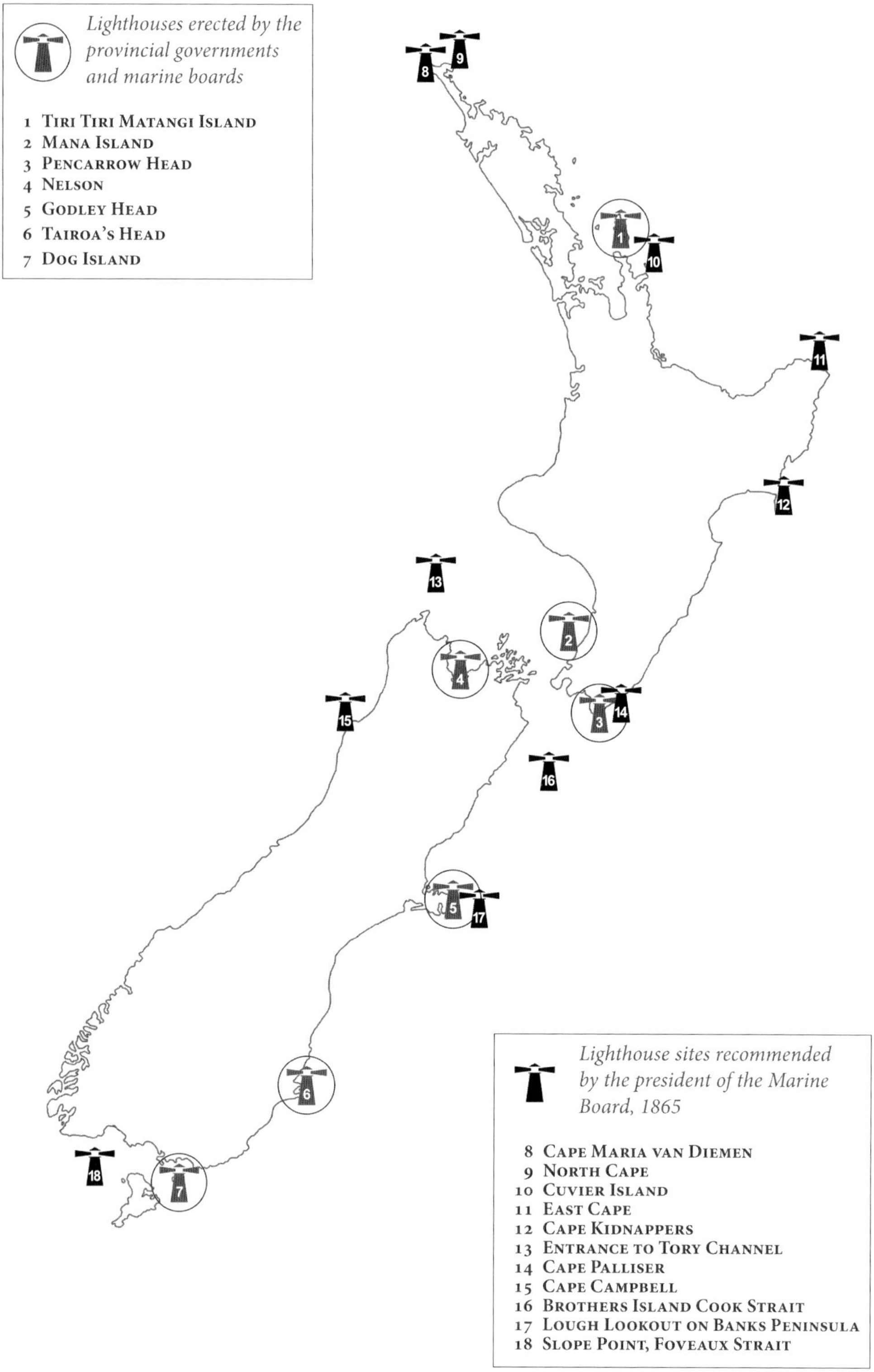

Figure 1.2: Lighthouses constructed or recommended by 1865

Assembly passed the Marine Act 1866, which repealed the Acts of 1863 and 1865. The 1866 Act transferred authority for lighthouses to the governor. The sections were identical to those in the 1863 Act with 'the Governor' substituted for 'the Board' thus ensuring that the governor 'shall have all the powers and jurisdiction' of Trinity House.

Although the Marine Department's official history gives 1866 as the Department's start date, the Act did not establish a Marine Department or specify any form of administration.[15] The department seems to have just emerged as a section of the Post Office. In the third report mentioned below, it was commented that following the passing of 1866 Act, 'the Marine Department has been created in the room of the former machinery'.

The report to the House of Representatives for 1865–66 was dated 25 June 1866 and was called the Second Report of the Marine Board Department of New Zealand, while that of the following year was the Third Report of the Marine Department. Between these two reports it was decided to appoint James Balfour as colonial marine engineer. The appointment was announced in the *New Zealand Gazette* along with an announcement of the devolution to him of the duties previously discharged by the Marine Board which had ceased to exist.[16] This of course was not strictly true. It was the governor's duties which had been devolved to Balfour.

Balfour's appointment ensured that the Scottish technology, which had been already introduced into the Otago Province, would now be applied throughout the colony. The Stevenson firms would be consulting engineers for the New Zealand Government for the next 80 years, and lighthouse lanterns and apparatus were primarily manufactured to Scottish designs by Scottish firms, usually Dove & Co or Milne & Son. Glass lenses were manufactured by Chance Brothers of Smethwick or French firms such as Barbier & Fenestre of Paris. This ensured that the lighthouse parts imported were the latest available, and it was not unknown for specifications to be altered after the initial order when new technology became available.

Stores and lighthouse fuel were also obtained from Scottish suppliers, one specification going as far as to order that stores 'be obtained from usual suppliers to Northern Lighthouse Board'.[17] The Lighthouse Service was also based on the Scottish model, although without the uniforms. By the end of the nineteenth century the New Zealand Lighthouse System resembled an overseas branch of the Northern Lighthouse Board.

The Marine Department after Balfour

The Marine Department remained under the general authority of the postmaster general until the death of Balfour in December 1869 led to a period of hasty reorganisation. Captain Robert Johnson acted as temporary marine engineer until February 1870, when the Department was placed under the commissioner of customs and operated as the Marine Branch of the Customs Department. This was headed by William Seed who, as the secretary of customs, was responsible for 'administrative business'. John Blackett, the acting engineer-in-chief of the Public Works Department, was appointed marine engineer.[18]

Johnson remained as nautical adviser. In 1873 he and Commander Edwin, captain of the survey ship HMS *Acheron*, were requested by William Reynolds, commissioner of customs, to

Figure 1.3: John Blackett (ATL 1/2-080821-F) © Alexander Turnbull Library, Wellington, New Zealand

> take into consideration the whole question of Lights on the Coast, and to prepare a comprehensive Plan and report, showing where they would recommend the erection of additional lights, and the transposition of those already existing; the class of each proposed light, and the probable cost of the same in detail.[19]

Their reports formed the basis of the eventual development of New Zealand's lighthouses.

Seed's was to become a familiar signature on letters to D & T Stevenson for about 18 years. He established a relationship with the firm both by mail and a visit to Europe in 1873. The relationship was sufficiently well developed for him to appeal to Thomas Stevenson to help Seed's son further his civil engineering career in the United Kingdom.[20]

The Marine Department had a brief period of independence from April 1878 but was reunited with the Customs Department in October 1880. However, the Marine Department continued to issue an independent annual report which was published in the Appendix to the Journal of the House of Representatives (AJHR). From 1892 the annual report was presented to the Minister of Marine, indicating that the Marine Department had been officially separated from the Customs Department. However, cooperation continued and local collectors of customs frequently acted for the Marine Department, particularly in lighthouse administration.

Seed retired from the civil service in 1887 and was succeeded by W T Glasgow as Secretary of Marine. However the increasing work load led to Marine Branch administration being delegated to L H B Wilson, who was promoted from chief clerk to assistant secretary. Wilson was another Stevenson connection. His full name was Lewis Henry Balfour Wilson, and he shared a common maternal grandfather with the similarly named Robert Lewis Balfour Stevenson; he was the son of Marion Balfour, the elder sister of both James Balfour and Margaret Isabella Stevenson. When he needed advice, he wrote letters to Edinburgh addressed to 'Dear Uncle Tom' concluding 'with love to Aunt Maggie'.[21] He also loyally supported his cousin, on one occasion commenting, 'I was glad to see Lewis' books so well received'.[22]

Seed, in a letter to Thomas Stevenson, says of Lewis Wilson 'he has a perfect mania for Lighthouse business'.[23] He was also William Seed's son-in-law.

Glasgow served as secretary for only eight months and was replaced by H S McKellar. Glasgow was reappointed in May 1892 and served until the end of March 1903. At the time of his reappointment he was directed to also take responsibility for the Marine Branch, and Wilson was required to resign. Subsequently he established an insurance agency.[24]

It was during Glasgow's secretaryship that a further division of responsibility was made. From 1894, lighthouses construction and repair was undertaken by the Public Works Department (later the Ministry of Works) under the engineer in chief who was also the marine engineer, both titles appearing in official communications. However all aspects of administration of the Lighthouse Service, including decisions on erection of new lighthouses and procurement of lighthouse materials, remained with the Marine Department. This arrangement remained through the period of this study.

The last secretaries to communicate with Edinburgh were L B Campbell and B W Millier, who was in his second period as acting secretary in 1944 when the equipment for the upgrading of Castlepoint lighthouse, procured through Charles and D Alan Stevenson (son and grandson respectively of David Stevenson), arrived. Eight years later the firm of Stevenson was wound up with the retirement of D Alan Stevenson. For more than half of its 150-year lifetime, the firm had been consulting engineers to the New Zealand governments.

Endnotes

1 Votes and Proceedings House of Representatives, Session 1, 1854.
2 *NZG Gazette* 1875 No 17 Friday 26 June 1857
3 *NZG Gazette* 1875 No 17 Friday 26 June 1857
4 Province of Canterbury, *Journal of the Proceedings of the Provincial Council* Vol II, Sessions XI-XX, 29 September 1859–29 September 1863, Minutes of 18 December 1861
5 *NZ Gazette* No 37 p 117
6 New Zealand Statutes 1862, No 20
7 AJHR 1864, D – No.1.D, Reports of the Officers of the Marine Board
8 AJHR 1864, D – No.1.D, Auckland, Reports of the Officers of the Marine Board
9 Local Marine Board Act, 1863, session XV, No 15 Auckland Provincial Legislature
10 AJHR 1863 D – No. 4No 2 Papers Relative to the Working of the Marine Boards Act, .
11 AJHR 1864, D – No.1.D, Reports of the Officers of the Marine Board
12 New Zealand Statutes 1863, No 28
13 NZG Archives BBAO 5544/170a 1865by Memo Sharp Pres Marine Board to Colonial Secretary 23 March 1865
14 AJHR 1865 , D-No1.C. Report on the Marine Board Department of New Zealand 1864–5
15 Martin E R Marine Department Centennial History 1866–1966, Marine Department, Wellington New Zealand 1969, printed by A R Shearer, Government Printer, Wellington, 1969
16 *NZG Gazette* 1866 No 55 15 Oct 1866, p 389

17 Acc 10706 vol 85 1876a requisition C.S 29 Lighthouse Stores Required by the Government of New Zealand requisition from Agent General to D & T Stevenson, 26 April 1876

18 AJHR, 1871 G-No.6 Sixth Report of the Marine Department, for the years 1869–70 and 1870–1.

19 AJHR, 1873 H-19 Eighth Report of the Marine Department, for the year ended 30th June 1873, Report on the Lighthouses on the Coasts of New Zealand

20 Acc10706 Vol 88a Letter William Seed to D & T Stevenson Wellington 22 September 1876

21 Wilson letters in Acc 10706 Vol 88a

22 Acc 10706 Vol 88a Letter Wilson to Thomas Stevenson Wellington 4 Jan 1879

23 Acc 10706 Vol 88a Letter Seed to Stevenson Wellington 22 September 1876

24 Wilson, L H B *The Cyclopedia of New Zealand* [Wellington Provincial District]. Pub The Cyclopedia Company , Wellington, 1897

2 'English' Provincial Lights

Introduction

The English provincial lights were those ordered from English manufacturers before the appointment of either of the Marine Boards.

Two lights, Pencarrow and Nelson, were erected by the provincial councils and the remainder by the marine boards. However, when the lighthouses were upgraded, it was with Scottish technology and not English.

Pencarrow and Baring Head

Pencarrow

Pencarrow Head marks the eastern entrance to Wellington Harbour, and the lighthouse was the first exhibited in New Zealand, on 1 January 1859.

The lighthouse was designed by Edward Roberts. He was a civilian member of the Royal Engineers who in January 1851 was seconded to the New Zealand Government as colonial engineer. The design was done in 1852[1] and in 1853 he sent plans and specifications to England to obtain quotes. Roberts designed a revolving screen system to cause the light to flash as there was thought to be a danger that the light would be confused with a bush fire. However, by July 1854 the only construction had been the lightkeeper's house with a light in the window.[2]

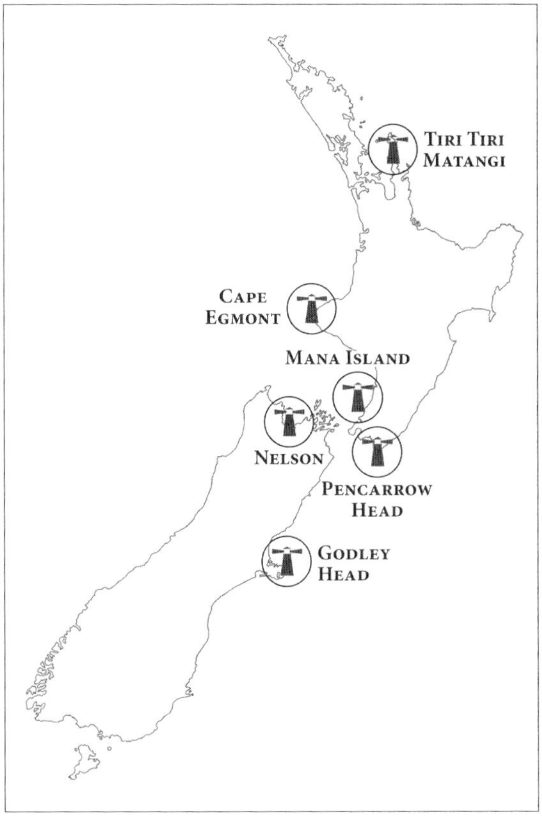

Figure 2.1: English provincial lights

Figure 2.2: Entrance to Wellington Harbour, 1877 (ATL-PUBL-0016-04-1)
© Alexander Turnbull Library, Wellington, New Zealand

Figure 2.3: Lighthouse cottage at Pencarrow (ATL A-032-012)
© Alexander Turnbull Library, Wellington, New Zealand

In 1857, the Wellington Provincial Council requested 'copies of the plans, specifications, and estimates' from the General Government.[3] In a later letter, Featherston, the provincial superintendent, reported that he had given 'the necessary instructions to have the lighthouse sent out from England'.[4] This was soon followed by the problem of financing mentioned in Chapter 1.

The order for the cast iron tower was sent to Cochrane & Co, whose Woodside Ironworks were in Dudley, in the Black Country near Birmingham.[5] A specification sent to Stevenson in 1879 described the apparatus as 'Second order dioptric apparatus by Chance Bros & Co, 1858',[6] although the lens was in fact catadioptric.

It appears therefore, that the lantern, optic, and revolving screens were built by Chance Bros, although this is difficult to check as the order books in the Chance Archive in Sandwell Archive in Smethwick only begin in 1858. Chance's premises at Smethwick (near Birmingham) were very close to those of Cochrane & Co in Dudley. A later record in the order book (see below) states that the lamps were 'made by Milne & Son', the Edinburgh firm which was to be very important in the establishment of New Zealand's lighthouses. The lamps burnt colza oil, a vegetable-based oil frequently used in lighthouses of the time.

The lighthouse tower, lantern, and apparatus were shipped on the *Ambrosire* and arrived in 1858. Construction was overseen by Edward G Wright who, with his family, immigrated to New Zealand in 1857.[7] On completion of the lighthouse in 1859 he moved to Hawkes Bay as Director of Harbour Improvements and Public Works. He also appears to have formed a short-lived partnership Wright and Griphook. In a letter dated 2 May 1862 the partnership offered 'to contract for providing and fixing (of Taiaroa Head Light) provided is substantial structure, 1st or 2nd order'.[8]

The offer was not accepted, and Wright moved to Christchurch and spent the remainder of his life in Canterbury.

The exhibition of the light on 1 January 1859 was announced in in the *New Zealand Gazette* in November 1858.

The light was described as

> second order (Catadioptric system) with eclipses at intervals of two minutes; it will be 420 feet above high water mark and visible in ordinary weather at a distance of thirty English miles.[9]

Within a very short period it was decided to change the light to a fixed one. An updated Notice to Mariners was published in the *Wellington Government Gazette* of 10 February 1859. However, the request for it to be published in the *New Zealand Government Gazette* led to an acidic exchange between Henry John Tancred, writing on behalf of the colonial secretary, and Featherston, the superintendent of Wellington Province.

Tancred arranged for the insertion of the notice but pointed out 'the very great danger which is likely to be the consequence of these repeated changes' and referred to 'the necessity of all lighthouses on the Coast being managed upon a uniform system'.[10]

Featherston in turn pointed out that[11] once the light was built, it was obvious that there was no danger of confusion with fire, that the revolving machinery was liable to derangement from earthquakes (Wellington had had a major earthquake in 1855), and would probably from time to time get out of order and no on-site repairs would be possible. He also pointed out that this did not constitute an example of 'repeated changes' and they were giving seven months' notice of the change, and as it was the only lighthouse in the colony, there was none to confuse it with.

There seems to have been some suggestion that the General Government take over the lighthouse. Featherston, in the same letter, indicated that the

> Provincial Government was prepared to hand lighthouse over to General Government on 'your redeeming the pledge given by Mr Stafford in the House of Representatives to repay its cost to the Province'. Until repayment or Act of General Assembly to bring light under your 'uniform system' will keep lighthouse.

This exchange of letters indicates the degree of independence shown by the provincial governments. The lighthouse and reserve were eventually purchased by government from Wellington Provincial Council for £6,422 in 1865.[12]

In August 1865 Chance Brothers received orders for stores[13] including goods for Pencarrow.

Four years later (on 31 July 1869), Chance Brothers received an order for 'Three Pressure Lamps having Second order bodies'.[14] The order record noted that

> [t]hese Lamps are to replace the Mechanical Lamps send [sic] by us for Pencarrow in 1858 (made by Milne & Son).

This order must have come direct from the New Zealand Marine Department, as the order is described as 'transmitted by James M Balfour, Colonial Marine Engineer'.

The first order for Pencarrow lighthouse under the new regime was in 1867 when stores were ordered for it along with stores for the other five then existing lighthouses.

Pencarrow and Tiri Tiri Matangi were the last two lights to use colza oil, and their Argand burners were upgraded in 1879. The specification (C/S120), sent on to D & T Stevenson by the Agent General in London,[15] ordered

> One set (3) four wicked [*sic*] paraffin burners to be fitted on pressure lamps, with regulating overflow tubes. (If fitting was not possible, two new holders were ordered). One gross of each size wick and one gross of lamp glasses.

These paraffin burners were Doty lamps, which were first installed at Manukau Heads (Chapter 4).

By 1892 some of the panes in the lantern had become yellow. Wilson sent samples of the discoloured glass to D & C Stevenson, (David had now been joined in partnership by his son, Charles) who commented:

> We are greatly interested with the specimen of glass from the lantern of the Pencarrow lighthouse which you have sent us. We are aware that some glass underwent a change in colour by long exposure, and for some time we were making experiments with different glasses fixed in a frame, but we have never come across so marked a change as what appears in the pieces sent. We have never seen any change to speak of on the French glass – We intend to try and test the loss of light in the two samples but as the specimen of clear glass is so small it will be difficult to do so. The glass is thicker than

Figure 2.4: Pencarrow Lighthouse (ATL 1/2-136029-F)
© Alexander Turnbull Library, Wellington, New Zealand

> what we send usually to you, although in special cases we have used ½ inch glass in very exposed situations. We notice that the specimen of clear glass sent is nearly ⅝" and the specimen of amber coloured glass ½ inch thick.

D & C Stevenson do not appear to have provided any explanation of why the glass had yellowed and may have regarded it as another example of the inferiority of the English product. The glass was replaced.[16]

About this same time the keepers were asked to try out a new form of oil, called 'Mineral Colza oil', which was expected to be cheaper than the imported paraffin[17], but the oil was not adopted by the Marine Department.

However, the Pencarrow Head was occasionally obscured by fog, and in 1899 local shipmasters petitioned the Minister of Marine to place a Norwegian oil lamp on the rocks below Pencarrow Head.[18] No immediate action was taken until the effect of the upgraded Matiu/Somes Lighthouse (see below) could be assessed. In 1906 a low-level light was placed at Pencarrow Point at the foot of the Head.[19] This light was 32 ft (10 metres) above sea-level at high water and had a Wigham patent three-wick lamp burning kerosene (Figure 2.6).

In 1907 the Pencarrow light was converted to a Chance Brothers incandescent light,[20] one of the first in New Zealand (see Chapter 8), but the Pencarrow keepers, like their fellow lightkeepers, had difficulty with them. Various types of incandescent burners were used over the next 28 years as the Pencarrow light was frequently used to test alternative burners. Although it was suggested in 1913 that the light should be changed to occulting[21] and some thought was given to automating the light with a Dalen Incandescent Lamp,[22] these incandescent lamps remained in place until the light was discontinued on the morning of 17 June 1935 with the exhibition of the Baring Head light nearby.[23,24] The lens was later dismantled in November 1941 and placed in the Wellington Harbour Board store.[25] The tower was retained as a daymark.

Baring Head light

While Pencarrow Head lighthouse was the first of New Zealand's lights, Baring Head, which replaced it, was the penultimate major coastal light established in New Zealand. Like Cape Reinga, the last lighthouse to be constructed, the Baring Head tower was of reinforced concrete.

On 14 April 1921, MacIndoe, the Secretary of the Merchant Service Guild of Australia, wrote to the Minister of Marine and expressed the guild's view that either the existing Pencarrow light should be changed to flashing or the light should be shifted to Baring Head.[26]

The marine engineer suggested that as a flashing light was required and Pencarrow was a fixed light and the tower in poor condition, the Castlepoint lighthouse be dismantled and re-erected at Baring Head. Captain Bollons, who as master of the Department steamer *Tutanekei* inspected all the lighthouses, agreed that a new flashing light was required at Baring Head, but pointed out that the tower at Pencarrow was in good condition and that as the range of the light depended on the height of its focal plane, moving the Castlepoint light was not a good idea.[27]

It should be remembered that by this time, the marine engineer was also the engineer-in-chief for the Public Works Department and thus while being a professional civil engineer was not necessarily an expert in lighthouses.

In September 1928 Captain MacIndoe and Captain Harris, master of the Union Steam Ship Company oil tanker *Otokia*, provided the Marine Department with a further guild proposal which formed the basis of the eventual proposal[28] i.e.:

- Pencarrow Head to be replaced by a flashing light with a minimum range of 25 nautical miles at Baring Head
- The fog horn at Pencarrow Head to be moved to Baring Head
- The low level light at Pencarrow to be upgraded by the Wellington Harbour Board to provide a range of 15 nautical miles.

Figure 2.5: Lower Pencarrow Light (ATL EP/1960/1588-F)
© Evening Post Collection, Alexander Turnbull Library, Wellington, New Zealand

In June 1929, Sharp, the Public Works Department lighthouse inspecting engineer, suggested that the new lighthouse have an acetone-acetylene apparatus such as that on order for Cape Egmont. He also suggested that the Pencarrow tower be reused – on the questionable grounds that 50-year-old cast iron was earthquake resistant. However, he also recommended that a final decision be delayed till the department had completed fitting a new 'outfit' into the existing tower at Cape Egmont and thus possible problems identified.[29]

Figure 2.6: Baring Head lighthouse under construction (ATL EP-Transport-lighthouses-02) © Dominion Post Collection, Alexander Turnbull Library, Wellington. New Zealand

The final decision was to install electric light and to build a new reinforced concrete tower and to place on it the lantern which had been ordered for North Cape in 1912. (The North Cape incandescent lamp, lens, and machine were used to upgrade Dog Island lighthouse – Chapter 3.) The lens which had been removed from Cape Egmont lighthouse was used at Baring Head, and the only new apparatus purchased were the electric lamps and the accompanying diesel generators. These were supplied by Chance Brothers.[30] The final choice of characteristic was light 5 sec, eclipse 2 sec, light 2 sec, eclipse 2 sec, light 2 sec, eclipse 2 sec.

The tender to build the tower, keepers' cottages, and service buildings was let to J Jackson and Sons for £2029/4/9.[31]

The light was officially lit on the afternoon of 17 June 1935 by Mrs Riddiford, whose husband Eric Riddiford had donated the lighthouse site to the government.

Local conditions forced a modification to the lighthouse in 1943. The site was particularly windy and the lighthouse housing had to be constructed within a pentangle of brush windbreaks.

Figure 2.7: Baring Head lighthouse station with brushwood windbreaks (ATL PA Coll-7327-1-146-0) © Evening Post Collection, Alexander Turnbull Library, Wellington, New Zealand

The principal keeper described the problem as follows:

> I have to inform you that the moveable cowl attached to the Lighthouse dome is not suitable owing to it being subject to contrary winds during Southerlies. The South Winds strike the cliff & shoot up creating what is known as a Welly-War. And owing to this the cowl does not face a true wind causing the cowl to swing in all directions violently.[32]

As the cowl was now revolving continuously instead of swinging out of the wind, there was considerably wear and very heavy vibration in the lantern. As the electric lamp did not require the degree of ventilation required by the incandescent burner for which the lantern had originally been designed, the marine engineer agreed with the principal keeper's suggestion that a new fixed cowl be substituted.[33] This appears to have been done the following year and the new cowl was identical to that designed for Manukau Head lighthouse (See Figure 5.6)

The light was automated in 1989 and the keeper withdrawn.

Nelson lighthouse

Buoys had been placed on Nelson Harbour entrance by the pre-provincial government in 1850. In April 1857 the Provincial Government established a select committee 'to enquire and report upon an efficient light for the Harbour of Nelson'.[34] This committee reported on 14 May 1857,[35] and recommended that a 'light should be set up on the Boulder Bank, on the spot indicated by the Harbour Master and others'.

(Boulder Bank is a barrier island consisting of large cobbles, which protects Nelson Harbour.)

The committee indicated that they had 'received material assistance from a work entitled *Description and Plans of Lights and Lighthouses on the Catadioptric System*; this appears to have been a Chance Brothers catalogue. They therefore recommended the purchase from Chance Brothers of a 'fixed light, fourth order, with a range of 15 miles illuminating 180 degrees of the horizon'.

Two years later, the report had not been acted upon and a second select committee, chaired by Alfred Dommet, who was to become provincial secretary, urged the construction of the harbour light.[36]

In 1860 orders were sent to England for the tower and for the lantern and apparatus. The cast iron tower was designed by John Blackett who was then the provincial engineer. Blackett had started his career with the Great Western Railway, so it is not surprising that the plans and specification were sent to Slaughter, Gruning & Company of Bristol.[37] Sometime after construction the tower acquired a plaque with the legend 'Stothard and Pitt, Bath', and some authorities have attributed the tower to them. (Stothard and Pitt later built cranes for Wellington Harbour.)

The order for lantern, burner and lens was entered by Chance Brothers in their order book on 13 April 1860. It was described as being sent by Edmund Buxton who was a member of Nelson's business community although not a member of the Provincial Government. In 1855 he established E Buxton & Co, a firm of general merchants in Nelson. He may have been given the responsibility to place the initial order because of his business contacts in the United Kingdom; however, the actual procurement was supervised by James Stuart Tytler of Edinburgh, who was agent for the Provincial Government. Like Messrs Auld & Crawford, who acted as agents for the Otago Provincial Government, Tytler was a Writer to the Signet (WS), a solicitor who specialised in drawing up legal documents. Hence the Nelson lighthouse was not without a Scottish contribution at the time of construction.

The lens was a fourth-order catadioptric consisting of 180 degree fixed Fresnel lens and a silvered metal mirror in halves of 180 degrees.

Figure 2.8: Nelson lighthouse on Boulder Bank from sea (ATL 1/4-055864-G)
© E R Williams Collection, Alexander Turnbull Library, Wellington, New Zealand

Figure 2.9: Nelson lighthouse with attached workshops (ATL PA2-2720)n © Alexander Turnbull Library, Wellington, New Zealand

The lamps were described by William Seed as 'ordinary 2 wick moderators made by Nelson, Chance & Brothers 1861/1862'.[38] These burned colza oil, and the Provincial Government was able to obtain tenders from local suppliers for the oil.[39]

The aim was to send the lantern etc in the same shipment as the lighthouse sections. It appears that the complete lighthouse was shipped to Nelson on the *Glenshee*.[40] The light was exhibited on 4 August 1862,[41] and in 1864 the lighthouse was purchased by government for £2,824/8/9.[42]

The original lighthouse had workshops on both sides of the tower as well as lightkeepers' accommodation, but these were subsequently removed.

Nelson had part of the August 1865 order to Chance Brothers for stores,[43] but from 1867 all successive orders for stores were sent to Scotland.

Nelson lighthouse was converted to paraffin oil earlier than Pencarrow. Doty burners were ordered in 1876.[44] Sometime before 1888 the light acquired a red sector which in July of 1888 was extended.[45] In 1914 it was decided to automate the lighthouse by replacing the Doty lamp with a Dalen acetone-acetylene lamp,[46] which was supplied directly by the Swedish manufacturer and installed the following year.[47] In 1982 the lighthouse was decommissioned and replaced by an automatic beacon although the lighthouse remains on the Boulder Bank.

Tiri Tiri Matangi

In 1860, the Auckland Provincial Government was urging the General Government to establish a lighthouse on Tiri Tiri Matangi Island. This light would guide ships coming from the east into the Hauraki Gulf and hence into Waitemata (Auckland) harbour.

On Tuesday 6 March 1860 Mr Harris moved a motion in Session 12 of the Auckland Provincial Council '[t]hat an address [be made] to His Honor the Superintendent, requesting him to ascertain from General Government if any steps have been taken to provide a Light-House on Tiri-Tiri-Ma-Tangi'.[48] Ten days later the superintendent replied, sending a letter from the colonial secretary which indicated that 'measures on the subject would be submitted to the House of Representatives'.[49]

Design and construction had to wait until the establishment of the Marine Board. The Board ordered identical lighthouses for Mana Island (referred to by Chances as 'Cook's Straits') and Tiri Tiri Matangi,[50,51] and this order also included apparatus for Port Cooper, now called Godley Head. The order was recorded by Chance Brothers on 25 September 1863.[52] The record specifies 'Three Second Order Fixed Lights'; these were paid for by the loan of £10,000 from the Bank of New Zealand to the General Government.[53]

The order was headed by reference to Simpson & Co, Engine Works, Grosvenor Road, Pimlico, London S.W. and For the Government of New Zealand Engineers McClean and Stileman, 23 Great George Street, Westminster, S.W.

This engineering firm was the partnership of John Robinson McClean and Francis C Stileman. The firm's other projects included docks at Barrow, railways in Cumbria and the Eastbourne water works. They also acted as engineers to Emperor Napoleon of France [III not Bonaparte]. Appropriately, the side of Great George Street where McClean and Stileman had their office is now taken up with the headquarters buildings for the Royal Institution of Chartered Surveyors and appropriately the Institution of Civil Engineers (I.C.E.). McClean was president of the I.C.E. 1864/5. Balfour indicated that the lighthouses of Tiri Tiri Matangi, Mana, and Godley Head, had been constructed 'mainly from the designs of Messers McLean [*sic*] and Stilman [*sic*], C.E., Great George-street, London'. Another source describes the Tiri Tiri Matangi and Mana Lights as being 'prefabricated by Simpson & Co. of Pimlico'.[54]

Balfour described the towers as

> being of cast iron, 70 feet high from base to vane, 15½ feet in diameter over all above the base course, and 11 feet in diameter below the gallery.

The lanterns were described as

> of the 'English' construction with inclined sash bars, and are glazed with massive panes of plate glass of great size and half an-inch [12.5 mm] thick.

The Chance order book describes the apparatus as

> Apparatus to light 360°, being
> 6 Lenses 60° each (not inclined,)
> 6 Upper Panels --"-- 12 prisms,
> 5 Lower Panels --"-- 5 prisms.
> Mounted and framed in gunmetal as usual
> Cast iron column 5' 10" high, 30" dia'r
> At base, minimum thickness 5/8 in

Service Table 5/8 In thick, 4'11½" dia'r

Balconies etc etc, Stay rods etc.

Two Pressure Lamps & burners duplicating parts complete

with one spare burner to each lamp

(four burners in all) on our system

(sheet brass barrel) with suitable stand

Two sheet iron dampers – copper ventilations

Tube 3½ In dia'r – all as see specif'n

of Lamp & also Copper Dish

Appurtenances for service of Light, as per

List given to us which includes 3 100 gals

oil cisterns and differs a little from ours.

Balfour's description of the apparatus was

> fixed dioptric apparatus of the second order [and the lamps] pressure or 'moderator' lamps of a very simple construction [with burners] made to carry three concentric wicks ... the largest wick being 2¾ inches diameter, and they produce a very brilliant flame, which when carefully tended, should be from 3 to 3¼ inches high.

The Chance order makes no reference to the lantern apart from a comment on balconies. It appears that the lantern was designed by McClean and Stileman and fabricated by Simpson.

The 1879 specification referred to below described the Tiri Tiri Matangi light as 'second order dioptric supplied by Chance Brothers & Co, 1863'.

The tower, lantern and apparatus for Tiri Tiri Matangi were shipped from London on 8 February 1864 on the *Queen of the Deep*, arriving in Auckland on 15 June 1864. Also aboard were Richard Aylmer and H B Rolton, who had been contracted by the Marine Board to supervise construction of the Tiri Tiri Matangi, Mana Island, and Godley Head lighthouses.[55] The Tiri Tiri Matangi light was exhibited on 1 January 1865.[56]

Figure 2.10: Tiri Tiri Matangi lighthouse (PH-ALB-98-p74n2) © Auckland War Memorial Museum – Tāmaki Paenga Hira

Stores were ordered from Chance Brothers for Tiri Tiri Matangi as part of the August 1865 order, and the first Scottish stores were those ordered in 1867 and supplied by Smiths.[57]

In 1879 the colza burners were replaced by Doty paraffin burners. These burners were ordered along with those for Pencarrow, and these lamps were the last ones to be converted to paraffin burners.

The Tiri Tiri Matangi specification sent to D & T Stevenson called for[58]

> One set (3) three wicked [*sic*] paraffin burners to be fitted on pressure lamps, with regulating overflow tubes, two new holders, one gross of each size wick and one gross of lamp glasses.

The next alteration to the light coincided with the exhibition of the Moko Hinau light (Chapter 7); this was to insert a red pane to show a red arc over Flat Rock to Kawau Island.[59]

The frame for this was manufactured by Luke Son & Williams at their Te Aro Foundry;[60] this same foundry also cast the iron tower for Castlepoint lighthouse (Chapter 11). The red glass remained from 1883 until 1922, when an automatic light was installed on Flat Rock.[61] A beacon had been recommended as early as 1867 by Balfour.[62]

The Doty paraffin burners remained in operation until 1916, when they were replaced by a Chance's incandescent burner.[63] In 1925 the light was automated by the installation of an acetone-acetylene light, and the characteristic was changed to group flashing. The light was electrified in 1955 and the keepers were withdrawn in 1984. The island became part of the Hauraki Gulf Maritime Park, and the last keeper became the first conservation officer.

Mana Island and Cape Egmont

Mana Island

The Pencarrow lighthouse was erected to light the eastern entrance to Cook's Strait, the Mana Island lighthouses the western end.

The site was recommended in an 1863 report from Captains Sharp and Johnson of the Chief Marine Board. While the report described a visit to The Brothers (a group of rocky islets further into the Cook Strait) the authors decided that the cost of erecting and maintaining a lighthouse there was too great, and they recommended one on Mana Island instead.[64] In February 1863 the secretary *pro tem* of the Chief Marine Board wrote to the colonial secretary asking if a 'proper site on that island can be placed at the disposal of the Board for the erection of a Light'.[65] The Marine Board report for 1865 does not state who won the contract to erect the lighthouses but that the construction was superintended by Messers Aylmer and Rolton. While the president of the Marine Board complimented their oversight, he made the particular point that

> [t]he completion of this work was somewhat retarded, owing to the imperfect and unfinished state of the iron tower, many borings which should have been made at the manufactory having to be done here with great labor and expense.

Mana Island lighthouse was first exhibited on 1 February 1865,[66] a month after its twin on Tiri Tiri Matangi, and earlier than expected.

In December 1864 the General Post Office issued a Notice to Mariners describing the four lighthouses which the Marine Board expected to have completed within the next six months. These were Mana Island, Dog Island, Godley Head, and 'Tairoa's Head'. The notification of Mana Island stated that:

> Mana Island Light-house is now in course of erection, on the North-western promontory of Mana Island, off the entrance to Porirua Harbour, on the Eastern shore of Cook's Straits, in 41°5' South latitude, and 174°48' East longitude, as measured on the Admiralty Chart.
>
> The light will be a Fixed Dioptric White Light of the second order, and will be seen in every direction where not intercepted by land.
>
> The light will be elevated about 450 feet above the sea, and will be visible in clear weather about 29 nautical miles, allowing fifteen feet for the height of the observer's eye, and at lesser distance according to the state of the atmosphere.
>
> The tower will be seventy feet in height, from the base to the vane
>
> The light is expected to be ready for permanent exhibition about March, 1865.

By 1874 problems had emerged with the light. Two ships, *City of Newcastle* and *Cyrus*, were wrecked when their masters mistook the Mana lighthouse for the Pencarrow Head lighthouse.[67] Captain Johnson also pointed out[68]

> [t]he altitude of the site on Mana Island has also proved too high, and the power of the light too small for the altitude; and the result is that while the rays can be seen by an observer whose eye is 15 feet [4.6m] above the sea at a distance of twenty-nine miles [47 km] in clear weather, the power is only sufficient to enable the light to be seen eighteen miles [29 km] in a mean state of the atmosphere; and at much less distance than this (owing to a large amount of the rays being overhead), the light is subject to be obscured at a time or in such weather when it ought to be seen.

Basically the resulting lighthouse was unsatisfactory because the Wellington Provincial Government had ordered a complete lighthouse from England, and had done it before the final site had been chosen. William Seed wrote in 1874[69] that

> [t]he iron tower that was sent out was 48 feet high (14.6m) when one less than half the height would have been more suitable; and the illuminating apparatus was made to show all around, when one with an apparatus to illuminated the arc to seaward of about 250° was all that was required.

Given the problems that had occurred in constructing the tower, the height problem must have been particularly galling.

In contrast, the Otago Provincial Government ordered lanterns, lens and apparatus from Scotland and designed the towers locally. Subsequently this was done for all lighthouses constructed by the Marine Department. Towers included the block masonry of Taiaroa Head, the distinctly New Zealand weatherboard towers designed by Blackett, and described in his paper, in the Minutes of the Proceedings of the Institution of Civil Engineers,[70] the cast iron towers such as Cape Brett constructed by New Zealand foundries, and ones built of reinforced concrete.

In the ninth report of the Marine Department, Seed also pointed out that 'the opinion of shipmasters who trade through the Strait is now, I believe, entirely in favour of a light on the Brothers'. If it was necessary to retain the Mana Island light, he suggested that its characteristic be changed to avoid confusion with Pencarrow.

The Marine Department opted to build a lighthouse on The Brothers, and Mana Island lighthouse remained operational until September 1877, when the Brothers light was exhibited (see Chapter 7). Four years later the Mana Island lighthouse was dismantled, and no photographs of it exist. The tower was moved to Cape Egmont in Taranaki.

Cape Egmont

A lighthouse was proposed for Cape Egmont as early as 1866, in that year's Marine Board report.

The comment was a little obscure, as Captain Sharp pointed out that

> [i]t has been suggested that, during the long winter nights, the Panama boats might have their passage expedited by lights being placed at Capes Egmont and Palliser, those two points forming the north-east and north-west extremity of Cook's [*sic*] Strait.[71]

It is not clear how a lighthouse on the Taranaki coast would be of value to boats coming from Panama; Sharp may have meant ships traversing the Cook Strait from the east.

In their 1873 report Johnson and Edwin observed that:

> [t]he land about Cape Egmont being low, vessels from off sea voyages seldom first make it; but as the position is valuable as a guide to the coasting trade of the Colony, we consider that a Second-order light will be sufficient.[72]

In 1874 it was suggested that if a new lighthouse were to be established on The Brothers, the light that had been ordered for Tory Channel could be diverted to Cape Egmont.[73] However, the Tory Channel light was eventually used at Moeraki (see Chapter 6). The following year the Statement on Marine Matters recommended the

Figure 2.11: Cape Egmont lighthouse from the sea with Mount Taranaki behind (ATL WA-21727-G) © Whites Aviation Collection, Alexander Turnbull Library, Wellington, New Zealand

> removal of Mana Island to Cape Egmont. The latter will have to await the completion of the light on The Brothers, and the further acquiring of a suitable site; but the land being Native property requires further action to be taken by my colleague the Native Minister.[74]

The Taranaki Land War had ended less than five years previously, and the government were aware of how close they had come to being completely beaten, and thus forced to return confiscated land.

The Marine Department report for 1878 indicated that it was

> proposed to use the Mana apparatus and tower for the Cape Egmont Light, and steps will be taken to move them to that place as soon as a site can be obtained from the Natives.[75]

It was 1881 before the Mana Island lighthouse was dismantled, shipped to New Plymouth and re-erected at Cape Egmont. It was first exhibited on 1 August of that year.[76]

The Marine Department report for 1881 pointed out that

> [o]wing to the disaffection of the Natives living near that part of the coast, it was found impracticable to proceed with its (Cape Egmont lighthouse) erection. On the establishment of military posts, and the occupation of the country in the vicinity of Cape Egmont, all chance of serious opposition by the Natives to the establishment of the light was removed, and the Government at once gave directions for its erection to be proceeded with.[77]

Figure 2.12: Cape Egmont lighthouse (ATL1/1-05693-G) © Northwood Collection, Alexander Turnbull Library, Wellington, New Zealand

In view of the 'disaffection of the Natives' the lighthouse was provided with a protective garrison of local armed constabulary. The troops had already contributed to lighthouse construction by building a ditch and bank fence around the lighthouse reserve and constructing a road about three miles long to the lighthouse.[78]

However the military–civil relationship was not smooth, and the 'disaffection of the Natives' was paralleled by the 'disaffection of the keepers'. The principal keeper frequently complained to the secretary of marine about the armed constabulary on the ground floor of the tower. Eventually, the constabulary were withdrawn in February 1882, having only once left the lighthouse, in November 1881, to join the forces which attacked and destroyed the nearby settlement at Parihaki, a centre of non-violent resistance to the confiscation of Maori land.

The opportunity was taken to upgrade the light and on 26 October 1880, the Agent General sent on to D & T Stevenson requisition CS 150/80 which included for Cape Egmont:

- 4, 3-wick paraffin burners for pressure lamps as supplied to Tiri Tiri with overflow tubes
- 2 burner holders as supplied to Tiri Tiri except that they are only 3 ¾" high.
- 1 Liverpool lamp for 3 wick burner
- Glasses and wicks for burners and Liverpool lamp
- 1 Paraffin tester complete.[79]

The agent general asked that the Cape Egmont lamps be sent 'as soon as possible by mail steamer, the other goods by sailing vessel'. The burners, holders, and Liverpool lamp were

[Opposite] Figure 2.13: Revolving Cowl for Cape Egmont ABVD 24480 W5528/7/[2] MD 4551 © Archives New Zealand, The Department of Internal Affairs, Te Tari Taiwhenua

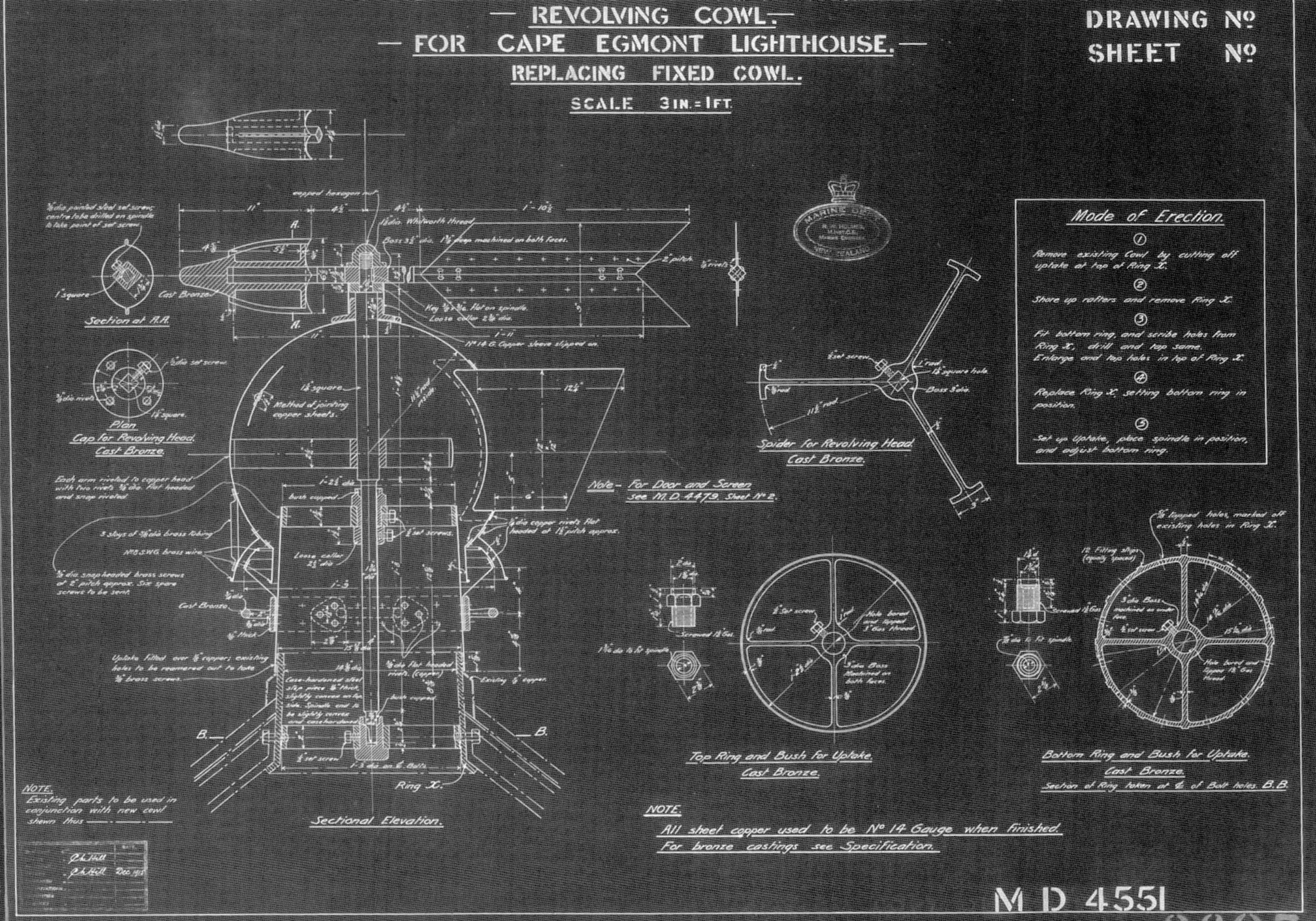

— REVOLVING COWL. —
— FOR CAPE EGMONT LIGHTHOUSE. —
REPLACING FIXED COWL.
SCALE 3IN.=1FT.
DRAWING No
SHEET No
Mode of Erection.
① Remove existing Cowl by cutting off uptake at top of Ring X.
② Shore up rafters and remove Ring X.
③ Fit bottom ring, and scribe holes from Ring X, drill and tap same. Enlarge and tap holes in top of Ring X.
④ Replace Ring X, setting bottom ring in position.
⑤ Set up Uptake, place spindle in position, and adjust bottom ring.
Section at A.A.
Plan
Cap for Revolving Head.
Cast Bronze.
Spider for Revolving Head.
Cast Bronze.
Note - For Door and Screen see M.D. 4479. Sheet No 2.
Top Ring and Bush for Uptake.
Cast Bronze.
Bottom Ring and Bush for Uptake.
Cast Bronze.
Section of Ring taken at ℄ of Bolt holes B.B.
Sectional Elevation.
Ring X.
NOTE.
Existing parts to be used in conjunction with new cowl shewn thus
NOTE.
All sheet copper used to be No 14 Gauge when finished.
For bronze castings see Specification.
MARINE DEPT.
NEW ZEALAND
M D 4551
3607

supplied by Milne & Son and the remainder by Smiths. Although the specification in the Stevenson Archives has a marginal note in pencil, *Mail leaves London 3 Nov 1880*, Milne & Son were unable to manufacture their order in time.[80] The order was despatched by rail from Edinburgh on 4 December 1880 to Orient Line of Steamers, South West India Docks, along with an invoice for £45.[81] On the same day a memo from Smith Morrison, the New Zealand Government shipping agent in London to the Agent General, pointed out that the Orient Line loaded at the Royal Albert Dock and not the South West India Docks,[82] but the package arrived safely nevertheless. The Agent General's memo of 7 December suggests that the unnamed mail steamer was the *Cuzeo*.[83] The upgraded light was exhibited on 1 August 1881 and described as 'a second order FIXED WHITE LIGHT visible between the bearings of N.23° W. and S. 5° W.[84]

In 1911 the Doty burners were replaced by incandescent light.[85] The lighthouse was further upgraded in 1916 when a new revolving cowl, manufactured to a Northern Lighthouse Board design, was installed.[86]

This was to become a familiar pattern: installation of incandescent burners followed closely by alterations to the cowl to allow for the extra heat produced by the new burners.

In 1926 it was proposed to automate the lighthouse with an acetone-acetylene lamp.[87] Two years later it was further proposed to change the light from fixed to flashing, due to the

> multiplicity of powerful electric lights in the locality, and the confusion caused by the headlights of motorcars on the main road between New Plymouth and Opunake.

The eventual upgrading was for a new triple revolving lens and an acetone-acetylene lamp, both controlled by a sun valve.[88] The lens removed was subsequently used in the Baring Head light.

The upgrading was carried out between 25 November and 8 December 1929. During this time three fifth order lights were arranged eight feet apart inside the existing lantern. The new characteristic was group flashing, 'flash 0.4 sec, eclipse 7.4 sec; flash 0.4 secs, eclipse 7.4. secs; flash 0.4 sec, eclipse 24 secs'.[89]

Electrification of the light began in 1950 and was completed the following year. The light was automated in 1986.

Godley Head

Godley Head (until 1900 also known as Cape Cachalot), is at the entrance to Lyttelton harbour, the main port for Canterbury Province. The area around Cape Cachalot was called Port Cooper and consequently the lighthouse was referred to as the Port Cooper light in early documents. (The bay opposite the lighthouse site is still called Little Port Cooper.) A lighthouse was suggested for this site as early as 1849, when Captain Thomas, Chief Surveyor for the Canterbury Association, visited the harbour.

One of the earliest actions of the newly established Provincial Council was to request the superintendent (James Edward Fitzgerald) to 'add to the estimates the sum of £80 for the Erection of a Beacon at the entrance of the Harbour'. The superintendent replied that he had referred the resolution in question to the Harbour Master for a report on the necessity of such a Beacon, and to the Legal Adviser of the Government for an opinion as to whether the Provincial Government are competent to appropriate money to such a purpose without a violation of the 19th Clause of the Constitution Act.[90]

The advice seems to have been negative, as in December 1861 the Provincial Council again addressed 'His Honor the Superintendent':

> The Provincial Council respectfully desire to bring to your Honor's notice the Inconvenience which has been felt by the Masters of Vessels arriving at Lyttelton from the Absence of a Lighthouse at the Heads of Lyttelton Harbour. The Council is aware that under the existing state of the Law, it is debarred from passing any Bill for that purpose. The Council therefore respectfully pray that your Honor will direct the attention of his Excellency's Government to the matter, and urge upon the Colonial Secretary the necessity of Legislation upon the subject.[91]

The punctiliousness of the Canterbury Provincial Council contrasts strongly with the aggression of the Wellington Provincial Council and the nonchalant laissez-faire approach of the Nelson and Otago councils.

The colonial secretary acknowledged receiving the address, in a letter dated 17 March 1862, and promised that the 'subject will receive consideration of Government'.[92] However, the government's consideration consisted of passing the matter to the newly created Chief Marine Board; the Godley Head light was one of the 'numerous applications' received by the Chief Marine Board at its first meeting in January 1863.

In the interim, E J Wright, who had constructed the Pencarrow lighthouse, offered to contract for providing and fixing of a lighthouse at Godley Head. The offer included provision of drawings, specification, and costing; if the coast light was not proceeded with he would accept £25 for the drawings and specification.[93] Although not taken up, the offer was seriously considered as a handwritten note on the document suggested buying Wright's drawings and sending to England for the lighthouse.

Wright also offered to contract for the Otago lights (see Chapter 3).

In a letter dated 30 June 1863, Captain Sharp of the Marine Board informed the Superintendent of Canterbury Province that an

> order has been sent to England to procure 3 lighthouses for the New Zealand Service of a similar construction and one will be placed at the disposal of Canterbury province for Lyttelton Harbour if lighthouse is suitable to the position upon viewing.[94]

This was the 25 September 1863 order to Chance Brothers. The Port Cooper order was for[95]

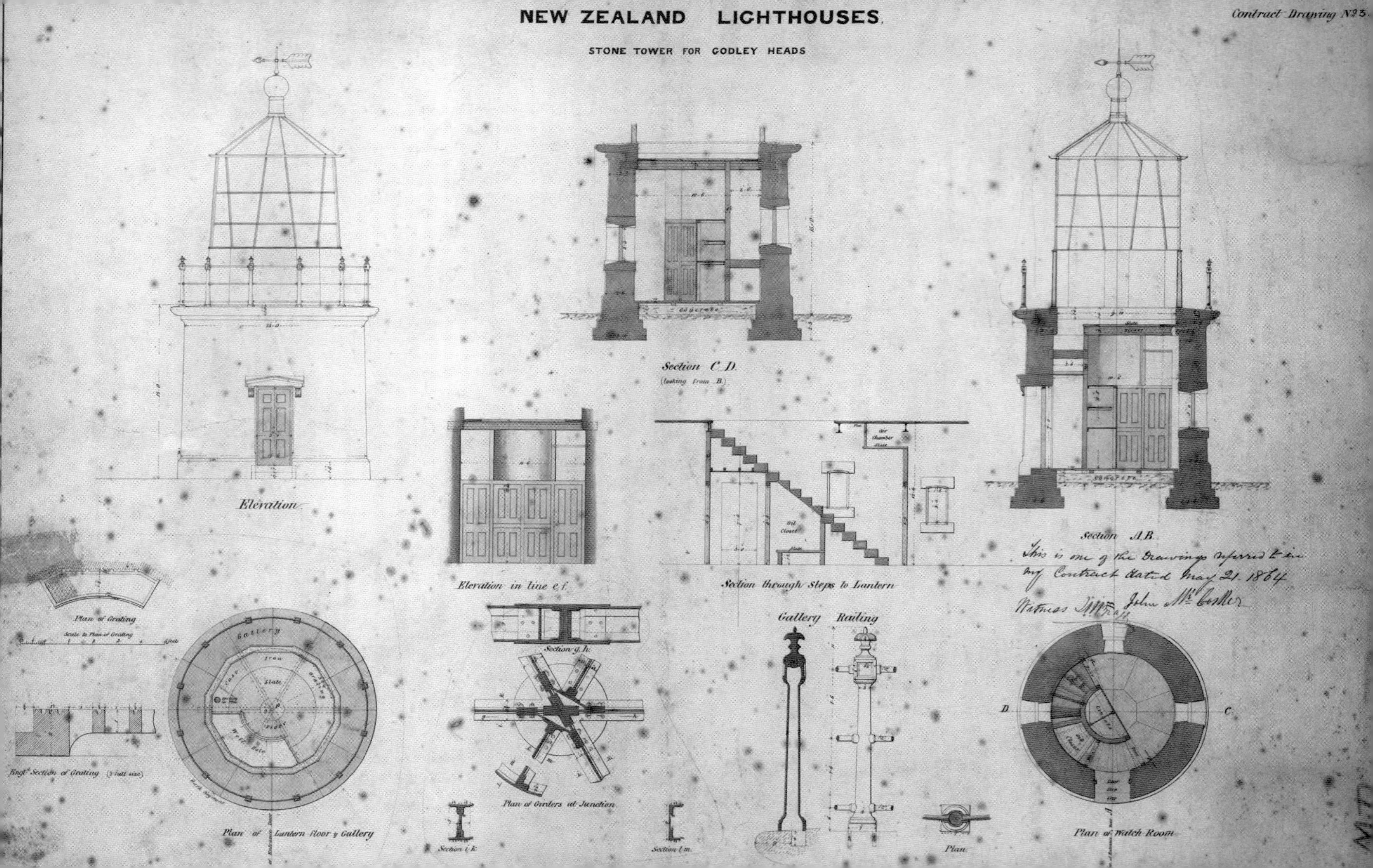
NEW ZEALAND LIGHTHOUSES.
STONE TOWER FOR GODLEY HEADS
Contract Drawing No. 3.
Elevation
Section C.D.
(looking from B.)
Section A.B.
This is one of the Drawings referred to in my Contract dated May 21. 1864
John McCosker
Witness
Elevation in line e.f.
Section through Steps to Lantern
Oil Closet
Plan of Grating
Engt. Section of Grating (½ full size)
Gallery
Plan of Lantern floor & Gallery
Section g.h.
Plan of Girders at Junction
Section i.k.
Section l.m.
Gallery Railing
Plan
D
C
Plan of Watch Room

Apparatus to light 200° being

4 Lenses (not inclined)

4 Upper Panels – 12 prisms

4 Lower Panels – 5 prisms

Armature, Lamps, Accessories etc as for

the preceding [*sic*] Lights, Tiri Tiri & Cook's Strait

It appears that Simpson & Co manufactured the lantern to the McClean and Stileman design. It is not clear who designed the 20-foot-high [6.1m] masonry tower. George Thornton, a Yorkshireman who had been appointed Assistant Provincial Engineer for Canterbury Province on 17 November 1863,[96] prepared the drawings for the contract and may have designed the tower, although the drawing does carry the a reference to McClean and Stileman.

The following year, the provincial secretary again approached the Marine Board requesting reconsideration of a site for the Port Lyttelton lighthouse. The Marine Board agreed to review the situation 'in course of a week'.[97]

In his 1864 report Sharp noted that the provincial assembly had voted a sum of £5,000 to enable the board to erect the lighthouse.[98] However a review of the Proceedings of the Provincial Government shows no reference to such a vote. Perhaps the Provincial Council's scruples about Clause 19 of the Constitution Act led them to 'hide' the grant. Thornton let the contract, and superintended construction of the tower. The local firm McCaster were contractors for the construction, and the erection of the lantern and apparatus was supervised by Aylmer and Rolton.

Balfour in his report[99] commented that the lighthouse's

> power at night would be very greatly increased were the light on the land side, which is at present wasted (*more than half of the whole light from the lamps*) [italics his] reflected seawards, as at Tairoa's Head and as intended to be done at Cape Saunders.

This is probably a reference to the incorporation into the Otago lights of the new spherical mirror invented by Thomas Stevenson (see Chapter 3).

The anticipated light was included in the four light Notice to Mariners of 8 December 1864 mentioned above. This part of the notice stated that:

> Godley Head Lighthouse is situated on Godley (or Cachlot) head, forming the north-western entrance of Port Lyttelton, Bank's Peninsula, in 43°35'32"south latitude, and 172°49'30"east longitude, as measured on the Admiralty Chart.

[Opposite] Figure 2.14: Stone Tower for Godley Head (ABPL 8607 W5520/34 MD421) © Archives New Zealand, Department of Internal Affairs, Te Tari Taiwhenua

Figure 2.15: Original Godley Head lighthouse (ATL 1/4-023421-G)
© Adkin Collection, Alexander Turnbull Library, Wellington, New Zealand

> The light will be a Fixed Dioptric White Light. Of the second order, is elevated about 450 feet above the level of the sea, and illuminating an arc of 200°. It will be visible from seaward, in clear weather, about 20 nautical miles, and at lesser distances, according to the state of the atmosphere, from E.¼S., round by the north to about N.N.W. ¼W.
>
> The tower will be thirty feet in height from base to vane.
>
> This light is expected to be ready for permanent exhibition about March 1865.

However, the light was exhibited on 1 April 1865.[100]

As with the other English provincial lighthouses, the first Scottish technology incorporated was a three-wick Liverpool lamp ordered in 1871. In 1872 wicks were supplied by Smiths.[101]

The colza lamps were upgraded with Doty burners in 1877.[102] These were manufactured by Milne & Son and despatched to New Zealand on the *Crownthorpe* from the West India Docks.[103] Their safe arrival was reported by Lewis Wilson in a letter to 'My dear Uncle Tom'.[104] A new Liverpool lamp with glasses and wicks was ordered in 1880 along with the paraffin burners needed comparatively urgently for Cape Egmont.[105] The Liverpool lamps for Godley Head and the remaining three English lighthouses followed at a leisurely rate by the sailing ship *Pareora*.[106]

By 1882 the lantern needed repairs and the old revolving cowl was replaced by 'a top of the pattern used in the Northern (Scotch) Board of Lighthouses'.[107]

Although in 1913 Captain Bollons (captain of the *Hinemoa*, the government steamer at the time which was used as the lighthouse tender) had suggested that the light be changed to flashing,[108] the matter was left in abeyance until 1929. Then it was decided to change the light, as

> universal electric lighting and the use of powerful headlights on motor cars ... [make it] necessary to provide a light which has a distinct flashing characteristic ... It is mainly for this reason that the conversion is being made.[109]

The conversion was completed by the middle of 1932 and the new light exhibited on 12 July. The character of the new light was 'group flashing white, having two flashes every 27 sec thus; Flash 2.5 sec, eclipse 4 sec., flash 2.5 sec, eclipse 18 sec'.[110]

In 1941 guns were placed on the headland and the lighthouse had to be moved to a new site down the cliff.

The existing light remained operating as the new site was excavated, and the lantern and lens which had been removed from Cape Foulwind in 1926 were installed.[111] This report also indicated that equipment for electrification of the light had been ordered.

Figure 2.16: Resited Godley Head lighthouse (ABPL 8848 W5221/19 17)
© Archives New Zealand, The Department of Internal Affairs, Te Tari Taiwhenua

Electrification was finally completed in 1954.[112] In 1975 it was decided that the light should be automated and handed over to the Lyttelton Harbour Board.[113]

The Christchurch earthquakes of 2010 and 2011 undermined the lighthouse's foundation, and in 2013 the lighthouse was lifted by helicopter off its ledge for restoration and possible relocation.

Matiu-Somes Island

Matiu/Somes is an island in Wellington harbour. Originally named Matiu, it was renamed by the European settlers after Joseph Somes of the New Zealand Company.

The lighthouse on it was thus a harbour light, and from its erection to 1874 it was the responsibility of the Wellington Provincial Government. It was managed by the Marine Department from 1874 to 1924 on behalf of the Wellington Harbour Board, which contributed to the cost of maintenance.

This lighthouse order to Chance Brothers was dated 28 December 1864,[114] and contained references to McClean and Stilman

The order was for a fourth-order fixed light for 270 degrees with a 90 degree silver-plated spherical reflector with fittings, to be supported on a cast iron column. A marginal note suggested that a 270° light was in stock. Two moderator lamps 'of improved construction' were specified; these were to burn colza oil. As the light was required to mark the main harbour channel, ruby glass screens in brass framing and a screen of green glass were specified. The ruby screens were described as being 'for the arcs indicated in the forementioned instructions as explained by McClean & Stileman'.

Figure 2.17: Matiu/Somes Island: original fourth order lighthouse (1886) and upgraded second order (1900) lighthouse (PH-ALB-98 -p72n2) © Auckland War Memorial Museum – Tāmaki Paenga Hira

The lantern, 'of suitable dimensions' was to be cast iron and gunmetal and glazed with 3/8 inch polished plate glass. It was apparently described as being 'as per the design supplied by us' and there was a marginal note 'as per our own design'. This part of the order had the interesting final paragraph:

> Sufficient ventilation will be ensured and the standards will present the minimum obstruction to the light.

Two years earlier Chance Brothers had received an order for the lights only for Otago (Chapter 3). The Otago lanterns were to be of Alan Stevenson design with its triangular astragals and thinner glass. Chance Brothers' comments could be interpreted as a reply to the opposition.

The cast iron tower was to be 10 feet high with an external diameter of 6 feet 8 inches, and designed to be bolted to a brick foundation. The lightroom floor was apparently wrought iron. Although this part of the order also bore the marginal note 'as per our own design', it concluded with: 'The Tower to be octagonal, according to the recommendation of McClean & Stileman – otherwise as detailed above'.

The lighthouse commenced operations on 17 February 1866.

After 14 months the light was reported to be working well but the initial adjustment of the screens which directed a beam of white light down the centre of the channel had been found to be defective. Balfour was asked to visit and superintend the fitting and adjusting of a permanent screen which separated the fixed white beam from its coloured neighbours.[115]

The lamps were upgraded to Doty paraffin lamps in 1878, the lamps having been part of an order that included Doty lamps for Nelson and Godley Head.[116] New lamp glasses were ordered in 1897.[117]

The 1895 Marine Department report stated that

> Representations have been made to Department that light on Somes Island is inadequate for the requirements of the Port of Wellington, and it has been suggested that it should be removed to Ward Island (nearby) or a more powerful light placed on the latter site and the light on Somes Island removed.[118]

Although soundings were taken and borings made on Hope Shoal off Ward Island,[119] the eventual decision was to upgrade the Somes Island light from fourth order to second order, with the Wellington Harbour Board contributing the sum of £800.[120]

The order for the new lighthouse apparatus was sent on to C & A Stevenson by the Agent General on 13 April 1898.[121] Although C & A Stevenson required further information from the Marine Department, they were able to report on 6 September 1898 that they had 'prepared the working plans and specifications of what is required for Somes Island Lighthouse' and hoped 'to be able to send you the offers shortly with our Report'.[122] The design was for a second-order dioptric with a four-wick Doty burner.

C & A Stevenson did have reservations about the design, and pointed out to both the Agent General[123] and the Marine Department:[124]

> We have designed the apparatus according to the instructions given to us, but as already reported we are of opinion that the new light, although better than the old will not be so strong as the authorities in New Zealand anticipate as to make the cut off sharp enough, the shades will require to be arranged as greatly impair the power of the light. We think it a mistake not to put the cut off on a stand in front of the light as we suggested – a method which we use in several places in Scotland with no evil effects from snow [*sic*] and so as to take full advantage of the strength of the light and secure a satisfactory cut off.

No lantern was ordered, as that sent out for Farewell Spit in 1868 had now become available with the erection of the new lighthouse there in 1897 (see Chapter 4).

Requests for tenders for the dioptric apparatus were sent to Barbier & Bernard[125] (successors to Barbier & Fenestre), Henri Lapaute,[126] and Chance Brothers.[127] Milne & Son[128] and James Dove & Co[129] were invited to tender for the four-wick fountain lamps and their supporting pillar and table. Kennaway, the Agent General, apparently asked why Chance Brothers were not invited to tender for the lamps as well as the optic. C & A Stevenson indicated that

> [w]e did not get an offer from Messers Chance and Co as it would not have done to fit up their lamps in another contractor's apparatus, moreover the apparatus requires to be tested by ourselves and the lamps etc. are made under our supervision in Edinburgh.[130]

Kennaway's comments were not forgotten, and five months later C & A Stevenson indicated that

> [w]e have tried and thoroughly tested the apparatus etc for Somes Island Lighthouse and the packages will be ready to be forwarded for shipment next week.

The tenders went to Barbier & Bernard and James Dove & Co, as their tenders of £358 and £174 respectively were the cheapest.[131]

As this was not a completely new lighthouse the only stores Smiths were requested to tender for were wicks, glasses, and ancillary equipment (including the usual Liverpool lamps and glasses) for the new lamps.[132] Smith's tender of £21/1/- was described as 'reasonable'. Coloured glass was also supplied.

The apparatus and stores were despatched by steamer from Leith on 18 March 1899 for shipment by SS *Ionic* from Royal Albert Docks.[133]

The existing lighthouse was dismantled and eventually sent south to Tuhariki (Jack's) Point, Timaru (see Chapter 8). The new tower was a brick one which was subsequently mortared.

The Notice to Mariners gave the following description:[134]

> Notice given new light exhibited about Wednesday 14th Feb1900, from circular brick tower painted white, 2nd order dioptric fixed. Height 95ft asl, visibility 16 nautic miles. Colour for harbour entrance same, white mid-channel, red western shore, green eastern shore. Lambton Harbour, white western side over Lambton Harbour & Evans Bay over an arc 80° between Gordon Point and Ngahauranga.

The new light was exhibited on 21 February 1900, thus completing the metamorphosis from English lighthouse to Scottish.[135]

The Doty paraffin burner was replaced by an incandescent burner in 1910. In 1924 this was replaced by an automatic acetone-acetylene lamp and the lighthouse was transferred to the Wellington Harbour Board.[136] It was converted to electricity after 1945.

Endnotes

1 http://www.historic.org.nz/corporate/registersearch/ProfessionalBio/Professional.aspx?ID=268

2 Vote & Proc H of R Session1 1854 p 415(?), *Report of Beacon & Lighthouse Committee*

3 NZG Archives, LE1/1860/220 Letter Featherston (Superintendent Wellington) to Stafford (Colonial Secretary) 5 Feb 1857

4 NZG Archives, LE1/1860/220 Letter Featherston (Superintendent Wellington) to Stafford (Colonial Secretary) 1 April 1857

5 Churchman Geoffrey B *New Zealand Lighthouses,* Wellington GP Books, 1989 p 12

6 Acc 10706 vol 85 Agent General for New Zealand to D & T Stevenson5 Sept 1879

7 http://www.historic.org.nz/corporate/registersearch/ProfessionalBio/Professional.aspx?ID=104218

8 NZG Archives AAAC/108/D500 142c/636 Letter Griphook & Wright, Napier to Superintendent, Otago dated 2 May 1862

9 *NZG Gazette* 1858 No 31 Notice to Mariners Pg 156 18 November 1858

10 NZG Archives LE/1860/220 No 10 Letter Tancred for Colonial Secretary to Superintendent Wellington 23 March 1859

11 NZG Archives LE/1860/220 No 11Letter Featherstone to Colonial Secretary 12 April 1859

12 New Zealand Marine Dept, Centennial History, Wellington 1969 p 55

13 Chance Archives Smethwick, BS6/12/1/1/1/ p 308 4 August 1865

14 Chance Archive Smethwick BS6/12/1/1/2 p 177 31 July 1869

15 Acc 10706 vol 85, letter from Agent General includes copy specification Pencarrow and Tiri Tiri Matangi, c/s120

16 AJHR 1892 , H-29 Marine Department (Annual Report for 1891-92)

17 AJHR 1894 H-18 Marine Department (Annual Report for 1893–94)

18 New Zealand Herald vol XXXIV Issue 11079 2 June 1899 pg 5.

19 *NZG Gazette,*1906 No 82 Notice to Mariners No 73 of 1906, pg 2529 27 September 1906

20 AJHR 1907, H-15 Marine Department (Annual report for 1906–)07

21 NZG Archives ADOE 16612 M1/762 8/52/3 note on Taiaroa Head file, signed Holmes, Marine Engineer

22 NZG Archives AAPR W3282 box 3 M.8/45/2

23 *NZG Gazette* 1934 No.79 Notice to Mariners No 42 of 1934, p 3386 23 October 1934

24 *NZG Gazette* 1935 No 41Notice to Mariners No 21 of 1935 p 1589 4 June 1935

25 NZG Archives AAPR W 3282 Box 3 letter WL Newnham Engineer-in-Chief, Public Works Department Wellington to Harbour Master, Wellington Harbour Board
26 NZG Archives AAPR W3282 Box 3 8/45/14 Letter F A Mac Indoe, Merchant Service Guild of Australasia to Minister of Marine April 14 1921,
27 NZG Archives ADOE 16612 M1 769 8/62/1 memo Bollons to Secretary Marine Department 27 March 1928
28 NZG Archives ADOE 16612 M1 769 8/62/1 Memo Godfrey Secretary Marine Department to Minister of Marine 27 Sept 1928
29 NZG Archives ADOE 16612 M1 769 8/62/1 Letter Sharp (Inspecting Engineer) to Marine Engineer 21 June 1929
30 NZG Archives ADOE 16612 M1 769 8/62/1 Report Sharp to Fukert, Marine Engineer 19 June 1931
31 NZG Archives ADOE 16612 M1 769 8/62/1 Schedule of Tenders 13 December 1933
32 NZGArchives ABPL W5221 7457 Box 1 45/11/2 letter Wilson Principal Keeper to Secretary Marine Department 26 August 1943
33 NZG Archives ABPL W5221 7457 Box 1 45/11/2 memo Newnham Marine Engineer to District Engineer Public Works Department, Wellington dated 9 Sept 1943
34 Votes and Proceedings, Nelson Provincial Council, Session IV, 1857
35 ibid
36 Votes and Proceedings, Nelson Provincial Council, Session VI, 1859
37 NZG Archives ACHZ 16465 NP 29 1Letter Blackett to Slaughter, Gruning,& Co, Engineers, Bristol 1 February 1860,
38 Acc 10706 Vol 85 Letter William Seed to D & T Stevenson 19 July 1876
39 NZG Archives ACHZ 16464 NP 7 7 1862/367 Specification for Supply Colza (Nelson catalogue 62/377)
40 http://www.historic.org.nz/Register/ListingDetail.asp?RID=41&rm=Brief&p=prints
41 www.tera.govt.nz/en/photography/28828/nelson-lighthouse
42 NZG Archives LE/1/58/1868/117 Accounts & Papers including Statement of Sums Expended in the Purchase and repairs of that on Boulder Bank
43 Chance Archives BS6/12/1/1/1 p 310
44 Acc 10706 Vol 85 Letter William Seed to D & T Stevenson 19 July 1876
45 *NZG Gazette*, No 38, , Notice to Mariners No 12 of 1888 p 748 , 28 June 1888
46 AJHR 1914 H-15 Marine Department Annual Report 1913–14
47 AJHR 1916 H-15 Marine Department Annual report for 1915–1916
48 Votes and Proceedings. Auckland Provincial Council Session XII, 1860, Address No XVII, Tuesday 6 March
49 Votes and Proceedings. Auckland Provincial Council Session XII, 1860, Message No 30, Friday 16 March
50 New Zealand Lighthouses – Report by Mr Balfour *Daily Southern Cross*, Volume XXI, Issue 2556, 27 September 1865, p 6
51 AJHR, 1865 D-No1C, Report of the Marine Board Department of New Zealand 1864-5 Report of Mr Balfour
52 Chance Archives BS6/12/1/1/1 p 210, 25 Sept 1863
53 AJHR1864 D-No1.D Reports of the Officers of the Marine Board
54 http://www.newzealandlighthouses.com/mana_island.htm
55 MS660, Auckland Museum and Institute Library, typescript 71/32
56 AJHR, 1865 D-No1C, Report of the Marine Board Department of New Zealand 1864-5 Report of Mr Balfour
57 Ac10706 Vol 35 p 929 letter to Messrs Smith Blair St
58 Ac 10706 vol 85, letter AG to D&T Stevenson . Cover note & Specification for Pencarrow and TTM. Specification copy S175/78 5 Sept 1879
59 *NZ Gazette* 1883 No 7 Notice to Mariners 17 of 1883, p 757 26 May 1883,
60 MS660 Library Auckland Museum and Institute
61 Maritime New Zealand, *Lighthouses of New Zealand, Tiritiri Matangi*
62 AJHR 1867 E-No. 6 Third Report of the Marine Department for year 1867
63 MS660, Auckland Museum and Institute Library, Typescript, 71/22
64 AJHR 1874 *Selection of Sites for Projected Coastal Lighthouses* (Correspondence relative to) H-23 Appendix No 5
65 NZG Archives ACIA 16199 WP7/1/7 letter Chief Marine Board to Colonial Secretary 3 Feb 1863
66 in Auckland Provincial Government Gazette, 22 Feb 1865, No 3 p 20 Notice to Mariners Wellington 27 Jan 1865,
67 AJHR 1874 H-23Ninth Report of the Marine Department for the year ended 30th June 1874
68 AJHR 1874 H-23 Appendix No 4 *Selection of Sites for Projected Coastal Lighthouses* (Correspondence relative to)

69 AJHR 1874, H-22 Ninth Report of the Marine Department, for the year ended 30th June 1874
70 Blackett J. 1880 *New Zealand Lighthouses*, Paper 1671, Minutes of the Proceedings of the Institution of Civil Engineers, Vol. 60, pp 334–337
71 AJHR 1866E – No 3 2nd Report of the Marine Board Department 1865–6 Enclosure No 1 sgd Sharp
72 AJHR, 1873 H-19 Eighth Report of The Marine Department, for the Year end 30th June 1873, *Report on the Lighthouses on the Coasts of New Zealand*
73 AJHR, 1874 H-22 Ninth Report of The Marine Department, for the Year ended 30th June 1874,
74 AJHR 1875 H-12 Statement on Marine Matters
75 AJHR.1878 H-12 Annual Report of the Marine Department
76 *NZG Gazette 1881*, Notice to Mariners No 28 of 1881, Marine Dept, Wgtn, 14 June 1881
77 AJHR 1881 H-27 Annual Report of the Marine Department –
78 AJHR 1881, H-27 Annual Report of the Marine Department –*Annual Report on Lighthouse Works by the Marine Engineer*
79 Acc 10706 vol 51. p 540 Letter to Milne & Son re New Zealand Lighthouses
80 Acc 10706 vol51. p 553 Letter to AG London re New Zealand Lighthouses
81 Acc 10706 vol 51. p 638 Letter to AG London re New Zealand Lighthouses
82 Acc 10706 vol 85 pg 266 copy memo Smith Morrison to AG, 4 Dec 1880
83 Acc 10706 vol 85 pg 270 memo Agent General, 7 Dec 1880,
84 *NZG Gazette*1881 No 25 Notice to Mariners, 11 of 1881, p 398 6 April 1881, and No 47 Notice to Mariners, 23 of 1881, 14 June 1881
85 NZG Archives ADOS 17056 ML-Cape Egmont 3/2/2Principal Keeper to Sec Marine, 2 August 1911
86 AJHR 1917, H-15 Marine Department Annual report for 1916–17, also NZG Archives ADOE 16612 M1 772 8/15/6
87 AJHR 1926 H-15 Marine Department Annual report for 1925–26
88 AJHR 1928, H-15 Marine Department Annual report for 1927–28
89 *NZG Gazette* 1929 No 77 Notice to Mariners,64 of 1929, p 2888 , 12 November 1929,
90 Journal Proceeding Canterbury Provincial Council Session II, March 1854
91 Journal Proceedings Canterbury Provincial Council, Session XVII, December 1861, also New Zealand Government Archives, CAAR/19936/CH287/CP26/ICPS 2018/1861 Copy of Minutes Provincial Council (18/12/1861),
92 NZG Archives, CAAR/19936/CH287/CP27/ICPS 295/1862 letter Colonial Secretary to Provincial Secretary
93 NZG Archives, CAAR/19936/CH287/CP30/ICPS 1180/1862. letter E J Wright to Provincial Secretary
94 NZGA CAAR/19936/CH287/CP40/ICPS 1309/1863 Memorandum of Interview Captains Sharp and Petersen and Superintendent Canterbury Province
95 Chance Archives BS6/12/1/1/1 p 211, Order dated 25 Sept 1863
96 Fukert FW Early New Zealand Engineers (rev Newnham), AH & AW Reed, Wellington 1953 p 279 *Thomson, John Turnbull*
97 NZG Archives CAAR/19936/CH287/CP54/ICPS 1154/1864 Letter Sharp to Provincial Secretary
98 AJHR 1864 Session 4 D-No1.D Reports of Officers of the Marine Board
99 AJHR 1865 D No 1 C (Appendix) No 2 Report on the Marine Board Department of New Zealand 1864–5. Report of Mr Balfour
100 ibid
101 Acc 10706 vol 40 p 941, letter to Agent General re New Zealand Lighthouse, 29 May 1872
102 Acc 10706, vol 85 copy letter Colonial Secretary's office to Agent General, 9 April 1877
103 Acc 10706 vol 85 Letter from Agent General 23 July 1877
104 Acc 10706 vol 88a letter Lewis Wilson to T Stevenson 4 January 1878
105 Acc 10706 vol 85 Cover letter from Agent General & Requisition CS150/88 26 October 1880
106 Acc 10706. vol 85 letter Agent General to D & T Stevenson 7 December 1880
107 AJHR 1882 Session I, Vol 2, H-12 Annual Report Marine Department report dated 20 June 1882
108 NZG Archives Letter from Bollons on "Hinemoa" reply to memo 1/91 of 20th inst. M1/762/8/52/3 26 November 1913
109 AJHR 1929 H-15 Marine Department Annual report for1928–29.
110 *NZG Gazette* 1932 No 42 Notice to Mariners 28 of 1932 p 1501 21 June 19323

111 AJHR 1941 H-15 Marine Department Annual report for the year1940–41,
112 AJHR 1954, H-15 Marine Department Annual report for the year 1953-54
113 NZG Archives AADX 889 W3148 / 40 63/190 letter dated 5 September 1975
114 Chance Brothers Archives BS6/12/1/1/1 p 268 28 December 1864
115 NZG Archives Letter John Halliday, Harbour office to W Hickson, (S'dent gen inwards letter) ACIA 16195WP3/21 194
116 AJHR 1878 H-12 the Marine Department Annual report of
117 Acc 10706 vol 66 p 108 circular to Smiths 26 June 1897
118 AJHR 1895 H-29 Marine Department Annual report for 1894–95
119 AJHR 1896, H-15 Marine Department Annual report for 1895–96
120 AJHR 1898, H-15 Marine Department Annual report for 1897–98
121 Acc 10706 vol 66 p 854 letter to Agent General 21 April 1898
122 Acc 10706 vol 67 p 168 letter to Agent General 6 September 1898
123 Acc 10706 vol 67 p 198 letter to Agent General 20 September 1898
124 Acc 10706 vol 67 p 226 letter to Secretary Marine Department 30 September 1898
125 Acc 10706 vol 67 p 172 letter to Barbier et Bernard 8 September 1898
126 Acc 10706 vol 67 p 172 letter to Henri Lepaute 8 September 1898
127 Acc 10706 vol 67 p 176 letter to Chance Brothers 9 September 1898
128 Acc 10706 vol 67 p 184 letter to Milne & Son 15 September 1898
129 Acc 10706 vol 67 p 184 letter to James Dove & Co 15 September 1898
130 Acc 10706 vol 67 p 228 Letter to Kennaway 30 September 1898
131 Acc 10706 vol 67 p 198 Letter to Agent General 20 September 1898
132 Acc 10706 vol 67 p 179 letter to Smiths 14 September 1898
133 Acc 10706 vol 67 p 729 letter to Agent General 18 March 1899
134 *NZ Gazette* 1900 No 1 Notice to Mariners 52 of 1889, p 19 29 December 1899
135 *NZ Gazette* 1900 No. 11 Notice to Mariners 8 of 1900, 5 February 1900 Jan, p 19
136 http://www.historic.org.nz/corporate/registersearch/ProfessionalBio/Professional.aspx?ID=268

3 The Otago and Southland Lights

The Otago settlement

The Otago settlement was planned in Edinburgh and was thus known as 'The New Edinburgh Scheme'. It had originated with a proposal for a Scottish settlement organised by the New Zealand Company, but this was gradually taken over by the Lay Association of the Free Church of Scotland, later known as the Otago Association. The earliest surveying was done in 1844 by Tuckett, the New Zealand Company's surveyor. Detailed surveying for the settlement was carried out by survey parties under Charles Kettle in 1846 and 1847. The first 278 immigrants arrived in March and April 1848, but by 1853 the number of immigrants had risen to only 2,300.[1]

In contrast to the Auckland, Wellington, and Nelson harbours, Otago Harbour is a sound stretching 21 km from the entrance on the Pacific Coast (the Heads) to the wharves at the city of Dunedin itself. Halfway along is the deep-water port at Port Chalmers.

In 1854 a first attempt was made to place a light at the Heads. In June of that year, C Logie, the harbourmaster at Port Chalmers, reported to the superintendent of the province that he had 'had a lamp constructed with two burners and reflectors', and stated that the pilot had 'expressed the opinion that with good oil, they would burn through the night'.[2]

It was planned to place the lamps in a stone cottage but they appear to have been placed in a wooden structure, as on 21 November 1855 Logie reported that the lighthouse at the Heads had been burnt down at about 3 a.m.[3]

Taiaroa Head lighthouse and leading lights

Taiaroa Head lighthouse

In 1860, gold was discovered at Gabriel's Gully, 92 km north-west of Dunedin. The resulting gold rush led to a major increase in shipping into Otago, and the ad hoc arrangement at the Heads which had been put in place after the fire of 1855 proved inadequate.

In September 1861, William Thomson, the harbourmaster at Port Chalmers, urged the 'appointment of an officer to combine the duties of signal man and lighthouse keeper at the Heads'.[4]

Subsequently, William Thomson carried out a site visit to the Heads and Cape Saunders along with John Turnball Thomson, who had been appointed chief surveyor of the province in May 1856.[5]

The site visit report recommended an ocean light at Cape Saunders, a harbour light at 'Tairoa Head' * and a pair of leading lights to take ships around the bar at the harbour entrance.[6]

D & T Stevenson had already been consulted on the question of lighthouses for the province as Crawford & Auld, the province's agents in Edinburgh, had sent on a letter from Thomas Stevenson dated 24 February 1862 which contained the following list of queries:[7]

1. The Characteristics, viz whether fixed, white, red, and time of revolution of nearest adjacent lighthouses.
2. The height of the site of the proposed lighthouse above the sea
3. Is the light required to show all around the horizon, and if not between what azimuths is it necessary that the lights show?
4. What sum is proposed to be spent on the works?
5. Is the light to be regarded as a harbour light or a sea light?
6. If the object of the light be to mark any special danger or show any particular passage or fairway, this should be carefully explained and if possible shown on a chart on an enlarged scale.
7. The sort of oil intended to be used.
8. A chart showing the position of the proposed lighthouse would also be helpful.

The letter also says of the list of questions: 'I know no one better qualified to fill up than Mr Thomson, the Government Surveyor, if his other avocations enable him to devote the time required'.

J T Thomson, in his addendum to the site report[8] modestly stated that he would 'make some remarks suggested by my former experience in Lighthouse Engineering for submission to Mr Stevenson's better judgement'.

His remarks included the following:

> Cape Saunders: advised:
>
> fixed natural light of first order as there are no Light Houses nearer than Wellington, climate same as South of Great Britain both with regard to temperature and haziness of atmosphere
>
> Prefabrication of lantern and light room at home [*sic*], flanges and bolts to be fixed to a plinth constructed of stone, no tower required.
>
> Store rooms and Light Keepers rooms constructed of wood or stone procurable on site, all packages sledged in.

* Although the current spelling is 'Taiaroa', documents to and from Scotland often used the mis-spelling 'Tairoa'.

> Tairoa: fixed red light, 3rd order, lantern and lightroom constructed at home – plinth, no tower required.
>
> Leading Lights: somewhat better than common lanterns

Thomson also suggested that D & T Stevenson arrange for the supply of one year's stores and the recruitment of an experienced lighthouse keeper on a three-year contract. This marks the start of New Zealand's professional lighthouse service, based on the Scottish model.

Richardson, the superintendent, seems to have had considerable confidence in the engineering advice received and moved quickly to order the required lanterns and apparatus. In his letter to the Colonial Secretary of 22 July 1862, he announced that

> Home agents of Provincial Government had been in correspondence with eminent engineers Messer Stephenson [*sic*] of Edinburgh' ... And that he had already ordered the 'necessary apparatus due to importance.[9]

Richardson appears to have been sliding over the question of Section 19. Earlier (in May 1862) he had sent a message to the Provincial Council, suggesting the expediency of authorising the expenditure of a sufficient sum for erection of a lighthouse. The government could then proceed if it found that it had authority 'as some doubt exists on question as to the party in whom the power of ordering such an erection is vested'.[10] His July 1862 letter to the Colonial Secretary has a strong element of *fait accompli*.

It is not clear when the order arrived in Edinburgh, but in a meeting on 22 September, D & T Stevenson suggested that lighthouses should be substituted for the proposed plinths and the lantern should be placed on a locally constructed tower for which they would provide suggested plans. In the end Balfour designed a 7.9-metre-high masonry tower. The brackets supporting the balcony are of the same style as those supporting the balcony on the Kinnard Head lighthouse at Fraserburgh, which Balfour would have known from his period of employment with D & T Stevenson.

D & T Stevenson also commented on the placement of the leading lights.[11,12] Ten days later D & T Stevenson wrote to the Commissioners for the Northern Lights, asking permission to use existing Scottish designs for the proposed New Zealand lighthouses.[13] This seems to have been given, as some of the New Zealand drawings in the Stevenson collection in the National Library of Scotland carry the NLB stamp. In contrast to the lighthouses ordered by the other provinces, those for Otago would have the lanterns designed by Alan Stevenson with their triangular panes of thinner glass and no vertical light-obscuring astragals.

Chances were invited to tender for the optical works for both Taiaroa Head and Cape Saunders,[14] and Milne & Sons for the first-class mechanical lamp for Cape Saunders and lamps and metal mirror for Taiaroa Head.[15] An advertisement calling for tenders for the lantern was also placed in the *North British Advertiser*.[16] Although there is no documentation relating to letting the contract for the lantern, the letter from D & T Stevenson to Milne & Son inviting them to tender for the Dog Island lantern (see below)

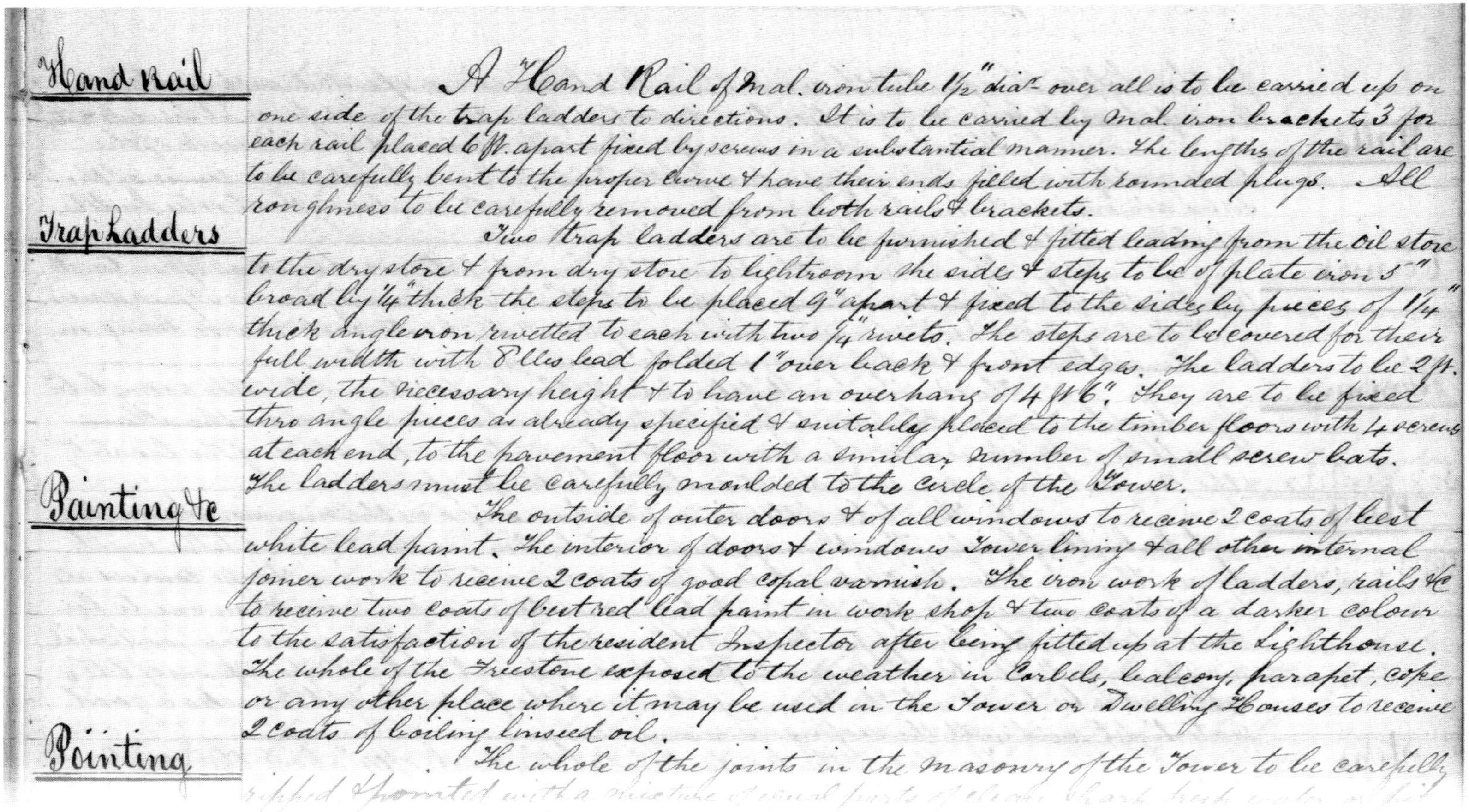

Hand Rail

A Hand Rail of mal. iron tube 1½" diar over all is to be carried up on one side of the trap ladders to directions. It is to be carried by mal. iron brackets 3 for each rail placed 6 ft. apart fixed by screws in a substantial manner. The lengths of the rail are to be carefully bent to the proper curve & have their ends filled with rounded plugs. All roughness to be carefully removed from both rails & brackets.

Trap Ladders

Two trap ladders are to be furnished & fitted leading from the oil store to the dry store & from dry store to lightroom the sides & steps to be of plate iron 5" broad by ¼" thick the steps to be placed 9" apart & fixed to the sides by pieces of 1¼" thick angle iron, rivetted to each with two ¼" rivets. The steps are to be covered for their full width with 8 lbs lead folded 1" over back & front edges. The ladders to be 2 ft. wide, the necessary height & to have an overhang of 4 ft 6". They are to be fixed thro angle pieces as already specified & suitably placed to the timber floors with 4 screws at each end, to the pavement floor with a similar number of small screw bats. The ladders must be carefully moulded to the circle of the Tower.

Painting &c

The outside of outer doors & of all windows to receive 2 coats of best white lead paint. The interior of doors & windows Tower lining & all other internal joiner work to receive 2 coats of good copal varnish. The iron work of ladders, rails &c to receive two coats of best red lead paint in work shop & two coats of a darker colour to the satisfaction of the resident Inspector after being fitted up at the Lighthouse. The whole of the Freestone exposed to the weather in Corbels, balcony, parapet, cope or any other place where it may be used in the Tower or Dwelling Houses to receive 2 coats of boiling linseed oil.

Pointing

The whole of the joints in the masonry of the Tower to be carefully

Figure 3.1: Part specification for Taiaroa Head lighthouse tower (ATL- MS-0126-114)
© James Melville Balfour Collection, Alexander Turnbull Library, Wellington, New Zealand

indicates that Milne & Son constructed the Cape Saunders/Nugget lantern, and suggests that they also constructed that for Taiaroa Head.

On D & T Stevenson's advice, Auld & Crawford accepted Chance's tender of £920 and Milne & Son's tender of £315/15/-[17] (see below). In the Chance Brothers records, the order for both Taiaroa Head and Cape Saunders is dated October 1862,[18] and is more detailed than those for the English lighthouses.

The order was:

> 3rd Order Fixed Light for Tairoa's Head
>
> 2 upper Panels of 11 prisms
>
> 2 middle panels
>
> 2 lower Panels of 4 prisms
>
> To be constructed like the first order (Cape Saunders) with inclined frames, as see specification tracing.

Very soon after the tenders were placed, it was decided to change the design for both Taiaroa Head and Cape Saunders. In the original design, both lights had metal mirrors to reflect light that would be otherwise wasted over the landward side. The light at Nelson had this system. The modification was to replace the metal mirror by an arrangement of vertical prisms which would completely reflect the light.

In his 1865 report Balfour described this refinement as follows:

> The light which at this station would naturally be wasted on the landward side, is reflected back again by a "Dioptric Spherical Mirror", an entirely new invention, by which the physical properties of glass are so skilfully taken advantage of so as to cause it to reflect the light back to the point from which it emanated (and hence, of course forward seaward) as perfectly as if it were an opaque substance; so that this combination of prisms, though apparently perfectly transparent, does not allow a single ray of light to pass through it which could be beneficially reflected, and actually casts a shadow on the wall behind. This kind of mirror was invented by Mr Thomas Stevenson, C.E., Edinburgh, some years ago, and a small one exhibited in London in 1862 attracted great notice, but this and the larger one intended for Cape Saunders are the *first* which have been made for actual lighthouse work, and they were found on trial to give highly satisfactory results.[19]

So in addition to the improved Scottish lanterns, the first Otago lights had the most up-to-date lenses available at the time.

The extra orders for the glass spherical mirrors were recorded by Chance Brothers in December 1862.[20] (The exhibition model mentioned by Balfour had also been manufactured by Chance Brothers.) The metal which had been ordered by Milne & Son and Son for the metallic spherical mirrors was utilised in the leading lights.[21] Milne &

Son were invited to re-tender, and their price of £122/8/- was accepted on 12 January 1863.[22]

On the same day, James Balfour, writing on behalf of Thomas Stevenson, sent instructions to Chance Brothers, apparently with regard to the glass work for the Otago lights.[23] This indicates that Balfour had a hand in the design and procurement of the lighthouses whose erection he was to supervise. On 23 April 1863 he was offered a two-year contract as provincial marine engineer for Otago Province, which he accepted.[24]

The specification for the lantern[25] described it as a '12 sided polygon ... not less than 9 feet 1 inch in diameter to the inner edge of the sole plate, and 6 feet 2 inches high from the upper surface of the sole plate to the under surface of the cornice plate'. The lantern was to be fitted with '26 frames of triangular glass to have a reserve of 12 panes' and '2 storm panes and spare glass for each'.

A letter from D & T Stevenson dated 9 February 1863[26] informed Milne & Son that the lanterns were to be forwarded on the *Wave Queen* about the middle of March.

There was also a separate specification for the 'Pillar Stand and manner of framing the Apparatus'.[27] The opening paragraph of the specification stated:

> Tairoa's Head Lighthouse is to exhibit a fixed red light. The apparatus is to be Dioptric & of the third order and to light $^{2}/_{5}$ of the circle and the light is to be surrounded at the back by 145° of a glass spherical mirror. The glass-work will be supplied in panels and the Contractor under this Specification will furnish the Pillar Stand and fit up the apparatus thereon complete.

The pillar stand was to be '6 feet 10 inches high over all when finished and the Table 4 feet in diameter'.

Other tenders were let to:

- Lloyd and Summerfield of Birmingham for '6 dozen red lamp glasses to fit the Northern Lights 2nd order pattern gauge' for Taiaroa Head;[28]
- John Ford of Edinburgh for 'Tairoa Head – 3 doz glasses for Liverpool lamp. Leading lights, 2 gross argand lamp glasses';[29]
- Smith & Company, Blair Street, Edinburgh for stores including wicks for the three-wick lamps to be installed at Taiaroa Head.[30]

The lanterns and stores for both lighthouses and the leading lights (see below) were despatched to Dunedin from the Clyde on the *Wave Queen* on 4 April 1863. The lamps and apparatus for Taiaroa Head followed on the *Pania*, * which sailed from the Clyde on 20 January 1864. The last equipment to be sent for Taiaroa Head consisted of boxes containing wire for lightning conductors and bell mountings. They came on the *City of Dunedin*, which sailed from the Clyde on 25 June 1864.

* The Pania was the first ship with a Māori name to carry lighthouse materials. It is significant that this was as early as 1863. Māori names for ships trading between New Zealand and the United Kingdom became more frequent as the century progressed.

In early 1864 the Marine Board called for tenders to be submitted to the Port Chalmers office for the lighthouse towers and dwelling houses to be erected at Dog Island in Foveaux Strait and on Taiaroa Head.[31,32] (The call for tenders also optimistically stated that the plans and specifications for Cape Saunders were in preparation and that it was anticipated that it would soon be proceeded with.) The contract was let to Hugh Calder of Dunedin, who laid the foundation stone at Taiaroa Head on 29 June 1864. The stone was quarried locally, and tower and houses were lined with kauri, a timber native to northern New Zealand. Construction was completed in six months.[33]

The Taiaroa Head light was exhibited on 2 January 1865.[34] The Notice to Mariners, with its alternative spelling, stated that:

> Tairoa's Head Lighthouse is situated on the headland bearing that name, on the east or seaward side of the entrance to Otago Harbour, in 45°47' south latitude and 170°45' east longitude, as measured on the Admiralty charts.
>
> The light will be a Fixed Dioptric Red Light of the 3rd order, and will be seen to seaward from about E.¾N., round by the North to about N.W.; and up the Harbour it will show from about W.½S. to about S.W. ¼W., the latter line indicating the turning point of the Cross Channel.
>
> The light will be elevated about 100 feet above the level of the sea, and will be visible in clear weather about 20 nautical miles, allowing 15 feet for the height of the observers eye, and at lesser distances according to the state of the atmosphere.
>
> The tower, which, together with the dwelling houses, will be painted white, will be 39½ feet high from the ground to the top of the lantern.

Alterations were suggested eight and nine years later. The 1873 Marine Department report suggested that when the proposed light for Cape Saunders was erected (see below), the Taiaroa light could be removed to Tory Channel entrance, and the entrance to Otago Harbour marked by a sixth order white light.[35] The 1874 report suggested that the light be changed to a fifth order white light.[36] However, neither suggestion was actioned, and the next actual change was to upgrade the lamp from colza oil burners to Doty paraffin lamps, in early 1877.[37] These were ordered in 1875, along with Doty lamps for Nugget Point, Farewell Spit and Cape Campbell. Tenders were invited from Milne & Son[38] and James Dove & Co.[39] The tender was given to Milne & Son, who supplied 3 three-wick burners with couplings and a Liverpool lamp with 12 red glasses, 1 gross wicks and mandril.[40] They were shipped to Dunedin on the *Penshaw*.[41]

The Taiaroa Head lighthouse was never converted to incandescent burners. The next change was the result of the wreck of the *Tyrone*.

On 27 September 1913, she hit Wahine Point, two miles from Taiaroa Head. Witnesses at the subsequent Court of Inquiry emphasised that the existing red light was inadequate, as

Figure 3.2: Completed Taiaroa Head lighthouse Hocken (ID 2614 file S14-276a)
© Hocken Collections, Uare Taoka o Hakena, University of Otago

red light had a far less penetrating power in foggy weather than white light. Consequently the court recommended 'that the present red light be replaced by some kind of high powered white light so as to make the light as effective as possible in thick and foggy weather'.

In 1920 it was decided to automate the light with an AGA acetone-acetylene flashing unit[42] which was supplied by the manufacturer's agent in Napier. This new light was exhibited in September 1921.[43] The light was converted to electricity, and a radio beacon installed, in 1955.[44] It was transferred to the then Otago Harbour Board in 1976.

Leading lights

As already mentioned, J T Thomson recommended that the leading lights be 'somewhat better than common lanterns'.

The eventual design was for small lanterns '4 feet 7 inches in daylight height of glass'[45] based on a small tower. These lanterns were made of riveted boiler plate with upright astragals, probably of cast iron.

The glass was to be, as specified for the lighthouses, plate glass with thickness between 1/4 and 5/16 inch thick. The lanterns were topped by small copper ventilators, corresponding to the upper portion of the ventilators designed for the other two lights.

The original design for the apparatus was for an Argand burner with two paraboloidal reflectors and a hemispherical reflector placed on a revolving carriage. The original specification prescribed zinc sheet No 12 Birmingham wire gauge thick, but as shown above the lamps were eventually made with the metal ordered for the Taiaroa Head and Cape Saunders lights.[46] A marginal note on p 462 of the specification states that the apparatus was made with silver plate. The lanterns, apparatus and lamps were all provided by Milne & Son; total cost £228/6/6.[47]

The specification required the lenses 'to be pressed in a mould (which will be lent to the Contractor) and carefully annealed'. These were provided by Sautier et Cie of Paris,[48] at a cost of £12/-/-. Ford provided the lamp glasses (£5/-/-), and Smiths the required stores (£12/18/3).

The apparatus for the leading lights was sent to New Zealand on the *St Kilda*.[49] She was not a commercial steamer. She was the small vessel purchased on behalf of the Otago Provincial Government. After modification, including alterations to the rigging, she sailed to New Zealand. (The Otago Government had initially wished to commission a new vessel from one of the Clyde shipyards. The shipyards however, all had full order books, building American Civil War blockade runners, and it became necessary to purchase an existing steamer and refit her)

Cape Saunders and Nuggets

The lantern and apparatus for Cape Saunders were subsequently used at Nugget Point.

The specification for the Otago lanterns takes up nine pages in Volume 110 of the Stevenson Archive in the National Library of Scotland.[50] Most of the detail required for both lanterns in included in the specification for the Cape Saunders lantern, with only one page being given for the Taiaroa Head lantern.

The 'first class Lantern' was to be 10 feet 3 inches [3.12 m] from 'the under bed of the sole plate to the upper edge of the cornice plate'. The eight-section sole plate was to form a 'polygon 12 feet 3¼ inches [3.74 m] in its least internal diameter'. The specification called for 63 panes of plate glass and a reserve of 18 panes 'of not less than ¼ inch in thickness nor more than 5/16 inch'. The lantern had a double dome, with a copper outer over a cast iron inner dome.

The apparatus specification stated that:

> The 1st Order apparatus for Cape Saunders is to illuminate ⅝ of the circle and is to consist of 5 panels of Upper prisms of 13 prisms each, 5 segments of Cylindric belt, and 5 panels of lower prisms with 6 prisms in each.[51]

It had a correspondingly sized spherical mirror.

The specification for the 'Pillar Stand and manner of framing' expanded the description with

> Cape Saunders Lighthouse is to exhibit a fixed light of the first order. The apparatus is to be Dioptric and to light ⅝ths of the circle; the remaining ⅜ths of the circle are to be enclosed by a glass spherical mirror.

The glass was to be set in a framework of gunmetal. The pillar stand was to be 9 feet high overall.[52]

When the lantern and apparatus arrived safely they were placed in store and the 1865 Marine Department report pointed out that

> The optical apparatus imported by the Provincial Government of Otago for placing on this headland has not been brought into use, the money voted by the Assembly for lighthouse purposes being insufficient to defray the cost of more works than the Board has already undertaken.[53]

The following year an alternative use was suggested:

> Further information obtained from shipmasters has caused the Board to think that the 'Nuggets', south of Cape Saunders, would, for the purpose of navigation be a far better position, both for directing vessels on their course, going either north or south, as also pointing out a safe place of anchoring during the height of south-west gales which prevail with great violence on that part of the coast.[54]

Nugget Point overlooks Molyneaux Bay, where the Clutha River (formerly the Molyneaux River) enters the Pacific Ocean.

The following year, James MacAndrew, the new superintendent for Otago, wrote to the colonial secretary, asking

> [i]f the Provincial Government provides necessary lighting apparatus would His Excellency's government provide buildings for Cape Saunders & Nugget Point.

Further the province suggested that timber buildings would be appropriate and offered to 'complete whole thing and hand lights over in full working order'.[55] However, after going through several Marine Acts, the General Government was not going to hand lighthouse construction back to the provinces.

In his report for the year 1867,[56] Balfour reported that an approximate estimate had been made for construction of the lighthouse at Nuggets and suggested that all the

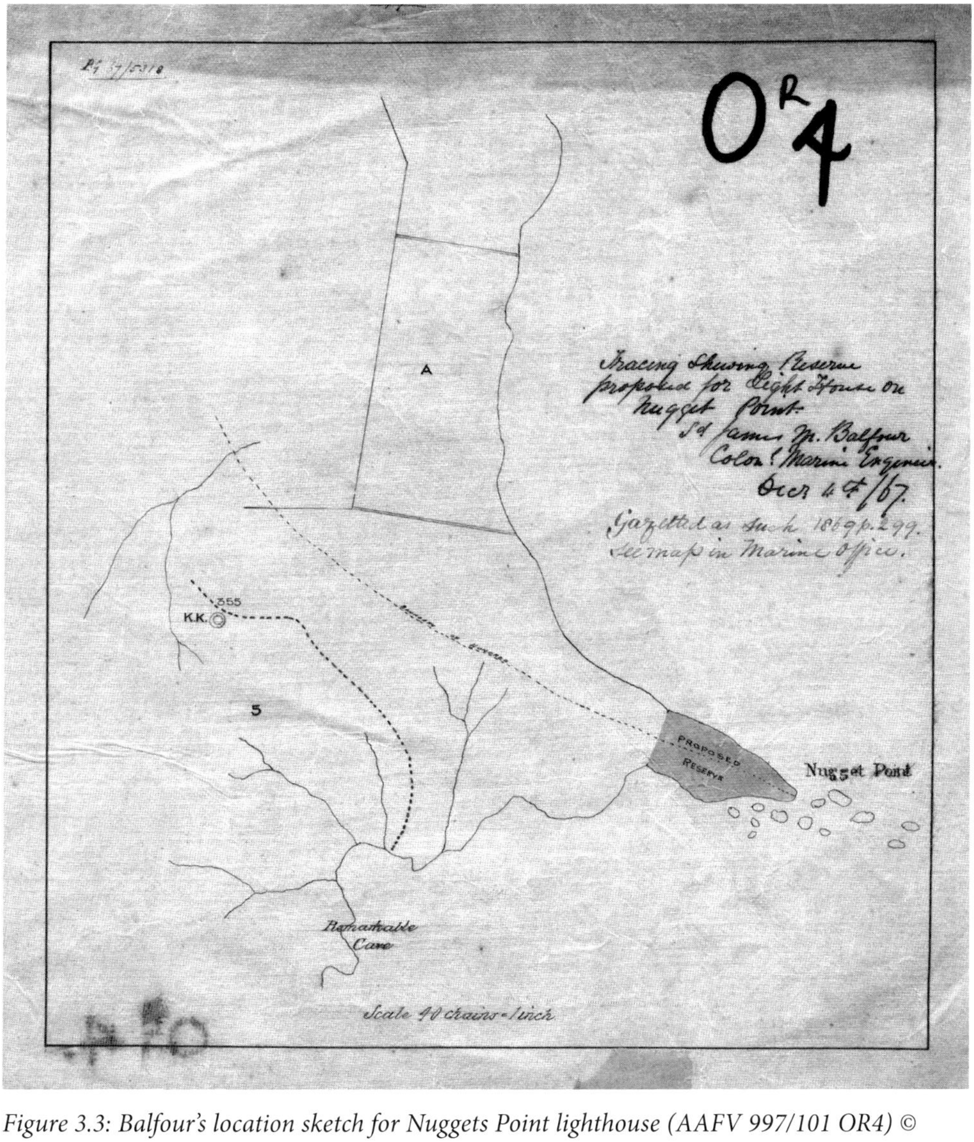

Figure 3.3: Balfour's location sketch for Nuggets Point lighthouse (AAFV 997/101 OR4) © Archives New Zealand, The Department of Internal Affairs Te Tari Taiwhenua

Figure 3.4 The Nuggets (ATL WA-06926-F)
© Whites Avaiation Collection, Alexander Turnbull Library, Wellington, New Zealand

lighthouses being designed or under construction should have wooden towers. (See Chapter 4 – Farewell Spit.)

However, the final design for Nugget Point was a stone tower, 31 feet high from ground to top of lantern and painted white. Lightkeepers later complained about the cold stone floor.

The tenders for the building were announced on 11 March 1869.[57] The accepted tender was that of Wm. P. Pearce of Dunedin for £2577/5/9. This tender included construction of a dray road to the lighthouse. Further details of the lighthouse were given the following month.[58]

While local contractors were competent to build the tower, houses etc, there was concern about the erection of the lantern and lighting apparatus, particularly as Balfour had been drowned in a surf boat accident at Timaru on18 December 1869. The erection of the lantern was supervised by Andrew Carnachan, who stayed on at the lighthouse as the first principal keeper. He had been a keeper with the Northern Lighthouse Board and was Principal Keeper Carswell Light when D & T Stevenson were asked to find a suitable candidate. The completed light was exhibited on 4 July 1870.[59] The Notice to Mariners described it as

> exhibited from a lighthouse erected on the outer extremity of Nugget Point, Lat 46° 27‘ S. and Long 160° 51‘ E. It is a fixed White Light of the first

Figure 3.5 Nugget Point Lighthouse (ATL 1/2-112291-F)
© R Hart Collection, Alexander Turnbull Library, Wellington, New Zealand

> order Dioptric, showing all around seaward between the bearings of N.E. by N¼.N. and South by East. It is elevated 250 feet above sea level, and may be seen in clear weather at a distance of 23 nautic miles, allowing 15 feet for the height of the observer's eye.
>
> The Tower is of stone painted white, and is 31 feet in height from base to top of lantern.

The first upgrade was in 1875 when the colza burners were replaced by three first-order (i.e. four-wick) Doty paraffin burners with couplings along with a Liverpool lamp, 12 glasses, 1 gross of wicks and mandril for the Liverpool lamp.[60] These were supplied by Milne & Son, who had supplied the original burners. This is probably why letters from D & T Stevenson referred to 'First order burners for Cape Saunders',[61] and later requested Milne & Son to forward the package containing burners for Cape Saunders etc to London to the order of Messrs Smith Morrison & Co, 117 Leadenhall Street for shipment per *Penshaw*, East India Docks.[62] The Doty lamps for Taiaroa Head formed part of the same shipment.

While the 1877 Marine Department report indicated that the burners had been installed,[63] the following year saw a requisition (CS 31) sent to D & T Stevenson to procure fountains and pillars to change the light from mechanical to fountain. These were also supplied by Milne & Son, and despatched to New Zealand in June 1878 on the *St Leonards* from East India Dock.[64] (The same shipment included the lantern for Hokitika, spelled in

Scottish documents as 'Ho Kitika'.) The 1879 Marine Department report indicated that the fountains had been installed.[65]

The 1909 annual report indicated that incandescent lamps were in the process of being installed at Nuggets. Their successful installation was reported in the 1910 report.

In the same year the cowl and ventilating inlet were upgraded, as the incandescent lamps produced more heat than the older Doty paraffin burners. The tender for supply was won by Messrs Jenkins and Mack.[66] It was no longer necessary to send to Scotland for these parts although the design was Scottish.

The light was converted to electricity in 1949 with the installation of diesel generator sets. In 1951 the light was connected to the national grid, and automated in 1989.

Dog Island lighthouse

Sailing ships coming to New Zealand frequently came around the south of Australia to pick up the Roaring Forties, and consequently made their first landfall at Bluff or Dunedin. Later steamers travelled the same route due to the commercial importance of Otago Province, beginning with the Otago gold rush. Ships were thus required to navigate the Foveaux Strait between the South Island (called at this time 'Middle Island') and Stewart Island

In 1861, Southland Province was created out of Otago Province, and one of the first actions of the new Provincial Council on was 16 August 1861 to pass a resolution:

> That a respectful Address be presented to His Honor the Superintendent, requesting him after he can procure the necessary information as to the best site for a Coast Light to impress on the General Government the necessity of erecting a Light-house on Dog Island, or some other convenient place in Foveaux Straits.[67]

J.A.R Menzies, the superintendent, canvassed local opinion as to a suitable site, and among those recommended were Centre Island and Dog Island.[68] (A lighthouse was built on Centre Island 18 years later.) The following year a memorial, dated 5 August 1862, from masters of steamers engaged in intercolonial and interprovincial trade and merchant and shipping agents in Dunedin, requested the erection of lighthouse on Dog Island, and asked the superintendent to use his influence with the General Government.[69]

Following a question raised in the House by one of the Southland MPs, the colonial secretary wrote to the Superintendent of Otago Province in September 1862 suggesting that if the two superintendents should agree on immediate erection and temporarily advance construction cost out of provincial funds, 'His Excellency's Government would propose repayment in the next session of the House of Representatives'. Before final arrangements could be made the Otago superintendent was asked to communicate full particulars of site, estimated cost, plans, and specifications for the proposed lighthouse.[70]

In the end the choice of site was left to the Superintendent of Southland, who was urged to arrange for surveys so that the General Government could be informed and building commenced, so as to be ready for the ordered lantern when it came.[71] In May 1863 a party visited Dog Island and assessed possible landing places and water supply, and the quality of the underlying rock. Consequently T Heale, the Provincial Surveyor for Southland, submitted a suggested design as an alternative to Thomson's (see below).[72]

Heale suggested a masonry tower 150 feet above sea level 22 feet 6 inches in diameter at the base and 15 feet 6 inches under the lantern. The lantern was to contain a first-order catadioptric light, the light to be visible at a distance of 19 miles and through at least 210° of the circle and preferable about 300°. This would fulfil two roles:

1. Sea light to guide vessels into the fairway of the Foveaux Strait
2. Leading light for the entrance to Bluff Harbour, which was now becoming an important port of call and was the only port of refuge for vessels 'caught in the Foveaux Strait by the furious gales so common in there'.

However, by this time the order had already been placed. It would seem that in typical Otago fashion, the lighthouse had already had some initial design work done by the Otago provincial authorities, and the lantern and apparatus ordered. In the letter of 10 March 1862, it was announced that 'the necessary light had already been ordered from England'[73] i.e. the order had been sent to the Agent General in London, who sent it straight on to D & T Stevenson. The Colonial Secretary could only acquiesce with a good grace and so wrote to the Otago superintendent that 'you were quite right to lose no time in ordering the revolving lights through Mr Stevenson C.E.'.[74]

In contrast to the cutting edge technology of Taiaroa Head and Cape Saunders lights, that for Dog Island was slightly older. The apparatus consisted of 16 small Argand lamps with individual parabolic mirrors and small holophotal lenses in front.

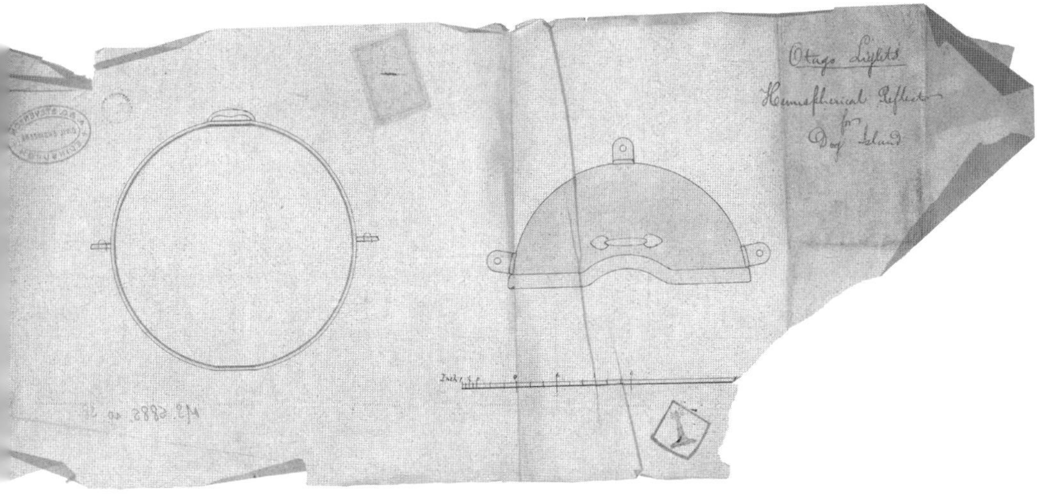

Figure 3.6: Holophotal lamps for Dog Island (MS5885) © National Library of Scotland

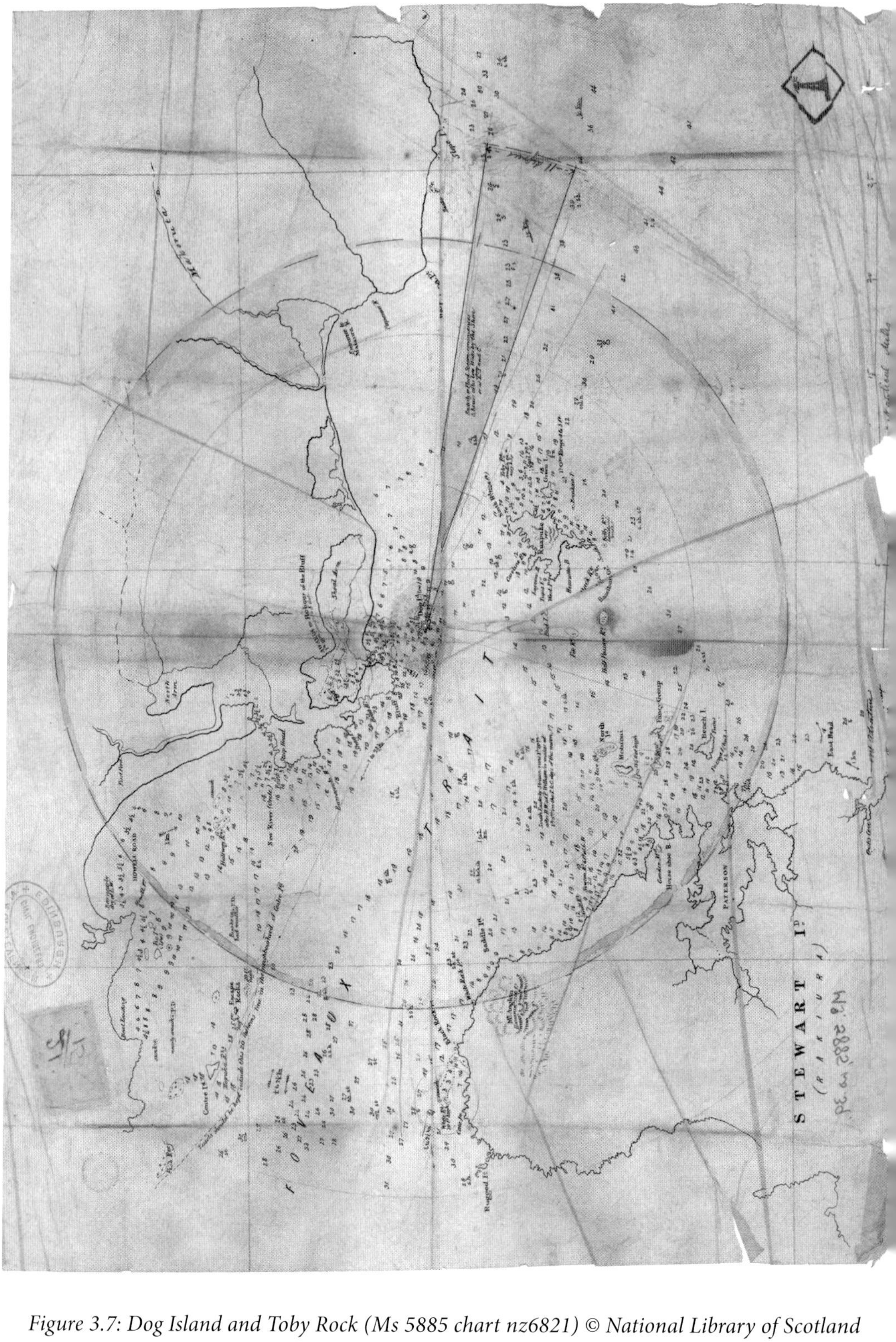

Figure 3.7: Dog Island and Toby Rock (Ms 5885 chart nz6821) © National Library of Scotland

The small lamps were supported on a four-faced frame or chandelier, with four lamps on each face. The chandelier was rotated by a clockwork mechanism. A similar chandelier system had been installed in the Bell Rock lighthouse 50 years previously, and this type of lamp was supplied to Newfoundland for the first time in 1858.[75] Balfour justified this design in his 1865 report.[76]

> This arrangement, designed by Messrs Stephenson [*sic*] C.E., of Edinburgh, is peculiarly applicable to inaccsssible [*sic*] localities like Dog Island, from the simplicity and durability of the argand lamps, and from the ease with which they can be repaired as well as from the fact that even should one or more of the lamps be temporarily disabled the light, though weakened, will not be totally extinguished.

Dog Island was the only New Zealand lighthouse to have this arrangement.

The initial design for the tower was done by J T Thomson, as the provincial engineer for Southland had no experience in lighthouse construction.[77] (As seen above, this did not stop the provincial surveyor from putting forward a suggested design the following year.) Thomson's proposed design was for a tower of 160 ft elevation, capable of being seen from a ship's deck at a distance of 20 miles. The lantern was to be 21 feet in diameter, revolving every half-minute and seen in all directions except 11 degrees, towards eastern channel which marked out Toby Rock. It was described in a letter to Major Richardson, the superintendent of Southland, with the request that it be forwarded to Mr Stevenson for his comments.[78]

Among the information sent to Edinburgh was a tracing of Foveaux Strait showing depths and the position of Dog Island and the proposed 11 degree arc to mark the area between Toby Rock to the south, and the rocks off Slope and Waipapa Points to the north.

Stevensons replied in a letter to the Otago provincial agent dated 5 January 1863 that they had no difficulty in understanding Thomson's questions and would proceed to procure the lantern and light apparatus. However, they also wrote:

> There is only one point in Mr Thomson's letter on which we have any difficulty and that is the suggestion to mask the revolving light over the small arc of 11° in the direction of Toby Rock to the South, and the Rocks off Slope and Waipapa Points to the North.
>
> From experiments we have recently made on cutting off revolving lights so as to mark distant dangers we are of the opinion that it would not be possible to cut off the revolving light in the manner proposed in a satisfactory way.
>
> We can send out a fixed light apparatus, to be exhibited in the required direction showing over an arc of 11° but in order to be visible at the requisite distance it would require to be a very powerful light and could not be exhibited from a 12 foot lantern, along with the revolving lights but would require to be shown from another lantern in the same tower at about 30 feet lower level.

> Even if this arrangement were carried out, we have our doubts whether owing to the distance to the dangers it would be found to work well. Toby Rock lies about 11 nautic miles and Slope Point about 25 nautic miles from Dog Island, both according to our experience too far off to be effectively guarded by a cut off, and we should therefore not be prepared to recommend its adoption.[79]

Hence, even though the Dog Island lighthouse used older technology, New Zealand was still benefiting from the latest Scottish research. The plan to mask Toby Rock was abandoned. Heale had also commented on the dangers provided by the rocks of the Otara Reef off Waipapa Point, but nothing was done until 1881 when the *Tararua* was wrecked with substantial loss of life. Consequently a lighthouse was placed at Waipapa Point and first lit on 1 January 1884.

In January 1863, D & T Stevenson approached Milne & Son to supply a lantern for Dog Island. They were asked to supply a lantern identical to the one for Cape Saunders and at the same price of £660.[80] Milne & Son's offer was formally accepted four days later.

The main specification described the light as

> a revolving light of the natural colour and is to attain its greatest brilliance every half minute. The apparatus is to consist of 16 glass holophotes arranged on a frame with four faces with silver plated hemispherical mirrors, lamps etc to be driven by a first class engine in a cast iron case which is to carry a first class revolving frame of the usual construction.[81]

This contract was to be for manufacture of machine and case, revolving frame, holophote frame, top steadying frame and spherical mirrors. The lamps and carriages were a separate contract and would be supplied to the machine contractor for assembling the apparatus. This contract was also let to Milne & Son for the sum of £908/1/6.[82] Additional instructions for these appear to have been supplied by James Balfour[83] although the document in the Stevenson Archives is not legible. There was a slight problem later when Milne & Son were found to have manufactured the revolving carriage with only 11 instead of 12 rollers and were requested firmly to 'make the apparatus in this respect according to the drawing and specification'.[84]

Smiths supplied lamps and stores for sum of £128/19/1,[85]James Dove & Co supplied 18 lamp carriages for £2/8/- each,[86] and Ford supplied lamp glasses.[87]

Chance Brothers and the French firm Sautier et Cie were invited to tender to supply 17 sixth order holophotes [150 mm focal distance].[88] The contract was let to Sautier et Cie for £34 each.[89] Although Sautier et Cie provided the small lenses for the leading lights into Otago Harbour, Dog Island was the first major New Zealand lighthouse to have lenses of French glass. From then on French manufacturers, particularly Barbier et Fenestre and its successor firms, figured prominently in the supply of lighthouse lenses to New Zealand.

The lantern was shipped from the Clyde on the *City of Dunedin* in May 1863,[90] and the remainder of the apparatus shipped on the *Beautiful Star* which left the Clyde in April 1864.[91] D & T Stevenson supplied two plans and the specification of the Dog Island lighthouse apparatus 'for the guidance of the workmen in fitting it up in the colony … and should any difficulty occur Mr Balfour will be able to give the necessary directions to the workmen'.[92] The copy received by the Otago Provincial Government included a margin note 'plan and instructions rec'd {drawings, instructions} referred to J M Balfour 26.6.62(?)'[93]

The specification for the contractor for the lighthouse and dwellings was signed by Balfour, Marine Engineer, Otago Province,[94] as was the surviving drawing. The tenders were called for in the *Otago Provincial Gazette* of February 1864,[95] and the specification includes a note, 'Contract signed George Garside 20 July 1864'.

Garside appears to have been in partnership with Walters, and the contract was not very successful. Three years later the Supreme Court, Otago and Southland District, was dealing with claims on the estates of both men and the registrar was concerning himself with 'the partnership debts and liabilities in respect of the Dog Island Contract'.[96] The fitting of the lantern and optical apparatus was superintended by Rolton.[97]

In his 1865 report, Balfour pointed out that as Dog Island was very low,

> [i]t was necessary to make the tower 100 feet high in order to give the light sufficient range and this, combined with the unworkable nature of the stone on the Island, and the great depth of foundation required for the dwellings, has made the undertaking comparatively a costly one but I have notwithstanding, been able to keep the cost within the average price of similar works in equally inaccessible situations at home (i.e. in Scotland). The tower is 21½ feet diameter at the base, 16½ feet diameter below the balcony, 100 feet high to the lantern, and 118 feet high over all. It is plain to a degree, economy having been looked upon as a paramount consideration, but its simple massiveness renders it a not unpleasing object. The external walls of the dwellings are also of stone, the interior partitions being of wood, and the whole has been satisfactorily completed by the contractor, Mr Garside, in the face of very considerable difficulties.

The 8 December 1864 Notice to Mariners described the lighthouse as follows:

> Dog Island Lighthouse is situated on a small island of that name in Foveaux Straits off the entrance to the Bluff Harbour, in 46°39'35" South latitude, and 169°26' East longitude, as measured on the Admiralty Chart.
>
> The light will show all around, and will be a Holophotal Dioptric Revolving White Light, of the first order, attaining its greatest brilliancy every half minute.

> It will be elevated about 150 feet above the level of the sea, and will be visible in clear weather about eighteen nautical miles, allowing fifteen feet for the height of the observer's eye, and at lesser distances according to the state of the atmosphere.
>
> The Tower is built of native stone of a grey colour, and will be 118 feet in height from the ground to the top of the lantern.
>
> The light is expected to be ready for permanent exhibition not later than April 1865.

The extent of Mr Garside's engineering difficulties can be seen in that light was not exhibited until 1 August 1865.[98]

Unfortunately there were to be continuing difficulties with the tower. Two years later, Balfour reported that the tower had been found to oscillate more than expected and had developed leaks which threatened the cohesion of the mortar. The tower was repaired by ripping and pointing with Portland cement, and in addition the middle third was repainted white and the upper and lower thirds black.[99]

By 1871 a crack had developed in the tower; its length was approximately 6 metres, beginning at the base.[100] Blackett's solution was a framework of durable hardwood encircled by wrought-iron bands.[101]

By 1917 the tower was again becoming unsafe due to perishing mortar, and the initial decision was made to build a new tower of reinforced concrete alongside the existing one.[102] The final decision was to encase the existing tower in a carapace of reinforced concrete. This was completed two years later[103] and it is still standing.

In 1871 18 new burners were ordered and manufactured by Milne & Son,[104] and in 1876 Doty burners were ordered as part of a major upgrade of New Zealand's lighthouses (see Chapter 7). A further 24 were ordered from Milne & Son in 1883.[105] In 1901 the principal keeper at Dog Island requested that a new set of 16 burners be ordered, pointing out that the current ones had been purchased from J Milne & Son of *Glasgow* [italics author's] on 4 May 1883.[106] Subsequently the Agent General was instructed to 'Procure through Messrs D & C Stevenson, one set of burners (16) for Dog Island Lighthouse. Forward to Secretary Marine Department Wellington.'[107] These were also supplied by Milne & Son at a tender cost of £17/-/-[108] and shipped on the *Wakanui* from the Royal Albert Dock. A final set of burners was ordered in December 1923. In this case D & C Stevenson recommended that the tender be given to R Laidlaw & Son (Edinburgh) Ltd.[109] These were despatched from Edinburgh on 4 January 1924 for shipment by *Remuera* from the Royal Albert Docks.[110] However it seems that the box was shipped on the *Calm.*[111]

Lamp glasses were ordered in 1877,[112] apparently from Smiths, and again in 1879. William Seed's order in 1883 was particularly detailed:

> To avoid confusion, 3 gross argand glasses for paraffin burners Dog Island supplied though you in 1877 for which you obtained 3 gross glasses from Messrs Smiths on 28/11/77. Put in 3 boxes, 1 gross each.[113]

Figure 3.8: Dog Island lighthouse with stiffening timbers (PH-ALB-98 p72n3c)
© Auckland War Memorial Museum – Tāmaki Paenga Hira

The last set of lamp glasses was supplied in December 1920, in this case by Butlers of Birmingham.[114] D & C Stevenson indicated that they had inspected them and found them satisfactory, but added that 'glass made in this country (UK) was not up to the standard of those made in Germany & Austria'.

In 1885 the lens of one of the holophotes fell out and apparently was broken. Stevenson approached Sautier, Lemmier et Cie for a replacement.[115] Although the letter asked if the French glassmakers had sufficient information to provide a replacement lens, the underlying assumption was that they would be able to reproduce a lens identical to that provided 22 years previously. Here the Scottish connection turned what might have been a complicated procurement into a straightforward one.

By 1923 the system had become antiquated. It had given 58 years' service. W A Fraser, the lighthouse artificer, pointed out that the cowl needed repairs and ventilation was inadequate. He suggested replacing the whole system.[116] Fukert, the then marine engineer, suggested replacing the apparatus with that which would be removed from the Cape Foulwind lighthouse when the latter was upgraded.[117]

However, the Marine Department still had in store the apparatus delivered in 1914 for the proposed North Cape lighthouse.[118] This had an incandescent burner, and the lens revolved in a mercury bath. (See Chapter 11.) It was decided to install this at Dog Island. Rebuilding at the lighthouse took about two months, beginning 1 October 1925. The lightroom floor had to be strengthened and a new hole cut for the driving weight.[119] The cowl was repaired and other maintenance done including repainting the white tower bands and erecting a new house for the principal keeper.[120]

The new light was exhibited on 2 December 1925 with a white group flashing characteristic – three flashes in quick succession every 30 seconds.[121]

This apparatus remained in place until the light was electrified in 1954.

The Otago lights marked the start of the Scottish contribution to New Zealand's lighthouses and set out the pattern which was, in general, to be followed for successive lighthouses. The decision to build was made in New Zealand, and the specification, usually accompanied by an official requisition, was sent to New Zealand's representative in London who sent it directly to D & T Stevenson (or D & C Stevenson) in Edinburgh. The procurement was usually divided into at least four separate contracts, one each for the lantern, the glass lenses, the apparatus (frequently rotating) which often included the lamps, although sometimes these were contracted for separately as was done for Farewell Spit and Cape Campbell lighthouses (Chapter 4). The fourth contract was for stores. The tower was always designed in New Zealand, thus avoiding a repetition of the Mana Island problem.

Figure 3.9: North Cape lens installed in Dog Island lighthouse © Voyager Maritime Museum, Auckland

Endnotes

1 The Free Church Project' from *An Encyclopaedia of New Zealand*, edited A H McLintock, originally published in 1966, downloaded from http://www.tera.govt.nz
2 NZG Archives, AAAC/707/D500/126a/76Letter C Logie Harbour Master to Superintendent 9 June 1854,
3 NZG Archives, AAAC/707/D500/127a/156Letter C Logie Harbour Master to Superintendent 21 Nov 1855
4 NZ Archives, AAAC/707/D500/135a/594 Letter M Thomson Harbour Master Port Chalmers to Superintendent 17 Sept 1861
5 NZ Archives, AAAC/706/D500/123d/30 Letter Colonial Secretary's Office, Auckland to Superintendent Otago 2 May 1856
6 NZG Archives, AEBE 18507 LE/1/34 1862/114-166 Letter W.Thomson and J T Thomson to Superintendent Otago Province, 17 June 1862
7 Acc 10706 Vol 30, p 957
8 NZG Archives, AEBE 18507 LE/1/34 1862/114-166 Letter J T Thomson to Superintendent Otago Province 18 June 1862
9 NZG Archives, AEBE 18507 LE/1/341862/114-166 Letter Superintendent Otago Province to Colonial Secretary, Auckland, 22 July 1862
10 NZG Archives, AAAC/701/D500/9f 65 Message No 11, Superintendent to Provincial Council, 6 May 1862 (Also Proceedings Otago Province?)
11 Acc 10706 vol 31, pp 366–368 letter to Agents Provincial Government of Otago
12 NNZG Archives LF/1/34/1862/114-66
13 Acc 10706 vol 31 p 389, letter D&T Stevenson to A Cunningham, Sec NLB 2 October 1862
14 Acc 10706 vol 31 p 396, letter D&T Stevenson to Chance & Co. 6 October 1862
15 Acc 10706 vol 31 p 399, letter D&T Stevenson to Milne & Son & Son 9 October 1862
16 Acc 10706 vol 31, p 405. letter Stevenson to Agents Provincial Government of Otago. 11 October 1862
17 NZG ArchivesAAAC D500 709 Box 340a 52 Letter Crawford & Auld to Otago Superintendent 18 Oct 1862
18 Chance Archives BS6/12/1/1/1 pp 167, 168, 172 order for Cape Saunders and Taiaroa Head
19 AJHR 1865 D-No 1.C Report of the Marine Board Department of New Zealand, report of Mr Balfour

20 Chance Archives BS6/12/1/1/1 p 173 Spherical Dioptric Mirrors

21 Acc 10706 vol 31, p 554 letter to Milne & Son re Otago Lights 12 December 1862

22 Acc 10706 vol 31, p 598 letter to Milne & Son re Otago Lights 12 January 1863

23 Acc 10706 vol 31, p 596 letter to Chance signed Balfour as 'Mr Stevenson left for south this morning' 12 January 1863

24 Acc 10706 vol 31 p 824, letter to J M Balfour dated 23 April 1863

25 Acc 10706 vol 110 p 364 part *Specification of Lanterns for Lighthouses to be erected by the Provincial Government of Otago New Zealand* dated 13 Oct 1862

26 Acc 10706 vol 31 p 677, letter to Milne & Son re Otago Lights 9 February 1863

27 Acc 10706 vol 110 p 427 *Otago Lights, Specification for Pillar Stand etc.*

28 Acc 10706 vol 31, p 612 letter to Lloyd & Summerfield, Birmingham headed Otago Lights

29 Acc 10706 vol 31 letter to John Ford glassmaker

30 Acc 10706 vol 31, letter to Smith & Co, Blair Street, re Dog Island and Otago lights

31 NZG Archives AAKJ D365 20d SP12/18 Marine Board Office, Port Chalmers, Letter to Provincial Sec and Southland

32 *NZG Gazette*, 1864 TO CONTRACTORS p 119, 8 March 1864

33 www.newzealandlighthouses.com/taiaroa_head.htm

34 *NZG Gazette* 1864, No 47 Notice To Mariners p 457 8 December 1864

35 AJHR 1873 H-19Eighthth Report of the Marine Departmen for the year ended 30th June 1873 Report by R Johnson and R A Edwin on Lighthouses on the Coasts of New Zealand, Enclosure in No 2

36 AJHR 1874 H-23 Nineth Report of the Marine Department for the year ended 30th June 1874 Correspondence Relative to Selection of Site for Projected Coastal Lighthouses No 2 Report Captain Johnson to Commissioner Customs

37 AJHR 1877) H-29 Twelfth Report of the Marine Department for year ended 30 June 1877

38 Acc 10706 vol 45 p 736 letter to Milne & Son & Son re New Zealand Lighthouses 5 July 1875

39 Acc 10706 vol 45 p 740 letter to James Dove & Co & Co re New Zealand Lighthouses 5 July 1875

40 Acc 10706 vol 45 p 849 letter to Milne & Son re New Zealand Lighthouses 4 August 1875

41 Acc 10706 vol 46 p 293, letter to Milne & Son.re New Zealand Lighthouses 17 December 1875

42 AJHR 1919, H-15 Marine Department Annual Report for 1918–1919

43 NZG Gazette, 1921, Notice to Mariners No 49

44 AJHR 1956 H-15 Marine Department Report for year ended 31 March 1956

45 Acc 10706 vol 110 p 365

46 Acc 10706 vol 110 p 462-464 Specification of Apparatus for Harbour Leading Lights for the Provincial Government of Otago

47 Acc 10706 vol 38 p 733 Letter Stevenson to Auld, 15 April 1864

48 Acc 10706 vol 31 p 814, letter to Sautier & Cie re Otago Lights 17 April 1863

49 Acc 10706 vol 32 p 102 Letter D&T Stevenson to the Agents of the Provincial Government of Otago, 15 July 1865

50 Acc 10706 vol 110 pp 359–367

51 Acc 10706 vol 110 p 343 Specification of Lighthouse Apparatus to be constructed for the Provincial Government of Otago

52 Acc 10706 vol 110 pp 397–400 Specification of the Pillar Stand and manner of framing the apparatus for Cape Saunders Lighthouse, New Zealand

53 AJHR 1865 D No 1 C (Appendix) No 2 Report on the Marine Board Department of New Zealand 1864–5. Report of President of Marine Board (Charles Sharp) Wellington

54 AJHR 1866 E – No 3 2nd report of the Marine Board Department 1865–66 Enclosure No 1 signed Sharp

55 NZG Archives AAAC/7/D500/27d/34 Copy letter MacAndrew, Superintendent Otago to Colonial Sec, Wellington, 12 April 1867

56 AJHR 1867 E – No 6.3rd Report on the Marine Board Department for the year 1867

57 *NZG Gazette* 1869 List of Tenders p 152 11 March 1869

58 *NZG Gazette* 1869 Notice to Mariners p 201 22 April 1869

59 N*NZG Gazette* 1870, Notice to Mariner No 15 of 1870 p 282 22 June 1870

60 Acc 10706 vol 45 p 849 letter to Milne & Son & Son re New Zealand Lighthouses 4 August 1875

61 Acc 10706 vol 45 p 736 letter to Milne & Son & Son re New Zealand Lighthouses 5 July 1875

62 Acc 10706 vol 46 p 293, letter to Milne & Son & Son.re New Zealand Lighthouses 17 December 1875
63 AJHR 1877 H-29 Twelfth Report of the Marine Department for year ended 30 June 1877
64 Ac 10706 vol 49, p 498 letter Agent General London re New Zealand Lighthouses 11 June 1878
65 AJHR 1879 H-10 Annual Report of the Marine Department 1879
66 *NZG Gazette* 1910, No 59, Tenderd p 1754 13 June 1910
67 AJHR 1862, D-15 Papers Relative to Lighthouses in Foveaux Straits
68 AJHR 1862, D-15 Papers Relative to Lighthouses in Foveaux Straits
69 NZG Archives AAAC 708 D500 144c 839 Dog Island Correspondence
70 NZG Archives ADOE 16612 M1/738 8/27/18 Copy(160) letter Colonial Secretary Wgtn to Superintendent Otago dated 13 Sept 1862 signed Domett
71 NZG Archives AAKJ D365 20d SP12/18 letter Richardson Superintendent Otago, to Menzies Superintendent Southland dated 15 Oct 1862
72 NZG Archives AAAC 708 D500 144c 839 Dog Island Correspondence Report Theoph Heales, Chief Surveyor, Southland Province dated June 1863
73 NZG Archives letter Superintendent Otago to Superintendent Southland 10 March 1862.AAKJ D365 20a SP12/18
74 NZG Archives AAAC 708 D500 144c 839 Dog Island Correspondence Letter Col Sec Office, Auckland dated 17 October 1862
75 Morrison-Low A.D. *Northern Lights,* National Museums Scotland, Edinburgh 2010 pp 188–189
76 AJHR 1865 D-No 1C. Report of the Marine Board Department: Report of Mr Balfour dated 18 July 1865
77 NZG Archives AAKJ D365 20d SP12/18 Letter Campbell, clerk General Assembly to Superintendent Southland dated 6 Nov 1862
78 NZG Archives 16618 M7/3/3 Report (copy?) I T Thomson to Superintendent re Dog Island 17 Sept
79 Acc 10706 vol 31 p 602, letter to Agents Otago Govt dated 5 Jan 1863, also NZG Archives AAAC/709/D500/3409a 66
80 Acc 10706 vol 31, p 600 letter to Milne & Son 3 Jan 1863
81 Acc 10706 vol 110 p 391 Specification of Holophote Framing, Revolving Carriage, Machine etc for Dog Island Lighthouse – New Zealand. 21 January 1863
82 Acc 10706 vol 31 p 656, letter to Milne & Son& Son.re Otago Lights 30 Jan 1863
83 Acc 10706 vol 31 p 775 Letter to Blackie, Milne & Son & Son, dated 31 March 1862, but probably 1863
84 Acc 10706 vol 32 p 502, letter to Milne & Son & Son re Dog Island Lighthouses 11 Jan 1864
85 Acc 10706 vol 31 p 657, letter to Smith & Co.re Otago Lights 30 Jan 1863
86 Acc 10706 vol 31 p 655, letter to James Dove & Co re Otago Lights 30 Jan 1863
87 Acc 10706 vol 31 p 655, letter to John Ford.re Otago Lights 30 Jan 1863
88 Acc 10706 vol 31 p 617, circulars to Chance and Sautier & Cie, Paris 19 Jan 1863
89 Acc 10706 vol 31 p 654, letter to Sautier & Cie. re Otago Lights 30 Jan 1863
90 NZG Archives AAAC/709/D500/340a/88 Letter Auld to Otago Superintendent stamped 24 July 24 1863 on enclosed Bill of Lading
91 NZG Archives AAAC/709/D500/340b/129 Copy letter Auld to Otago Superintendent stamped 20 June 1864
92 Acc 10706 vol 32 p 744, letter to Auld re Otago Lighthouses signed Dick
93 NZG Archives AAAC/709/D500/340b/130 Copy letter, Auld to Otago Superintendent stamped 20 Jun 1864
94 NZG Archives Documents inADOE 16612 M1/738 8/27/18
95 *Otago Provincial Gazette* IssueNo 291 24 February 1864 p 64
96 *Otago Provincial Gazette* Issue No 4646 February 1867 p 45
97 AJHR 1865 D No 1 C (Appendix) No 2 Report on the Marine Board Department of New Zealand 1864–5
98 *NZG Gazette* 1864, No 47 Notice to Mariners p 457 8 December 1864
99 AJHR 1867 E – No 36 3rd Report of the Marine Board Department for year 1867
100 NNZG Archives ADOE 16612 M1/738 8/27/18 Letter William Seed – Secretary of Customs Wellington
101 AJHR 1871 G – No 6 6th) Report of the Marine Department for Year 1869–70 and 1870–1
102 AJHR 1917, H-15 Marine Department Annual Report for 1916–17
103 AJHR 1919, H-15 Marine Department Annual Report 1918–19
104 Acc 10706 vol 39 p 812, letter to Milne & Son & Son re Dog Island dated 1 Mar 187
105 Acc 10706 vol 85, letter Agent General London to D&T Stevenson re Marine 210/35

106 NZG Archives ADOE 16612 M1/738 8/27/18 Letter Nicholson PK Dog Island to Sec MD 25/07/1901

107 NZG Archives ADOE 16612 M1/738 8/27/18 Memo to AG from Premier (R J Seddon)

108 NZG Archives ADOE 16612 M1/738 8/27/18 Memo AG to Premier (R J Seddon)

109 NZG Archives ADOE 16612 M1/738 8/27/18 Memo C6925 – B172/93 High Commissioner to acting dated 19 December 1923

110 NZG Archives ADOE 16612 M1/738 8/27/18 2 copies of invoices from Laidlaw (Simon Sq works) and Shipping specification dated 4 January 1924

111 NZG Archives ADOE 16612 M1/738 8/27/18 Letter Secretary Marine Department (Park) to Collector Customs.

112 Acc 10706 Vol 85 letter Agent General (24/10) incl requisition from Secretary of Customs 1877 also letter Stevenson to Agent General 24 November 1877

113 Acc 10706 vol 85, letter Wm Seed, to D & C Stevenson 327/36 dated 17 May 1883

114 NZG Archives ADOE 16612 M1/715 8/10/11 Memo High Commissioner to Prime Minister 8492 16 Dec 1920

115 Acc 10706 vol 56 p 143 letter to Sautier, Lemmier et Cie, Paris dated 3 September 1885

116 NZG Archive ADOE 16612 M1/738 8/27/18 Letter W A Fraser to Secretary Marine Department. 30 December 1923

117 NZG Archives ADOE 16612 M1/738 8/27/18 Copy Memo Furkert (ME) to Secretary Marine department 16 May 1924

118 NZG Archives ADOE 16612 M1/738 8/27/18 Letter Secretary Marine Department (Godfrey) to Bollons (Master of Lighthouse Tender).

119 NZG Archives ADOE 16612 M1/738 8/27/18 Letter Fraser to Sec MD 16 December 1925.

120 NZG Archives ADOE 16612 M1/738 8/27/18 Letter Fraser to Sec MD. 1925

121 *NZG Gazette* 1925 , Notice to Mariners No 83 of 1925, p 3217 19 November 1925

4 The Balfour Lighthouses

Balfour's first lighthouse responsibility when he was appointed colonial marine engineer was to complete the lighthouses ordered under the auspices of the Marine Boards. In the period following this he was responsible for the design and procurement of the lighthouses at Farewell Spit and Cape Campbell, and began the design and ordered the apparatus for the two wave-washed cottage lighthouses at Bean Rocks and Ponui Passage in Waitemata Harbour.

Farewell Spit

In 1865 gold was discovered on the west coast of the South Island.* Lighting Farewell Spit was suggested in the second report of the Marine Board of New Zealand, which commented that

> increasing amount of traffic from Cook's Strait, and Nelson in particular, to the Ports of Hokitika and the Grey renders it imperatively necessary to place a light near the extremity of Cape Farewell Spit.[1]

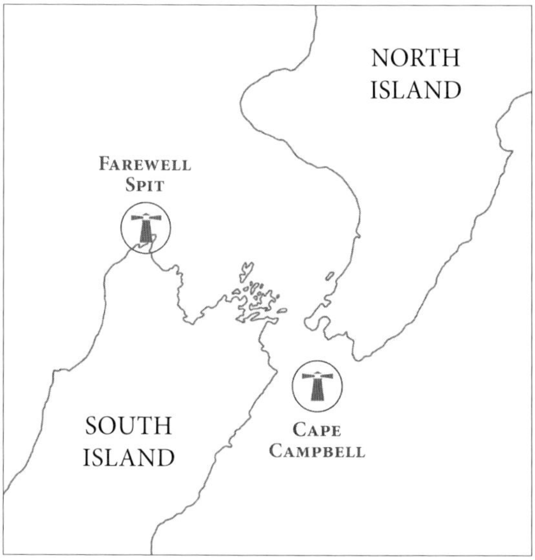

Figure 4.1: Locations of Farewell Spit and Cape Campbell lighthouses © Author

Richard Aylmer was retained to survey Farewell Spit and report on a suitable site for a lighthouse. In his first report, dated December 1865, he described the spit as comprising unstable sand hills and recommended that a lightship be moored off the spit.[2] Captain Sharp of the Marine Board disagreed in his part of the second report, for the following reasons:

*At this time, the area was referred to as the Gold Coast, and up to about 1900 what is now called the South Island or Te Wai Pounamou was referred to as the Middle Island.

1. There is no secure place of anchorage for a vessel where the light would be of service
2. The elevation of a floating light would be insufficient to be visible when approached from the south-west.
3. It would be liable to removal or entire loss, and might therefore lead vessels attempting to make it into danger
4. The cost of maintenance of a floating light would be greater than a shore establishment.

In January following, the government decided against a floating light and requested Aylmer to investigate the possibility of establishing a lighthouse on the spit.[3]

Aylmer's proposed design had foundations of cast-iron or wrought-iron screw piles screwed into a depth of 25 or 30 feet [7.6 to 9m]. Bolted to the tops of the piles would be a cast-iron framework (Aylmer called an 'entablature') to which the tower would be bolted. The proposed tower was to be cast-iron flanged plates with internal fittings of wrought iron approximately 60 feet (18.3m) high. The lantern was to have astragals and horizontal pieces of gunmetal, the former fixed diagonally, and roofed with sheet copper outside and wrought iron inside i.e. a Scottish lantern. The light was to be a second order dioptric revolving.[4]

The following year Balfour visited the site and his report to the postmaster general was reproduced in the third report of the Marine Department. In the early part of the report Balfour wrote

> nowhere in New Zealand is there such a crying need for light as on this low-lying and extensive danger.

While Balfour felt that the most efficient way to light the danger was to place a lighthouse on Cape Farewell and 'one on screw piles at the extremity of the Spit, in about two fathoms of water at low water', he recognised that the cost of erection and maintenance would be high.[5]

He considered that a 'useful light' could be erected near 'Bush End', the site which Aylmer had recommended at the end of the spit. This would 'indicate the dangers in a less perfect but still reasonably satisfactory manner', and this was what he proposed. It is of interest that he

> did contemplate the possibility of erecting a screw pile beacon on the outer end of the spit, and illuminating it by electricity generated at the lighthouse some four miles distant, but on close examination of the site, the difficulties appear so formidable that I cannot, for the present at least recommend the experiment.

At this time, Trinity House was experimenting with illumination by electricity at South Foreland lighthouse.[6] The system was designed by Frederick Hale Holmes, and used a carbon arc lamp powered by a generator driven by a steam engine. The Northern Lighthouse Board was also interested in this form of illumination and continued to experiment with it after the South Foreland experiments ceased in 1880. Once again the Scottish connection enabled New Zealand to keep up with the latest lighthouse technology, even if it proved to be a bit too advanced.

Balfour's suggested selecting a site 'about a-half mile further up the spit' from that selected by Aylmer and Aylmer's design for the tower be replaced with

> an open timber framework strongly braced, founded on sleepers mortised to the heads of piles driven not less than twelve feet [3.6 m] into the sand, and possibly guyed by wire ropes fixed to screw moorings, so as to diminish vibration.

The light was to be placed at least 100 feet [30 m] above ground level to give a range of not less than 16 nautical miles and was to be a revolving light with a 60-second or 30-second period. While the light was to be white in almost all directions, Balfour proposed showing a red light over the end of the spit, using a second light in the lantern.

It seems that this arrangement was submitted to D & T Stevenson, who demurred. In February 1868[7] they wrote to Balfour:

> We find in making the drawings that it will not be possible to put a light of the required power consisting of double holophotes within the lantern which you specified. We have tried several plans of adhering to the same arrangement but can make nothing of it. Moreover it is objectionable from the interrupting of the light which would always take place when the keeper had occasion to go inside of the frame. We also tried single holophotes but did not see how a light of proper power would be produced by employing holophotes of such small size.

As consulting engineers to the New Zealand Government, D & T Stevenson adopted

> what we consider to be the best alternative vizt a second order revolving light with single flame and mechanical lamp. We hope you may approve of this course. It is indeed the only one that we can think of in the circumstances. Even although any great excess of estimate might not have been an insuperable difficulty our great fear was that the towers may have been already erected and that the lanterns and apparatus found useless after their arrival.

The dangerous end of the spit was marked by placing red glass panes in the appropriate section of the lantern.

The lantern and light apparatus for Farewell Spit lighthouse and that for Cape Campbell (see below) were ordered at the same time. The specification[8] described the lanterns as being '7 feet 3 inches in height from the bottom to the top of the astragal'. The sole plate formed 'a polygon of 12 sides inscribed about a circle 9 feet 3 inches in diameter'.

Like the first order lantern for Cape Saunders, this was predominantly of cast iron manufactured using supplied patterns with some gunmetal (for sash frames) and glass of thickness between 1/4 and 5/16 inch. The cowl was again a double one with copper outer skin and iron inner.

In contrast to previous specifications, the Farewell Spit/Cape Campbell specification included instructions for lightning conductors of copper wire rope, with platinum points and copper plates at the termination point. (Other lighthouses were to have the copper wire spliced into a steel wire at ground level.) The length of wire for Farewell Spit was 125 feet [38 m].

There may have been problems with gunmetal in the past, as this specification included the following detailed requirements:

> The sash frames are to be of gun-metal consisting of 11 parts of tin and 84 of copper, the two metals to be mixed and run into ingots and afterwards remelted to form the castings. The metal is to be tested by the Inspector to a tensile strain of 33,000 lbs to the square inch at the expense of the Contractor.

In February 1868, James Dove & Company, Milne & Son and Slight were invited to tender for the two lanterns.[9] D & T Stevenson recommended that Dove's offer of £819/-/- be accepted.[10]

On 10 March Chance & Co were asked

> for what sum you will furnish two second order Dioptric Revolving apparatus. The top prism to be omitted and your offer to be for the apparatus settings complete down to a level of 3 feet under the centre of the lens.[11]

They were also asked to provide an offer for 130° of a fifth order Dioptric fixed apparatus 37½°. The apparatus was to be fitted with a hemispherical mirror in Edinburgh. This was for the proposed lighthouse at Manukau Head (Chapter 5).[12]

The description given in the specification was:

> The lighthouses of Farewell Spit and Cape Campbell are to exhibit Revolving lights showing once every minute. The apparatus to be of the Second Order of Dioptric holophotal Revolving Lights having a focal distance of 700 mm. It is to consist of 8 lenses, 8 panels of totally reflecting holophotal prisms above the lenses of 15 prisms in each, and 7 panels below the lenses of 6 prisms in each.[13]

Chance Brothers recorded the order for two second order lights on September 1868.[14] Their price for both was £1702/-/-. Subsequently they also supplied red glass for the cut-off over the end of the spit, cost £21/9/6.[15]

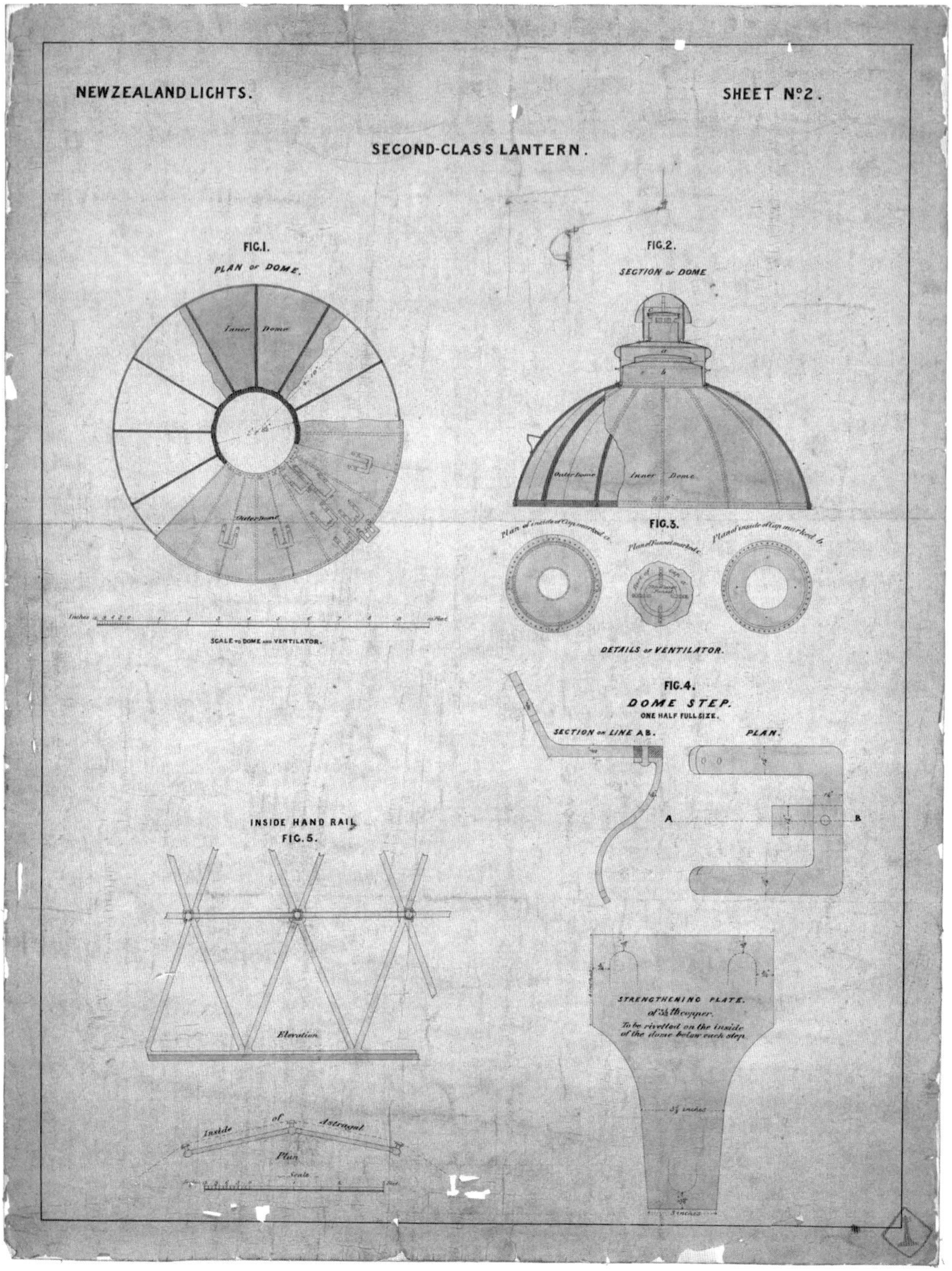

Figure 4.2: Typical second order lantern (Ms5885) © National Library of Scotland

The apparatus contract was for 'the machine, the machine case and revolving carriage and the top steadying frame, and to fit up the apparatus on the revolving carriage'. Although drawings were provided, the contractor had the option of supplying an alternative design, provided it would 'work smoothly and steadily at the required rate [one revolution in 8 minutes] and he could not change the 'large internal wheel and pinion and also [the] size of machine case'. The contractor was warned:

> nor shall the adoption of that arrangement liberate him from the responsibility of making a machine to perform its work to the satisfaction of the Engineers and the strength of the various parts must be sufficient for whatever weight may be required as a motive power.

The weight for Farewell Spit was designed to pass down the centre of the tower, but that for Cape Campbell had a system of pulleys.

The specification also required the supply of spare gearing including:

- A set of nine cast-steel bearing rollers,
- Two spare leaves for each lantern pinion,
- One spare worm wheel,
- One spare fly shaft with bushes for it
- 10 spare fly shaft footsteps
- One spare main driving pinion of 30 teeth
- One spare horizontal driving shaft
- Five spare footsteps for the upright shaft.

Offers were received from Milne & Son and Slight, and the contract was given to Milne & Son for £660/-/-.[16] The same letter which reported the offers for the apparatus also enclosed offers from Milne & Son for the mechanical lamps for £388, and from Smiths for the stores for £318.14/-.

On 25 May, a little more than two months after the granting of the tender for the lanterns, and apparently in response to the submission of tender information for the apparatus contract, John Morrison, the New Zealand Government agent, wrote to D & T Stevenson. He appears to have requested copies of the lantern specification, queried the way contractors were paid, and seems to have indicated that the New Zealand Government expected to receive the apparatus in October.

On 27 May D & T Stevenson answered as follows:

> In answer to your letter of the 26th instant we beg to send you herewith copy of the specification for the Lanterns. In all cases the Contractors receive instalments as the works proceed.

There has been no necessary delay with the work, which from its nature cannot be hurried without detriment, but no time will be lost in its speedy completion. We do not recall to have heard that the Government expects to receive the apparatus in the month of October otherwise we should have at once advised you that it was not possible to accomplish this. The lanterns will be sent off whenever they are ready so that fitting of them up may be proceeded with in the country and the apparatus will follow as soon thereafter as practicable.[17]

(On a later occasion, D & C Stevenson were forced to point out to the then Agent General that if one accepts a tender, then that is the sum one pays.)

It seems that the original specification may have been for pressure lamps as D & T Stevenson wrote to Balfour on 10 July 1868 saying:

We have received your letter and shall order the pressure lamps as requested.

In 1865 we made a series of experiments with pressure and mechanical lamps but we were not able to keep so high a flame with the pressure lamp as with the mechanical; the flame from the pressure lamp being always about 2/10ths of an inch lower than that of the mechanical one. There was also with the pressure lamp a greater tendency for charring of the wicks than in the mechanical lamp. We mention these facts for your information and guidance.[18]

This advice was taken, and Milne & Son also supplied mechanical lamps.

The lanterns were sent to London on 12 September 1868 and shipped to New Zealand from the East India Dock by the *Wild Duck*.[19] The stores for both lighthouses and the light apparatus for Farewell Spit were shipped to New Zealand by the *Melita*, which left the East India Dock in August 1869.[20]

In early 1869 the Marine Department called for tenders for construction of the lighthouse and allowed for the options of wooden pile or iron pile foundations. Only one tenderer gave a price for iron piles. The contract went to Samuel Brown of Wellington who quoted £2496/-/- for construction with wooden piles.[21]

A preliminary description of the light was given in a *Gazette* notice dated 22 April 1869[22] and signed by James Balfour. But before the light was first exhibited, on 17 June 1870, he died and hence the final Notice to Mariners[23] was issued over the signature of Julius Vogel, the commissioner of customs, later Agent General and premier of New Zealand. The light was described as

a Revolving White Light, of the Second Order Dioptric, attaining its greatest brilliancy every once a minute.

Over the Spit End (that is between the bearings of N.W.¾N to W. by N.¼N) the light will be Red, in order to caution Mariners of their approach to that danger

> The light is elevated about 120 feet above the sea level, and will be visible in clear weather about 17 nautic miles, allowing 15 feet for the height of the observer's eye.
>
> The Tower is an open-framed structure of timber, painted in alternate bands of red and white, and is 113 feet in height from the ground to the top of the lantern.

The Notice also cautioned against entering the area of the red light.

While the Sixth Report[24] stated that the value of Farewell Spit light 'is constantly and warmly acknowledged by shipmasters', the author acknowledged that

> [t]he light at Farewell Spit occupies the most isolated position … The ground … is composed of sand, not fit for cultivation, and almost destitute of vegetation. Owing to the barren and uninviting character of the station, it will be difficult to get keepers to stay at it for any length of time together, and they will therefore have to be frequently changed.

In their 1873 report Johnson and Edwin recommended that a first order light, revolving every two minutes, be placed on Cape Farewell and the existing light on Farewell Spit be replaced by a fourth order light with the same range. They anticipated that the dismantled light 'could be removed elsewhere'.[25] Although this dismantling eventually happened, a lighthouse was not placed on Cape Farewell.

In 1875 the colza lamps were replaced by three-wick Doty burners manufactured by Milne & Son[26] as part of a larger order. They were shipped to New Zealand on the *Penshaw*, which left East India Docks in December. However three of Farewell Spit's overflow tubes and couplings were used at Godley Head lighthouse and replacements were ordered three years later.[27] Unfortunately the ones sent were not satisfactory; D & T Stevenson, passing on the Marine Department's complaint, described them as 'useless if not positively dangerous'.[28] Milne & Son admitted that their foreman had sent off the tubes but also pointed out that '[t]hey were seen by your inspector before being packed'.[29] The replacement tubes were sent the following month.[30]

In the third report (1867), Balfour had recommended building timber structures for lighthouses:

> In all cases the approximate cost has been kept as low as possible, on the understanding that the buildings will be of timber, and consequently in so far of a temporary and perishable nature, as it has been considered better to get the coast more completely lighted, even in a temporary manner, than to incur a large expenditure in erecting a smaller number of more permanent structures.

New Zealand's wooden lighthouse towers were usually of kauri and totara (two local trees), and many of them, particularly the weatherboard structures designed by Blackett, lasted until automation. However, the Farewell Spit tower was to demonstrate how temporary these structures could be in an exposed location; 23 years after its erection the tower was found to be badly weakened by erosion from the windblown sand, and an iron replacement was suggested.[31]

It was also decided to replace the existing lantern, and the order was acknowledged by D & T Stevenson in October 1891 in a letter to Lewis Wilson.[32] The 1892 Marine Department report stated that 'it had been ordered from England'. The following year, Lewis Wilson decided to give credit where credit was due and reported that '[t]he new lantern for Farewell Spit has been procured from Scotland'.[33]

Figure 4.3: Original Farewell Spit lighthouse (1870) (ATL 10x8-0986-G) © Tyree Collection, Alexander Turnbull Library, Wellington, New Zealand

This was a second order lantern with a revolving cowl. It was topped by a standard Northern Lighthouse Board weather vane to keep the cowl in the correct alignment. This was the design of cowl which was reduced to circulating impotence by the updrafts of Baring Head (Chapter 2). The lantern was of the same design as had been used nine years previously for the lighthouse at Waipapa Point. With commendable frugality, D & T Stevenson recycled the Waipapa Point drawing, crossing out the existing title and replacing it with 'Farewell Spit'. Likewise the original signature and date (8 April 1882) were crossed out and a new signature, dated 28 October 1891, added beside it (see Figure 9.11 a,b). A pencil note also indicated that the drawing was to be used for Cape Palliser.

This time saving enabled Stevenson to invite tenders from both Milne & Son and Dove & Co on 28 October 1891, and to request that they by lodged on or by 30th. In a letter dated 2 November, D & T Stevenson were able to inform the Agent General that Dove & Co's price was £642 and Milne & Son's was £612.15.-, and to recommend the acceptance of Milne & Son's. The letter of acceptance to Milne & Son was dated 4 November. This exchange shows how fast communication between London and Edinburgh was in a pre-electronic age. It was made possible by overnight mail train service between the two capitals. *

* The 1936 John Grierson Post Office documentary 'Night Mail' showed that this service was still available 40 years later.

D & T Stevenson appear to have sent two tracings of the sole plate, one on 26 November 1891[34] and another one on 19 December,[35] apparently to show the fitting stops.

The letter of 26 November also indicated that the gutter-plate was to be made of gunmetal, as suggested by the Marine Department. Also the lightning conductor was only being brought down to the solc plate as it was assumed that the lightning conductor from the old tower was being transferred to the new one. Nevertheless this shorter conductor was to have a new system, with the upper terminal of the conductor consisting of pieces of platinum wire inserted into the copper rod rather than have it wrapped around a copper cone as previously.

Although the existing apparatus was to be used in the new lantern, D & T Stevenson arranged to include a new top steadying frame for it.

Milne & Son tried a new method of lantern construction, namely to cast the sole plate sections and trimming path sections in single pieces. Volume 61 p 122 of the Stevenson Archives contains a letter to Milne & Son, dated 4 February 1892, that refers to a conversation relative to the trimming path plates, but the remainder of the letter is illegible and one cannot be certain whether this was agreement for a new method of construction.

This experiment led to a delay in completion[36] and finally D & T Stevenson instructed

> With reference to our conversation yesterday – as the two castings have again cracked, there is no other alternative but to revert to the old and well tried plan of casting the trimming plate separately from the sole plate. This will cause delay but there is no help for it.[37]

Not all Scottish innovations were advantageous.

Two months later the Agent General was again raising the question of delivery and was informed:

> The Lantern for Farewell Spit Lighthouse will be packed and ready to be forwarded to London this day fortnight. It has been delayed in consequence of difficulty with some of the castings which we had to reject. Will you kindly send us the address for packages and let us know where they are to be forwarded for shipment.[38]

Finally on 4 May 1892, the packages containing the lantern were despatched from Leith by steamer to Royal Albert Dock for shipment on the SS *Duke of Westminster*. On arrival the packages were stored and, in a situation reminiscent of the fate of the original Cape Saunders light, remained there until a decision could be made about the tower.

The 1893 report, which announced the arrival of the lantern from Scotland, also expressed the hope that:

> a sum of money will be provided for the erection of a new tower There is a convenient site for the new lighthouse a little further towards the spit end,

> and the tower need only be 60 ft in height, instead of 100 ft as at present; this reduction can be made as the light will not be of so much importance after the exhibition of Stephens Island Light.

Some progress was made in the following year, and the 1894 report of the Marine Department announced that:

> Plans and specifications have, therefore, been prepared for a tower to be constructed of wrought-iron open frame-work, on concrete foundations; the height of the proposed tower being 75 ft from the top of foundations to the sole plate of the lantern.[39]

The tender for the lighthouse tower was awarded to J & A Anderson of the Canterbury Foundry, Christchurch, at a cost of £828 6s 0d.[40]

Although the notice of acceptance of the tender was dated 23 January 1895, the 1896 report[41] (dated 30 May) apologetically informed the House that:

> It was anticipated that the new tower at Farewell Spit would have been erected during the past summer, but, owing to the contractors [*sic*] failing to supply the iron tower within the prescribed time, the work of erection has not yet put in hand.

The tower sections and lantern packages were transferred to the site by government steamer, and the tower was eventually erected and the new lantern assembled by the Public Works Department in late 1896. The light in the old tower was discontinued on the morning of 17 January 1897. The machine and apparatus were taken down and erected in the new tower so promptly that the light was shown again on the night of the 19th.[42] Notice to Mariners No 36 of 1896[43] stated that:

> The light (which is a revolving white light of the second order dioptric, attaining its greatest brilliancy once a minutes) will show all round, except on the outer side of the western portion of the Spit from the lighthouse, where it is shut in by sandhills.
>
> Over the Spit end the light will be red, in order to warn mariners of their approach to that danger.
>
> The light will be elevated 97 ft above the sea-level, and, allowing 15 ft. for the height of the observer's eye, it will be visible 15¼ nautical miles in clear weather [i.e. as the new tower was shorter than the previous one, the range of the light was reduced].
>
> The tower is an open-framed iron structure, painted in alternate bands of red and white; and is 88 ft in height from ground to the top of the lantern.

Figure 4.4: Second Farewell Spit lighthouse (ATL 10x8-0993-G)
© Tyree Collection, Alexander Turnbull Library, Wellington, New Zealand

The redundant lantern was used for the new lighthouse on Matiu/Somes Island.[44]

Doty burners to replace the existing colza lamps were ordered for both Farewell Spit and Cape Campbell lighthouses in 1875, along with orders for Taiaroa Head and Nugget Point.[45] In both cases the order was for 3 three-wick burners with couplings and a Liverpool lamp with 12 glasses, 1 gross wicks, and a mandril. The whole order was shipped on the *Penshaw* in December 1875.[46]

The Doty burners were replaced by incandescent burners in 1912.[47]

The lighthouse was damaged in the 1929 Murchison earthquake. The principal keeper reported that severe earthquakes shifted whole machine bodily 6" [15 cm] from the bed plates and damaged the burner.[48] The machine was successfully jacked back into position.[49] Earthquakes were a major problem for New Zealand's lighthouses, with damage ranging from dislodging equipment and, in the case of newer lighthouses, spilling mercury (Chapter 11) to destruction of the lens and damage to the tower (Kahurangi Point lighthouse). However, no lighthouse has ever been destroyed.

In 1951 the Secretary of Marine recommended that the Farewell Spit light should be electrified[50] and this was finally completed in 1955.[51]

Cape Campbell

Cape Campbell is a promontory on the Pacific Coast in southern Marlborough, and thus lies approximately midway between Wellington and Banks Peninsula. A lighthouse was needed to fill the gap between Pencarrow Head and Godley Head.

The Cape Campbell lighthouse was first mentioned in the Appendix to the Journal of the House of Representatives in the third report of 1867[52] where Balfour stated that he had 'also furnished an approximate estimate of the cost of construction of a lighthouse at Cape Campbell'.

The lantern and light apparatus were ordered along with those for Farewell Spit. The Cape Campbell lantern was identical to that for Farewell Spit, except that the lightning conductor required only 55 feet [16.8m] of copper wire compared to 125 feet for Farewell Spit. The Cape Campbell lantern was sent out with that for Farewell Spit on the *Wild Duck*[53] in September 1868.

Although the glass work and mountings for the two lighthouses were identical there was a significant difference in the apparatus. In contrast to Farewell Spit, where the driving weight was designed to pass down the centre of the tower, that for Cape Campbell was placed down the side of the tower. The specification stated:

> the weight line is to be led to the side by pulleys and up to the top of the lantern at the dark side where a proper hook is to be fitted up for another pulley from which the weight is to hang. All these pulleys are to run in antifriction bushes.

The apparatus was despatched to New Zealand on the *Electra* in July 1869. Also in the shipment were the lighthouse stores, 10 oil cisterns and stands, and 1 gross of three-wick lamp glasses supplied by Smiths.[54] Three years later Smiths also provided 10 fathoms of machine rope for Cape Campbell, which was also shipped on the *Electra*.[55]

For Cape Campbell, Balfour designed a timber structure to support the lantern. The tender for its supply and construction was awarded to H Carter & Co, Wellington, and the tender price was £1960/-/-.[56]

Shortly before the light apparatus was despatched, a preliminary announcement of the lighthouse's erection was made.[57]

The lighthouse was finally completed in July 1870 and the light was exhibited on 1 August 1870.[58]

The Notice to Mariners described the lighthouse as

> erected on a knoll at the extreme end of Cape Campbell, in Lat 41°43'15"S., and Long.174°18'30"E.
>
> The Tower is an open-framed structure of timber, 73 feet in height from the base to the top of the lantern, and is painted in alternate bands of red and white.

The Light is elevated about 155 feet from the sea level, and may be seen in clear weather at a distance of 19 nautic miles, allowing 15 feet for the height of the observer's eye.

It is a revolving White Light, of the second order Dioptric, attaining its greatest brilliancy once a minute, and will be seen all around from seaward between the magnetic bearings of N.¼W. and E. by N.½N.

Doty burners to replace the existing colza oil burners were ordered for Cape Campbell in 1875, along with those for Farewell Spit. As with Farewell Spit, the order consisted of 3 three-wick burners with couplings and a Liverpool lamp with 12 glasses. 1 gross wicks, and a mandril and was shipped on the *Penshaw* in December 1875.

The timber tower lasted longer than that on Farewell Spit, but eventually had to be replaced although the process was not swift. In March 1903 it was announced that the tender for a new iron tower had been awarded to Charles Judd of Thames (who also cast the Cape Brett lighthouse tower) for a price of £1499 6s 0d.[59] Consequently, the 1903 Marine Department report[60] announced that:

A new iron tower is being constructed for Cape Campbell, to replace the present wooden structure which was erected in 1870.

Figure 4.5: Original Cape Campbell lighthouse (Hocken ID 19145)
© Hocken Collections, Uare Taoka o Hakena, University of Otago

The report also stated that a new lantern had been procured. Due to the date there is no correspondence on this in the Stevenson Archives, and the New Zealand files are silent as well. However, the National Library of Scotland contains a drawing originally done for Kahurangi lighthouse with Cape Campbell added in pencil (see Figure 10.17a). It seems that the new lantern was obtained in the usual fashion, and D & C Stevenson had thriftily recycled the drawing.

There was some delay in erection as the following year's report[61] stated:

> It is proposed to erect a new iron tower at Cape Campbell as soon as the steamer is available to convey the balance of the ironwork from Wellington.

The 1905 report[62] indicated that by May 1905 '[t]he erection of a new tower at Cape Campbell is nearly completed'. The tower, which was erected about 40 feet [12 m] from the earlier tower, was completed about four months later. The existing light apparatus was moved during the period from sunrise 13 October to sunset 15 October. During this period a fixed white light produced by four port lights was shown from old tower. The elevation of the light remained the same, but the range was now increased to 18 miles.[63]

By this time, New Zealand foundries were capable of producing the iron towers required for lighthouses. The first lighthouse tower cast in New Zealand was that for Cuvier Island lighthouse, which was first exhibited in September 1889 (Chapter 10).

The Marine Department report for 1911[64] reported that Chance incandescent burners had been installed at Cape Campbell. However those at Cape Campbell were as troublesome as those at the other lighthouses, and so in February 1933 they were replaced by Smith-Ford burners.[65]

In September 1914 Cape Campbell lighthouse was shaken by a major earthquake. The whole apparatus slid across the floor 6 inches east and 3 inches north.[66]

Figure 4.6: Second Cape Campbell lighthouse (ATL 1/2-092152-F) © Randle Collection, Alexander Turnbull Library, Wellington, New Zealand

The burner lifted out of its socket and was left hanging by the tubing. Mantles, lamp glasses and a burner tube were broken.[67]

The keepers were able to make temporary repairs and bring the light back into operation with only one night's darkness. A wooden framework was placed around the apparatus to prevent it moving until it could be repaired properly. Two years later, another earthquake dislodged one of the central portions (the bull's eye) from the lens,[68] although with some outside assistance it was replaced within a week.[69]

In August 1937 it was announced that Cape Campbell lighthouse would be electrified[70] although the required diesel generators did not arrive until June 1938.[71] The completion of the installation in both lighthouses and keepers dwellings was announced in the Marine Department report of 1939.[72] The light was automated in 1986.

Bean Rock and Ponui Passage

The City of Auckland was founded in February 1841 by Governor Hobson, to be the capital of the new colony after the first signings of the Treaty of Waitangi in 1840. Hobson maintained that his new site would be more 'centralised' although there was probably an argument for moving the nascent capital from Kororareka, an area colloquially known as 'the hellhole of the Pacific'. The city is sited on an isthmus dotted with volcanic cones and with natural harbours on both sides.

On the eastern is the deep-water Waitemata Harbour, and on the western the shallower Manukau with its fearsome bar. Balfour was to contribute to the lighting of both harbours.

As a result of the city's growth, the Provincial Government ordered the lighthouse for Tiri Tiri Matangi to guide ships towards the Waitemata Harbour. This was subsequently erected by the Marine Board (Chapter 2).

In 1868 gold was discovered near Thames on the Coromandel Peninsula.[*] As this was quartz gold and not alluvial, its extraction required heavy machinery, most of which was shipped from Auckland. On 25 August 1868, a memorial signed by 95 members of Auckland's shipping and mercantile community was presented to the Auckland Provincial Assembly, requesting that three lights be erected between Auckland and Thames especially to assist the greatly increased night traffic.[73] At the assembly's request, this was transmitted by Robert Crighton (on behalf of the deputy superintendent) to the colonial secretary on 31 August 1868.[74] On 17 September 1868, the colonial secretary replied. He pointed out that the lights were local ones, and were thus the responsibility of the provincial authorities.[75]

It was eventually decided to place lighthouses on Bean Rocks[4] in Waitemata Harbour and on the end of the sandspit branching out from the north side of Pakihi Island into the waterway between Pahiki and Ponui Islands. The waterway is shown on charts as Sandspit Pass, but the lighthouse was named Ponui Passage. D & T Stevenson usually referred to it as Sandspit Light.

* There had been a short-lived rush in 1852, but as most of the gold was in quartz strata rather than alluvial gold in streams, the rush soon petered out.

4 The Bean Rocks were named after P. C. D Bean, the captain of the survey ship HMS *Herald* used in the survey of Waitemata Harbour in 1840. The lighthouse was named Bean Rock.

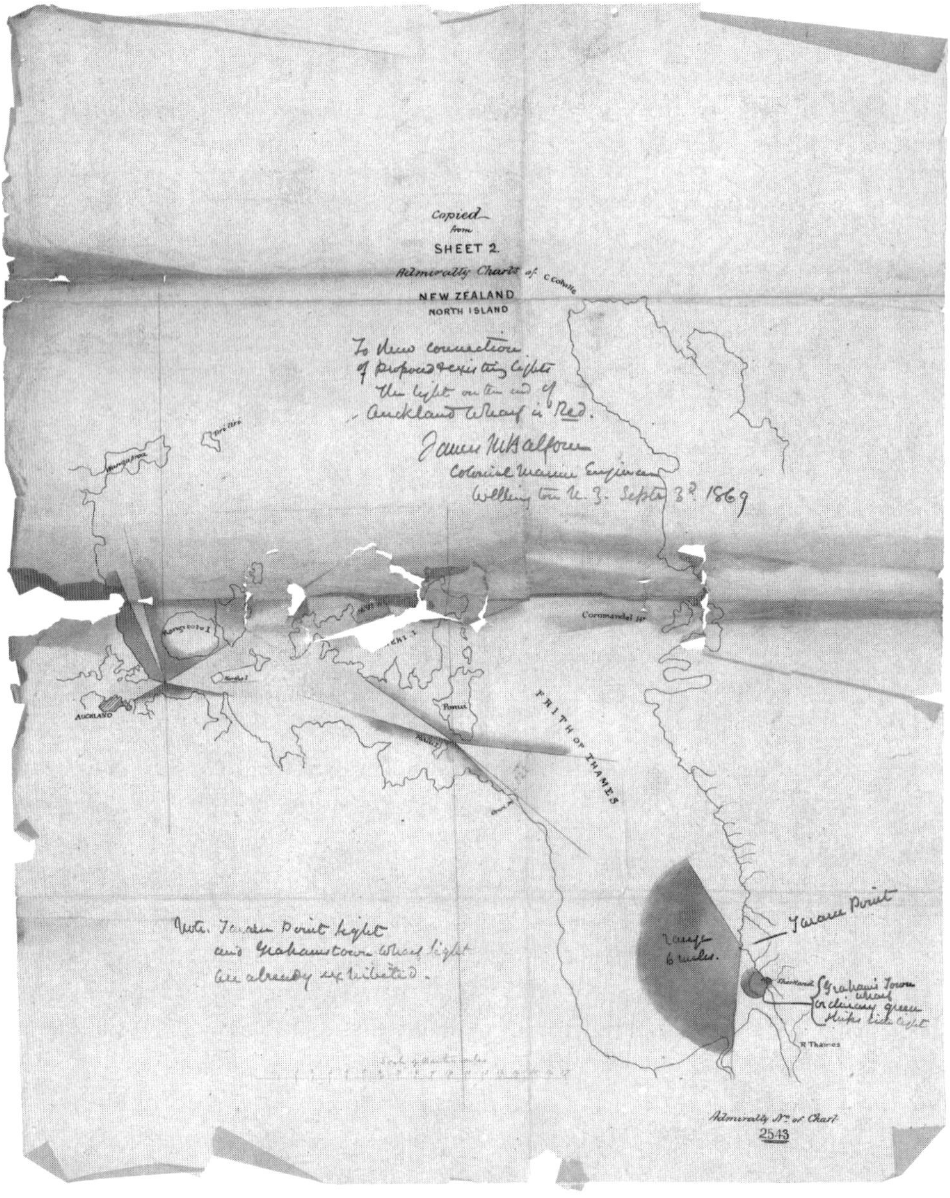

Figure 4.7: Location of Bean Rock and Ponui Passage lighthouses, as sent to D & T Stevenson (Ms 5885) © National Library of Scotland

As the provincial engineer (C O'Neill, another D & T Stevenson appointee) had no experience in lighthouses, the Provincial Government asked that Balfour's services be placed at their disposal. This was agreed to by the Marine Department.[76] Although the letter agreeing was dated 21 November 1868, Balfour was not able to officially start work until September 1869. In his letter to D & T Stevenson of 3 September he apologised for 'a very hurriedly written and I fear incoherent specification'. He pointed out that, having been warned that his skills would probably be needed, he:

> urged upon him (O'Neill) the necessity of giving me ample time before the mail left. However the Engineer … is also a Politician, and politics carried the day, so that it was only yesterday afternoon – not 36 hours before mail time – that I was authorised to prepare drawings and specification.[77]

His letter included a tracing based on Admiralty Chart 2453, which gave the positions of the two lighthouses.

The letter also indicated that the Bean Rock light

> is meant to answer a double purpose. To lead vessels through Rangitoto Passage clear of all dangers and also to benefit the local trade between Auckland and Shortland Gold Fields which are situated at the head of the Firth of Thames and are approached by Tehamaki [now Tamaki] Strait.

For the lanterns, Balfour contemplated 'using the lanterns of the size of which Dove made four in one order at a cost (by my notebook) of £377 for the four. They were made I think for the Indian Government'. Although he described them as 'nice little things', Balfour felt that the astragals had been 'unnecessarily clumsy' and requested that if cast iron was used it should be lighter but that gunmetal would be preferred.

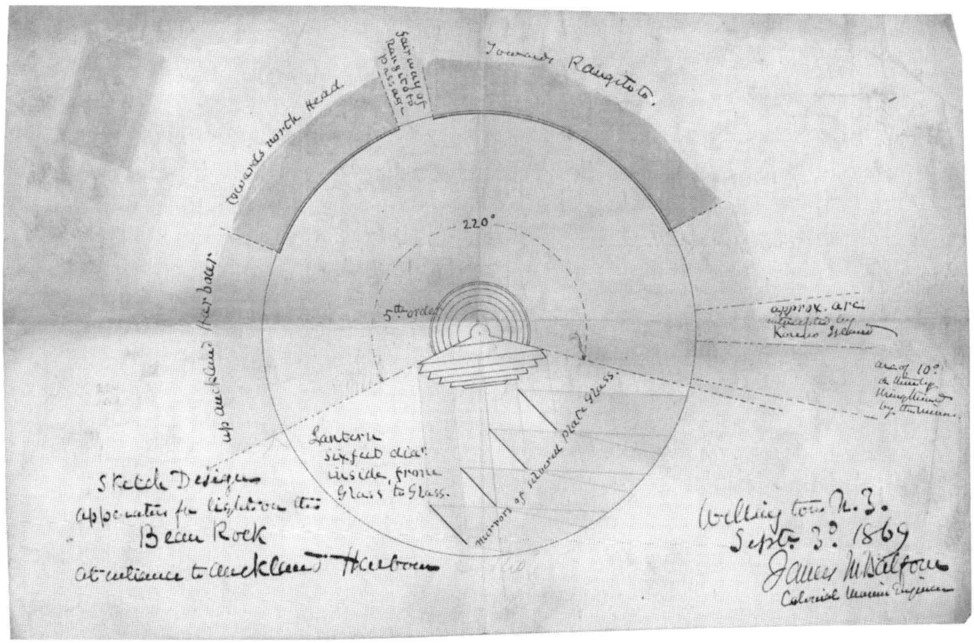

Figure 4.8: Sketch of the design apparatus, Bean Rock, dated Wellington 30 Sept 1869 (Ms5885) © National Library of Scotland

With this letter he provided an initial specification as follows:[78]

Lights between Auckland and Frith [*sic*] of Thames

429d

It is intended to erect a small Lighthouse on the Bream [*sic*] Rock at the entrance to Auckland Harbour also a light on piles on the extremity of the Sandspit between Pahiki and Ponui Islands as shown on the chart herewith.

Neither light requires to be seen for a great distance, and the lamps will not be more than 50 feet above the sea in either case.

The lights are intended to change color on certain azimuths somewhat as shown on the chart but the arrangement will be definitely described next mail after I have had time to consult the Chief Harbour Master of Auckland with regard to certain details.

Lanterns

The lanterns are to be six feet in internal diameter and one tier of panes high to be cast from the patterns made under the directions of Messrs D & T Stevenson C E 84 George St for four lanterns for the Government of India a few years ago. The Astragals to be inclined as usual with all lanterns made in Scotland, to be as thin as consistent with the requisite rigidity, and to be of gun metal by preference, if the funds available will permit. The sole plate, cornices etc to be cast iron; the dome to be copper, and as the climate of Auckland is very equable no inner dome will be required. An ample supply of spare glass, also storm panes to be furnished for each Lantern, also one set of dome riveting tools, and a set of taps and dies for all screws used in constructing the Lanterns to be supplied. As the Lanterns will probably be erected by the same workmen one set of tools will suffice for the two lights.

Apparatus

The lightrooms will be of Timber and the sole plates are to be cast without feathers. The Lantern bed will be of timber eight inches thick and holding down bolts are to be sent to suit.

The Lights are to be fixed of the fifth Dioptric, the lamps to be of the form introduced by Messrs Chance of Birmingham for burning Kerosine, which have been found to answer very well in Australia.

Each Lighthouse to have a set of three lamps complete, all parts being interchangeable.

The following tools and stores to be furnished for each Lighthouse;

1. A set of Copper oil measures, Quart, Pint & Gill
2. One oil measure tray
3. One oil funnel
4. Four pairs of lamp scissors
5. One dozen lamp glass brushes
6. Ten gross wicks
7. One dozen fine chamois skins. Two dozen linen rubbers. Two dozen woollen rubbers
8. 4 lbs rouge in japanned tin case. 10 lbs rotten stone in japanned tin case
9. 2 pieces of sponge. Four feather dusters
10. 2 pairs flat plyers – two pairs cutting plyers – two pairs nippers – two spirit levels – 1 glazier's diamond – 2 putty knives – two soldering bolts – 4 lbs tin solder – 4lb resin
11. One small hand lantern

All the stores to be carefully packed in white cotton waste of the best quality which will become part of the Lighthouse stores.

In addition to the above there are to be furnished for each light, Four hand brushes or dusters – two dust shovels, and six scrubbing brushes.

The goods for the two lighthouses to be separately packed in properly marked cases, all goods liable to be damaged by moisture to be soldered up in zinc.

They are to be shipped to Auckland direct, it will be hereafter arranged how they are to be addressed.

The whole cost of the each lantern and apparatus complete with all stores is not to exceed £3000 f.o.b. or £6000 for the two.

In September Balfour proposed red, green, and white sectors to define different channels and hazards, but before sending final instructions to Edinburgh he consulted Captain Burgess, the chief harbourmaster for the Port of Auckland, whose suggestions varied from Balfour's.

Hence in a later letter (dated 2 October 1869) Balfour sent for D & T Stevenson's comments a copy of Burgess' report and two tracings, one Balfour's original and 'one showing the several proposed arrangements of the lights'. This latter showed Balfour's original design, two suggestions from Burgess and the final proposed arc pattern.[79]

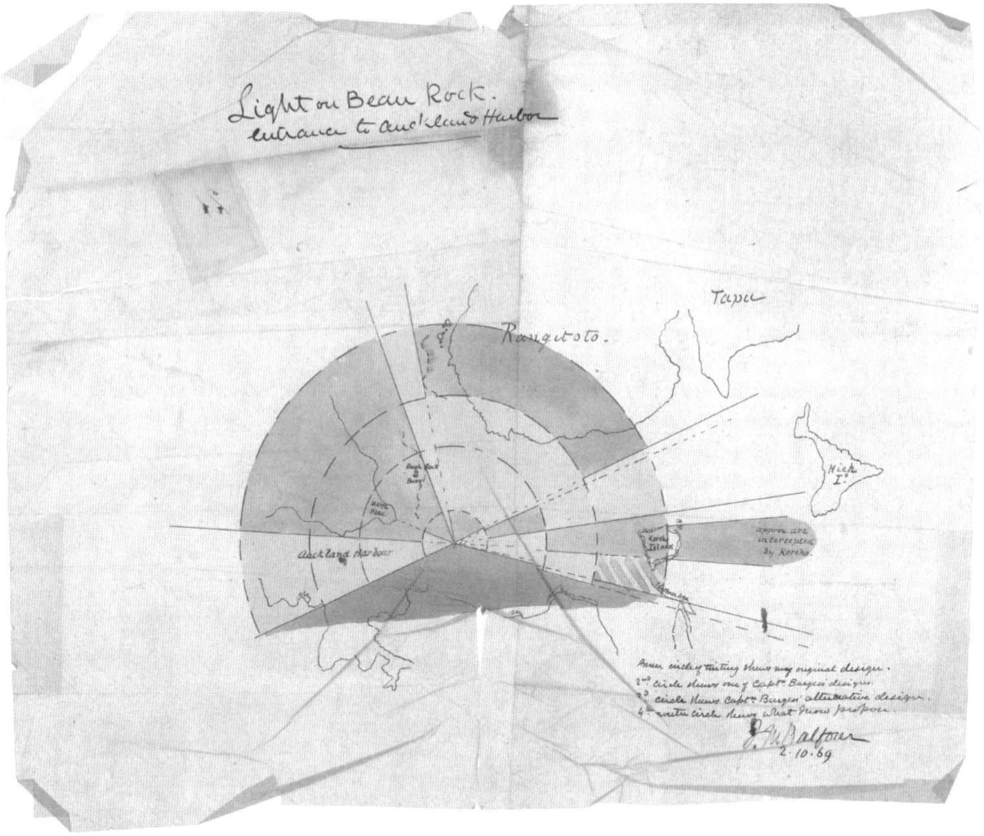

Figure 4.9: Light on Bean Rock, alternate arcs (Ms 3885) © National Library of Scotland

The current pattern is shown below.

He also sent a sketch of the design for the Ponui apparatus with the proposed sectors up the 'Tehmaki Strait' and the 'Frith of Thames' at Ponui Passage.

In his 1868 report, Balfour had commented:

> During the past year I have made some experiments with the view of ascertaining the possibility of burning kerosene in the lighthouse lamps, and am sanguine of ultimate success for Dog Island at least, where a number of small lamps are used. The introduction of kerosene would not only enable us to procure the necessary supplies in New Zealand, but would, I am satisfied, effect a considerable saving in the cost of maintenance of all the lights in which it could be used, and would also I believe make the light more powerful.[80]

When Balfour eventually incorporated kerosene burners into the design for the Bean Rock and Ponui Passage lighthouses, D & T Stevenson appear to have been a bit dubious about using kerosene burning lights. In his September letter Balfour quoted from a letter received from Lieutenant Heath, the portmaster of Brisbane and the chief nautical officer

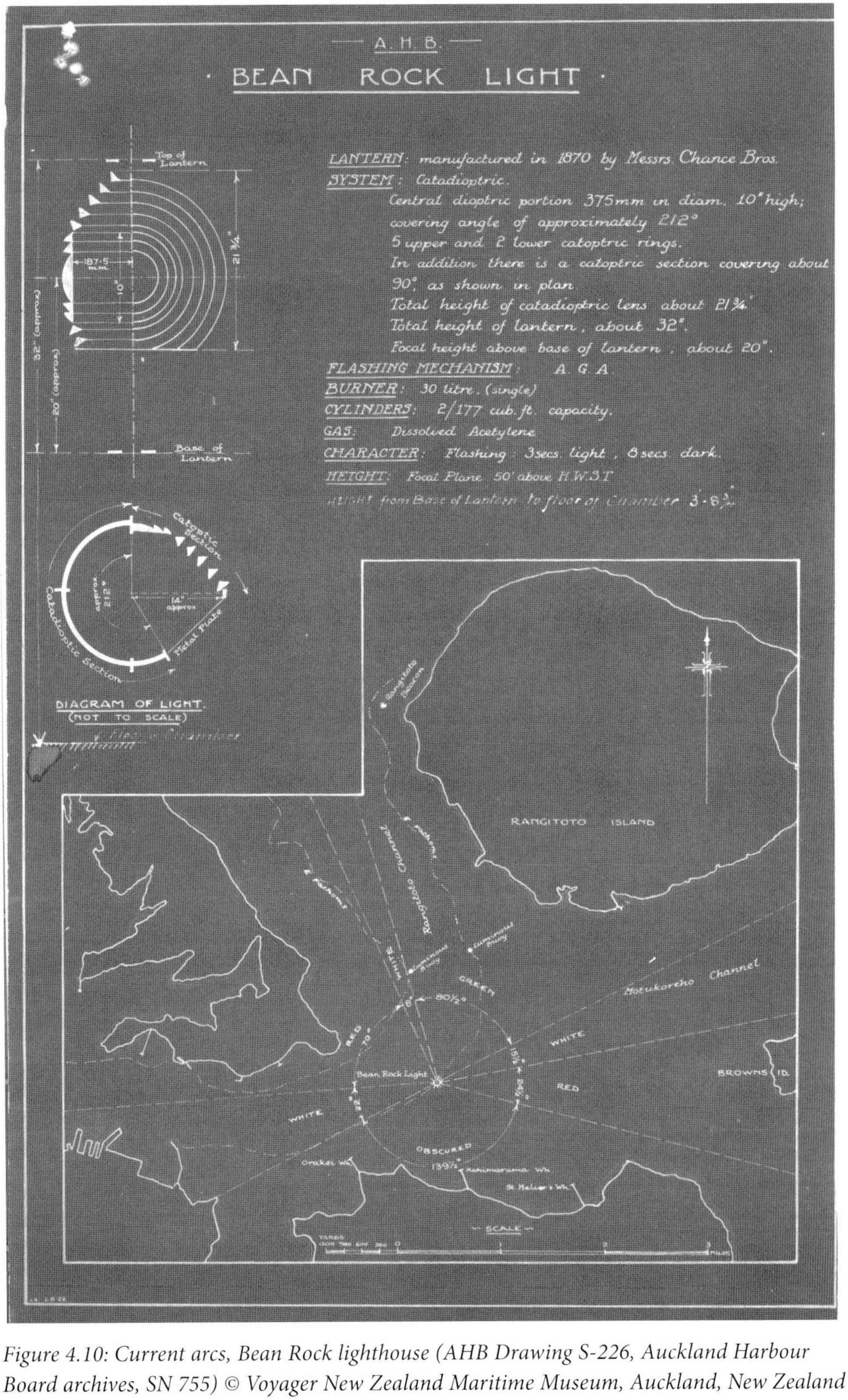

Figure 4.10: Current arcs, Bean Rock lighthouse (AHB Drawing S-226, Auckland Harbour Board archives, SN 755) © Voyager New Zealand Maritime Museum, Auckland, New Zealand

for the Colony of Queensland. Heath stated that for the previous four years all lights except first and second order had used kerosene. He spoke highly of a fourth order kerosene burner which had been designed jointly by James Hicks & Co of Birmingham and Chance & Co. Balfour also pointed out that kerosene was 'always available in the market & reasonable in price whereas the Colza must be specially imported'.

Balfour died two months after writing his October 1869 letter, and responsibility for the design and erection of the lighthouses was passed over to James Stewart, the district engineer for the Public Works Department in Auckland.[81]

It is not clear how much of the final design Balfour had accomplished before his death. As can be seen from his letters to D & T Stevenson, he had already designed the lighting apparatus and lantern. On the other hand, Stewart in his 1877 report to the collector of customs, Auckland,[82] stated:

> Prior to the lamentable death of Mr Balfour in Dec 1869, he was understood to have made some progress with the designs for the Lighthouses, but nothing was found among his records relating to such.

Charity suggests that Stewart was apparently referring to the lighthouse structures; they were not strictly towers. The final design for these was for octagonal cottage lighthouses on an open timber framework bolted to piles. The lanterns were set into the roofs of the 'cottages'. The construction drawings showed that the cottages for both lighthouses were identical. However when the Bean Rock Lighthouse was built, the balustrade around the lantern consisted of horizontal railings not the diagonal railings that were used for Ponui Passage.

In his September letter Balfour asked that Stevenson move quickly as 'local authorities are apt to credit general Government officers with all kinds of delays'. Requests for tenders for both lanterns were sent to Milne & Son and Dove & Co in November 1869,[8384] and the tender was given to Dove & Co for £280. However, the final price was higher, due to the need to install different coloured glass in different sections of the lanterns. This information came after the original tenders were called.[85]

In his letter, Balfour admitted to concern about the budget, but made a reference to Lough Foyle lighthouses, which was probably a D & T Stevenson in-joke. Unfortunately Balfour's reservation was justified and there had to be changes in the specification for the light apparatus during the tender period.

The original design was for a 220° of beehive and 140° of holophote for Bean Rock, and 210° of beehive and 150°of holophote for Ponui Passage. (In April 1859 D & T Stevenson had conducted experiments in Queens Park in Edinburgh of the effectiveness of the combination of beehive and holophotes to define a coloured arc.) Subsequently Chance Bros were asked to tender for:[86]

> **Sandspit Lights [i.e. Ponui Passage]**
>
> A beehive of 5th order full for 235° and the upper prisms only for 42° *more* Also a part of a holophote for 125° and stripe(?) of silvered mirror set in

frames and arranged as shown in drawings. Red glass triangular panes are required for ? the Lantern is 6 feet diameter and 3 feet ? high.

Bean Rock Lights

A beehive of the 5th order for 215° 91½° of red glass and 75° of green. Lantern same as formerly.

The coloured glass should be in triplicate to allow for risk of breakages

In a letter five days later, Chances were also asked to give a price for:

three lamps complete for each light (all parts being interchangeable) of the form introduced by you for burning Kerosene, and also 2 gross of lamp glasses, 1 dozen lamp glass brushes, and 10 gross of wicks for each lighthouse.[87]

Chance's reply came promptly, but as it was beyond the budget D & T Stevenson wrote again asking what the

cost would be supposing Sandspit to be furnished with 270 deg of a 4th order beehive and Bean Rock with 208 deg of a 5th order beehive and a half holophote arranged exactly the same as that for Newhaven; this will dispense with all mirrors and one of the holophotes.[88]

They also asked for the diameter of the proposed Kerosine [*sic*] burner. Chances replied on the 24th, but on the 29th D & T Stevenson apologetically requested[89]

another offer substituting a 5th order for the 4th order for Sandspit as the cost still exceeds the sum named as the expenditure on the two lights. All the other parts of the apparatus, glass etc. for both lights remain as in your offer. PS We have increased the arc of Bean Rock to 212 deg.

The final price was £265.[90] Chance Brothers recorded the order on January 1870.[91] Stores were also provided by Smiths.

For most New Zealand lighthouses, with their first and second order lights, lanterns were sent out ahead of the light apparatus. In contrast, those for the fifth order Bean Rock and Ponui Passage lighthouses could be sent as a single shipment along with the stores. They were shipped on the *Countess of Kintore* from East India Docks on 15 June 1870.[92]

Bean Rock lighthouse

The foundation for the Bean Rock lighthouse consisted of seven cast-iron piles; a central vertical pile surrounded by six others braced so as to have a slope of 1 in 5. The piles were placed on the rock and an artificial foundation created by heaping rocks around the piles up to a height of about five feet [1.5m] above the low water mark with a batter of 1 in 4. Concrete was forced down the centre of each hollow pile and flowing out the lower end displaced the water in the surrounding rock pile to form a rock/concrete mass.[93]

Concrete was placed on the upper rocks only where they surrounded the piles. It was anticipated that rock oysters would adhere to the rocks and strengthen the mass. The Southern Cross article commenting on this bio-adhesive referred the reader to

> the old breakwater, the stone of which was so firmly embedded together that when, some time ago, the idea was entertained of utilising the stone for another purpose, it was easier to cart stone from the quarries.

The Bean Rock light was exhibited on 1 December 1871.[94] The Notice to Mariners was extensive; it described the two white, two red, and one green sectors, and gave advice on entering the harbour from each of the three approach channels.

A heavy gale in February 1874 damaged the masonry foundation,[95] but the lighthouse continued in operation. Further gale damage to the masonry around the piers occurred in March 1874.[96] The repairs were undermined the following July when another gale attacked the structure. The 1886 report refers to extensive repairs but also shows that the previous breakwater experience had not been forgotten. It stated that a quantity of rock-oysters had 'from time to time been planted in the interstices of the stonework, in the expectation that they will, in the course of a few years, spread over the entire surface and bind it together'.[97]

When the provincial governments were abolished in 1876, the Bean Rock lighthouse seems to have come under the Marine Department by default. In 1877 the Marine Department endeavoured to transfer responsibility for the Bean Rock lighthouse to the Auckland Harbour Board, which had been established in 1871. This was stoutly resisted[98] and the Collector of Customs Auckland, who acted for the Marine Department in lighthouse matters, was informed that:

> At a meeting of the Board held on 23rd January last it was resolved, that the Board do not take over the Bean Rock Lighthouse unless some provision made for the maintenance of same.[99]

This was replied to in a letter from the acting Minister of Customs dated 19 February 1877,[100] which stated that the writer had been

> directed to inform you that light essentially [*sic*] harbour light, and as such will need to be maintained by Harbour Board. The Government would maintain it until 31st next month [i.e. 31 March].

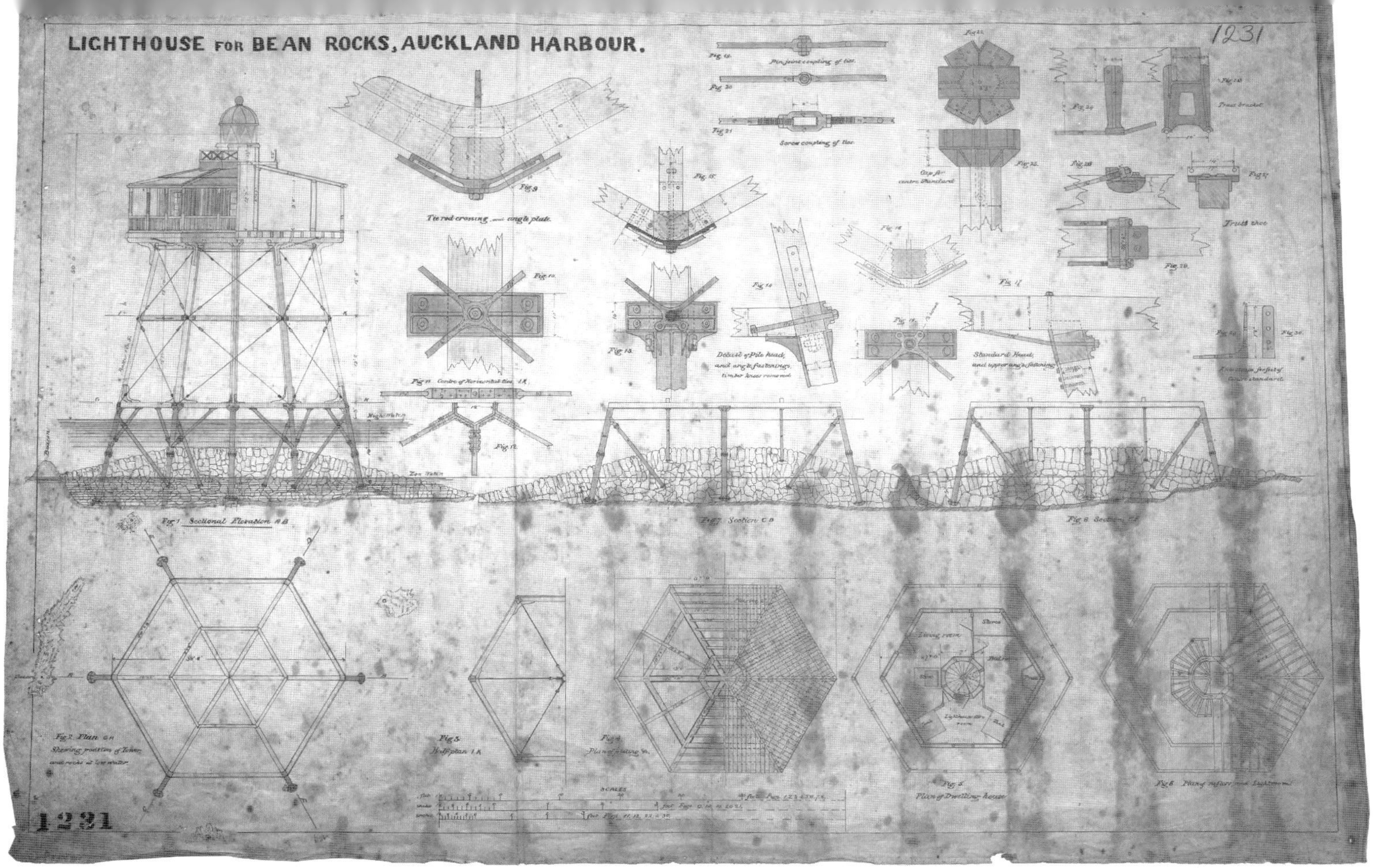

Figure 4.11: Design for Bean Rock lighthouse © Archives New Zealand (Auckland), the Department of Internal Affairs Te Tari Taiwhenua

Before this happened, the Works Committee met on 15 March and resolved:[101]

> That taking into consideration the fact that the General Government are receiving the whole of the light dues for the Colony and that this Board has not power to levy light dues the cost of maintaining the Bean Rock Lighthouse would become a burden on the present diminishing income of the Board, this Committee therefore recommend that the Bean Rock Light be not taken over by the Board.

The government made one last effort in reply. A telegram from William Seed to the collector in Auckland[102] stated that

> Bean Rock light is within jurisdiction of Auckland Harbour Board and will have to be maintained by that body. Have written to Chairman here on.

There is no record of a letter from Seed in the Auckland Harbour Board minute books and, as shown above, the Marine Department continued to maintain the lighthouse.

In 1912 it was decided to convert the light to an unwatched acetylene-acetone burner, and the Harbour Board agreed to take over the light from October. However, in a letter dated 22 May 1913,[103] it was pointed out that the Marine Department was still collecting light

[Left] Figure 4.12: Bean Rock Lighthouse after Restoration at AHB's workshops © Auckland Harbour Board collection, Voyager New Zealand Maritime Museum, Auckland, New Zealand

[Below] Figure 4.13: Resiting of Bean Rock lighthouse © Auckland Harbour Board collection, Voyager New Zealand Maritime Museum, Auckland, New Zealand

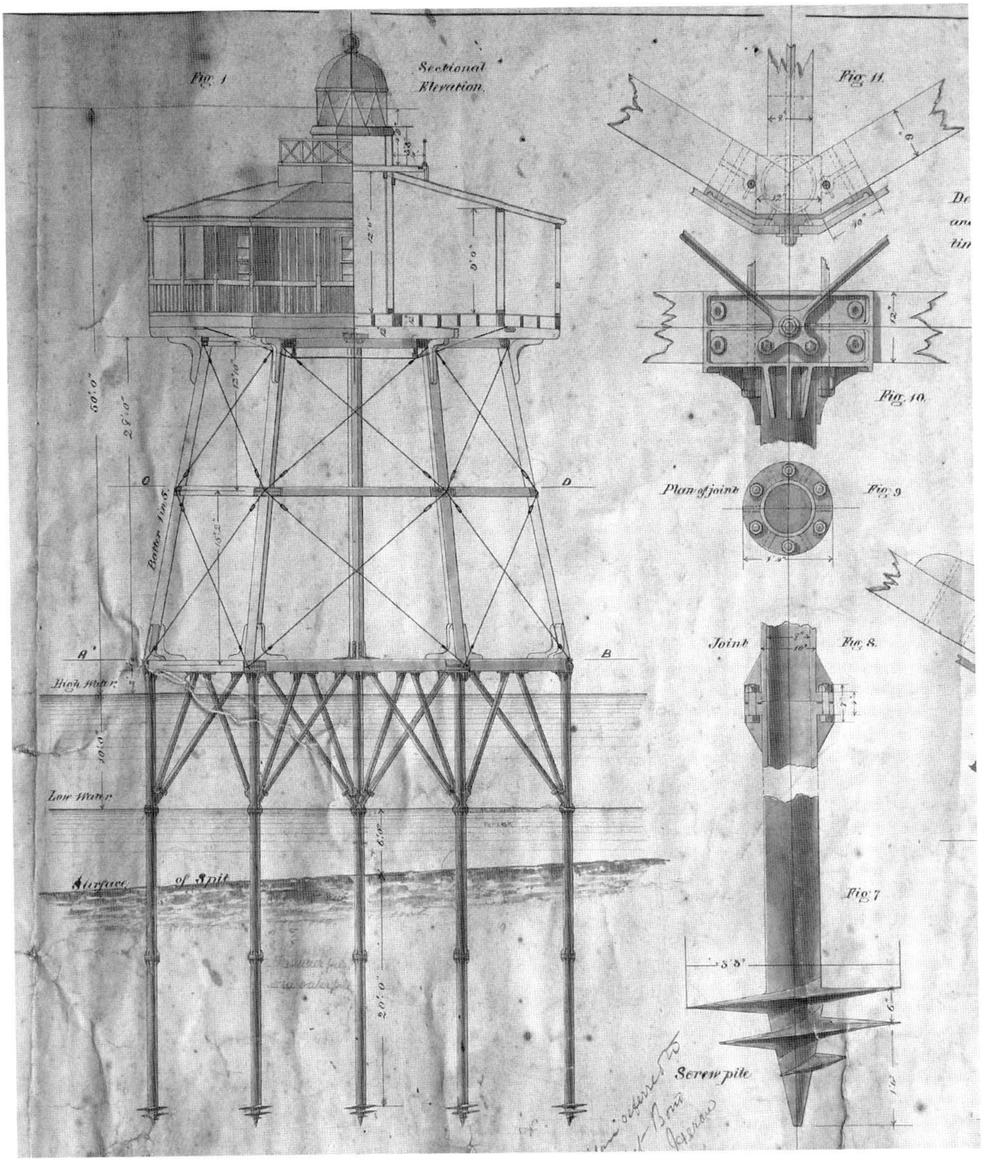

Figure 4.14: Design for Ponui Passage lighthouse
© Archives New Zealand (Auckland), the Department of Internal Affairs, Te Tai Taiwhenua

dues on vessels passing it on the way to islands in the Hauraki Gulf and coastal settlements. The matter was resolved by an Order in Council i.e. a directive from the governor.

The upgraded light was exhibited on 21 October, and the characteristic changed from fixed to flashing, with a flash of one second followed by four seconds of darkness.[104]

On 22 November 1915 the characteristic was changed to 3 seconds flash and 8 sec eclipse, and there was a change in dividing lines of coloured sectors.[105] A change to group flashing was made on 1 February 1924.[106] The new characteristic was 3 flashes every 8 seconds: 1 sec light, 1 sec eclipse,1 sec light, 1 sec eclipse,1 sec light, 3 sec eclipse; visible for 10 miles. The colours and sectors remained unaltered.

The light was electrified in 1936, when an undersea cable was laid from nearby Orakei Wharf.

In 1985 it was recognised that the lighthouse was in need of repair. While designs were put forward for a totally new lighthouse, public opinion convinced the Auckland Harbour Board to restore the existing lighthouse. This was particularly important from a heritage viewpoint, as the sister cottage on Ponui Passage had been demolished.

The lighthouse was lifted off its foundation and repaired at the Harbour Board's workshop in Auckland. The foundation was strengthened with concrete and the repaired lighthouse delivered to its site by the Harbour Board's floating crane *Hikunui.*

Ponui Passage lighthouse

The Ponui Passage light was to direct shipping southward to Thames through the narrow navigable channel at Ponui. Stewart, in his paper on the lighthouse foundations, estimated the width of the channel to be 'one-tenth of a mile' wide [approximately 160 m]. Test borings done by the chief harbourmaster showed that the sandspit consisted of

> loose sand and shells, with shingle, over-lying at a depth of seven to fourteen feet (2.1 to 4.2 m) a hard crust, the exact thickness of which was undetermined.
>
> Below this was a homogeneous sub-stratum of soft muddy and sandy clay; a very hard but not rocky bottom was reached a depth of thirty-one feet from the surface of the sand.[107]

For Ponui's foundation, Stewart opted to use screw piles driven vertically. These had been invented by Alexander Mitchell (1780–1868) an Irish civil engineer who built up a successful civil engineering and brickmaking business despite being blind. He was made AMICE in 1837. Screw piles were first used on the Maplin Sands lighthouse in 1838.[108]

Seven screw piles were driven by a capstan 'embracing the body of the pile, which moved upwards a few feet at a time as the pile descended'. (This was the method that Mitchell had used on the Maplin Sands.) Most of the weight of the structure was carried on the six outer piles. The seventh central pile was designed to take the thrust of the other piles through the bracing when winds were high

The Ponui Passage light was exhibited on 29 July 1871, although the Marine Department did not place a Notice to Mariners in the *New Zealand Government Gazette.*[109] On the abolition of the Auckland Provincial Government, Ponui Passage, like Bean Rock lighthouse, came under the Marine Department.

The Marine Department does not appear to have attempted to get the Auckland Harbour Board to take over the Ponui Passage light until it was automated. It was decided in 1914 to automate Ponui Passage lighthouse, along with the Nelson light,[110] with an acetylene-acetone burner. The successful installation at both lighthouses was announced in the 1916 report.[111] The new Ponui characteristic was 'a flash every 2 seconds, thus 1/3 second flash, 1 2/3 seconds eclipse'.[112]

Figure 4.15: Original Ponui Passage lighthouse (ATL PA Coll-007-0-4) © New Zealand Institute of Engineers Collection, Alexander Turnbull Library, Wellington, N.Z.

The next major change occurred in 1938 when the 'cottage' and the above-pile framework were removed and a new platform placed on the pile caps. The lantern was transferred to a new white wood and steel tower 24 feet [7.3m] high placed on this platform.[113] The tower had supporting struts, and contemporary photographs show a structure resembling a miniature Blackett lighthouse.

In 1984 the acetylene-acetone burner was replaced by a solar-powered light on the original screw pile foundation.

Endnotes

1 AJHR 1866 E – No 3 Second Report of the Marine Board Department 1865–66 Enclosure No 1
2 AJHR 1866 E – No 3 Second Report of the Marine Board Department 1865–66 Enclosure No 2
3 AJHR 1866 E – No 3 Second Report of the Marine Board Department 1865–66 Enclosure No 3
4 AJHR 1866 E – No 3 Second Report of the Marine Board Department 1865–66 Enclosure No 4
5 AJHR 1867, E-No 6 Third Report on the Marine Department for 1867 E-No 10, Report Upon the Proposed Lighthouse at Farewell Spit
6 Morrison-Law A.D. 2010, *Northern Lights, The Age of Scottish Lighthouses* p 209
7 Acc 10706, vol 36 pages 542 letter Stevenson to Balfour, 18 February 1868
8 Acc10706 vol 112 p 103–108 Specification of Two Lighthouse Lanterns for the New Zealand Government 14 May 1868
9 Acc 10706, vol 36 pp 592–594, letters to James Dove & Co, Milne & Son, and Slight
10 Acc10706 vol 36 p 595 letter to John Morrison Esq New Zealand Government Agents London dated 5 March 1868
11 Acc 10706, vol 36 p 644 letter to Chance Brothers, Birmingham
12 AJHR 1868 E – No 6 Fourth Report on the Marine Department for year 1868. Report Balfour (Colonial Marine Engineer) on working of Marine Department 1867–8
13 Acc 10706 vol 112 pp 121–123 Specification of Glass Works and mountings for Farewell Spit and Cape Campbell Lighthouses in New Zealand 4 April 1868
14 Chance Archives, BS6/12/1/1/2 pp 97, 98
15 Chance Archives, BS6/12/1/1/2 p 132 14 August 1868
16 Acc 10706, vol 36 p 903 letter to Morrison, New Zealand Government Agent, London 25 May 1868
17 Acc10706 vol 36 p 919 letter to Morrison, Agent for New Zealand Government 27 May 1868
18 Acc 10706 vol 37 p 44 letter D & T Stevenson to James Balfour 10 July 1868
19 Acc 10706 vol 37 p 163 letter to Morrison re New Zealand Lighthouses 18 Sept 1868
20 Acc 10706 vol 38 p 3 letter to Balfour Colonial Marine Engineer re New Zealand Lighthouses 5 August 1868
21 *NZG Gazette* 1869 No 15 Tenders p 152 11 March 1869
22 *NZG Gazette* 1869 No 22 p Notice to Mariners p 201 22 April 1869
23 *NZG Gazette* 1870, No 29, Notice to Mariners No 11 of 1870, p 257 8 June 1870
24 AJHR 1871, G-No 6 Sixth Report of the Marine Department 1869–70 and 1870–1
25 AJHR 1873 H-19 Eighth Report of the Marine Department Enclosure in No 2 Report on Lighthouses on the Coasts of New Zealand, by R Johnson and R A Edwin
26 Acc 10706 vol 45 p 849 letter to Milne & Son re New Zealand Lighthouses 4 August 1875
27 Acc 10706 vol 49 pp 740, 741 letters to Milne & Son and Agent General 11 September 1878 re New Zealand Lighthouses
28 Acc 10706 vol 50 p 200, letter to Milne & Son.re New Zealand Lighthouses February 1879
29 Acc 10706 vol 50 p 201, letter from Milne & Son.re New Zealand Lighthouses February 1879
30 Acc 10706, Vol 85 letter from Milne & Son, 13 March 1879
31 AJHR 1892, H-29 Marine Department Annual Report for 1891–92
32 Acc 10706 vol 79 p 962 letter to Wilson 24 October 1891
33 AJHR 1893, H-31 Marine Department Annual Report for 1892–93
34 Acc 10706 vol 61 p 23 letter to Wilson 26 November 1891
35 Acc 10706 vol 61 p 56 letter to Wilson 19 December 1891
36 Acc 10706 vol 61 p 157 letter to Milne & Son 24 February 1892
37 Acc 10706 vol 61 p 164 letter to Milne & Son 25 February 1892
38 Acc 10706 vol 61 p 342 letter to Agent General 23 April 1892
39 AJHR 1894, H-18 Marine Department Annual Report for 1893–94 Letter Marine Engineer to Secretary Marine Department 8 August 1894
40 *NZG Gazette 1895*, No 5, Tenders for Farewell Spit Lighthouse p 135 24 Jan 1895
41 AJHR 1896, H-15 Marine Department Annual Report for 1895–96
42 AJHR 1897, H-15 Marine Department Annual Report 1896–97
43 *NZG Gazette* 1896, No 94, Notice to Mariners No 36 of 1896, p 2025 4 December 1896
44 AJHR 1898 H-15 Marine Department Annual Report for 1897–98
45 Acc 10706 vol 45 p 849 letter to Milne & Son re New Zealand Lighthouses 4 August 1875

46 Acc 10706 vol 46 p 293, letter to Milne & Son.re New Zealand Lighthouses 17 December 1875
47 AJHR 1912 H-15 Marine Department Annual Report for 1911–12
48 NZG Archives ADOE 16612 M1/739 8/28/10 Telegram A Pearce Principal Keeper to Secretary Marine Department 17 June 1929
49 *Dominion* Wellington 20 June 1929
50 NZG Archives ADOE 16612M1/739 8/28/10 Memo Secretary Marine to Minister 2 March 1951
51 AJHR 1955, H-15 t Marine Department Report for the year ended 1955
52 AJHR, 1867, E-No 6 Third Report on the Marine Department for 1867
53 Acc 10706 vol 37 p 135 letter to Dove re New Zealand Lighthouses 26 August 1868
54 Acc 10706 vol 37 p 904 letter to Morrison re New Zealand Lighthouses 29 June 1869
55 Acc 10706 vol 40 p 890, letter Agent General re New Zealand Lighthouse, 6 May 1872
56 *NZG Gazette* 1869 No.15 Tenders p 152 11 March 1869
57 *NZG Gazette* 1869 No.Notice to Mariners p 201 22 April 1869
58 *NZG Gazette* 1870, No 23 Notice to Mariners No 16 of 1870, p 377 18 July 1870
59 *NZG Gazette* 1903 No 20 Tender p 801 13 March 1903
60 AJHR 1903 H-15 Marine Department Annual Report for 1902–3
61 AJHR 1904 H-15 Marine Department Annual Report 1903–1904
62 AJHR 1905 H-15 Marine Department Annual Report 1904–5
63 *NZG Gazette*, p 2339 Notice to Mariners No 76 of 1905 Sept 28 1905
64 AJHR 1911, H-15 Marine Department Annual Report for 1910–1911
65 NZG Archives ADOE 16612 M1/721 8/14/10 Letter Fraser to Secretary Marine Department dated 10 March 1933
66 NZG Archives ADOE 16612 M1/721 8/14/10 memo Fraser to Secretary Marine Department, dated 8 October 1914
67 NZG Archives Letter George Perrit, Assistant keeper in charge to Secretary Marine Department dated 30 September 1914
68 NZG Archives telegram Creamer Principal Keeper to Secretary Marine Department dated 24 November 1916
69 NZG Archives ADOE 16612 M1/721 8/14/10 Memo Creamer to Secretary Marine Department dated 1 December 1916
70 AJHR 1937, H-15 Marine Department Annual Report for the year 1937
71 NZG Archives ADOE 16612 M1/721 8/14/10 Invoice No 5164 to Store Manager PWD from Levin & Co Ltd dated 15 June 1938
72 AJHR 1939, H-15 Marine Department Annual Report for the year 1938–9 .
73 NZG Archives AEBE 18507 LE/1/58 1868/117 145
74 NZG Archives AEBE 18507 LE/1/58 1868/117 145
75 NZG Archives AEBE 18507 LE/1/58 1868/117-145
76 NZG Archives WP/3/23 68/478 draft letter Superintendent dated 21 November 1868
77 Acc10706 uncatalogued letter, Balfour Marine Department Wellington to D & T Stevenson, 3 September 1869. ref 432.d
78 Acc 10706 uncatalogued specification for Bean Rock and Ponui Passage prepared by Balfour dated 'Wellington September 3rd'.
79 Acc 10706 uncatalogued letter, Balfour Marine Department Wellington to D & T Stevenson, dated 'October 2nd 1869. ref 88.f'
80 AJHR 1868, E-No 6 Fourth Report on the Marine Department for the year 1868
81 NZG Archives BBAO 5544 box 19a record 1877/278 Letter 305/77, Public Works Department Auckland, to Collector Customs
82 NZG Archives BBAO A78 5544 192 1877/278, Report Stewart J to Collector of Customs, Auckland 18 April 1877
83 Acc 10706 Vol 38 p 196 letter to Dove & Co re New Zealand Lighthouses 3 Nov 1869
84 Acc 10706 Vol 38 p 197 letter to Milne & Son re New Zealand Lighthouses 4 Nov 69
85 Acc 10706 Vol 38 p 984 letter to Morrison re New Zealand Lighthouses 23 June 70
86 Acc 10706 vol 38 p 265 letter to Chance re New Zealand Lights 16 Dec 69 (copy in vol 112)
87 Acc 10706 vol 38 p 299 letter to Chance re New Zealand Lighthouses 21 Dec 69
88 Acc 10706 vol 38 p 303 letter to Chance re New Zealand Lighthouses 23 Dec 69
89 Acc 10706 vol 38 p 320 letter to Chance re New Zealand Lighthouses 29 Dec 69
90 Acc 10706 vol 38,p 322 letter Mr John Morrison London re New Zealand Lighthouses 31 Dec 69

91 Chance Archive BS6/12/1/1/2 p 211
92 Acc 10706 vol 38 p 821 letter to Chance re New Zealand Lighthouses 26 May 70 also pp 822, 823
93 Daily Southern Cross Volume XXVI, Issue 4029, 21 July 1870, Page 2 Proposed Lighthouse at Bean Rock and the Sandspit
94 *NZG Gazette* 1872 No 4Notice to Mariner No 3 of 1872 p 53 17 January 1872
95 NZG Archives ACFM 8180 14/ 564/74 Letter Burgess Harbourmaster to Superintendent
96 AJHR 1885, H-13 Annual Report of the Marine Department
97 AJHR 1886, H-24 Annual Report of the Marine Department
98 Auckland Harbour Board Archives, Committee Minute Books SN2/3 p 188, Minutes of Works Committee, 15 January 1877 Voyager New Zealand Maritime Museum
99 NZG Archives BBAO 5544 Box 19a 1877/100 Memo from Secretary Auckland Harbour Board (James Brigham) to Collector of Customs Auckland, Feb 5 1877
100 NZG Archives BBAO 5544 Box 19a 1877/144 letter to Chairman Auckland Harbour Board 19 February 1877
101 Auckland Harbour Board Archives Committee Minute Books SN2/3 pp 206, 206, Minutes of Works Committee, 15 March 1877 Voyager New Zealand Maritime Museum
102 NZG Archives BBAO 5544 Box 19 1877/17 Telegram Seed to Collector Auckland, 24 May 1877
103 NZGA BBAO A133 5544 Box 149a 1913/1732 Letter Secretary Auckland Harbour Board to Collector of Customs, Auckland 22 May 1913
104 *NZG Gazette* 1912 No. 78 Notice to Mariners 124 of 1912, p 3003, 15 October 1912 also in NZGA BBAO 5544 233/a 1912/2488
105 *NZG Gazette* No 124 Notice to Mariners 89 of 1915, p 3700 28 October 1915
106 *NZG Gazette* No 6 Notice to Mariners 5 of 1924, p 303 29 January 1924
107 Stewart J Assoc. Inst. C.E. *A Description of the Foundation of the Lighthouse in the Ponui Passage.* Transactions and proceedings of the Royal Society of New Zealand, 1868 p 135 Art.XL
108 Lutenegger A J *Historical Development of Iron Screw-Pile Foundation:1836–1900* Int J for the History of Eng & Tech, vol. 81 No 1 Jan 2011, 108–28.
109 *NZG Gazette* 1871 Notice to Mariners of 1871
110 AJHR 1914, H-15 Marine Department Annual Report for1913–14
111 AJHR 1916, H-15 Marine Department Annual Report for 1915–16
112 *NZG Gazette* 1916 No 6 Notice to Mariners 5 of 1916, p 266 21 Jan 1916
113 *NZG Gazette* 1938 No 63 Notice to Mariners No 41 of 1938, p 105 23 Aug 1938

5 Manukau and Napier Bluff Lighthouses: the Doty Burners

In the seventh report of the Marine Department, William Seed wrote:

> From particulars given in correspondence printed and laid before the Imperial Parliament, [i.e. in London] it appears that a very considerable saving could be effected, and at the same time a more brilliant light produced by using kerosene instead of colza oil. I have therefore been making enquiries of the lighthouse authorities in various places where I understand mineral oil is used with the object of seeing whether the same kind of oil could not with advantage be made use of in New Zealand.

At the time he wrote this (on 28 August 1872), Bean Rock and Ponui Passage lights with their kerosene burners had been in operation for nine and thirteen months respectively, so it is surprising that he did not mention them. Seed had assumed responsibility for the Marine Department in February 1870, and probably the two small Auckland lights had not yet 'swanned into his ken'.

The lights which were built at Manukau South Head and Napier Bluff marked another step forward in lighthouse technology in New Zealand.

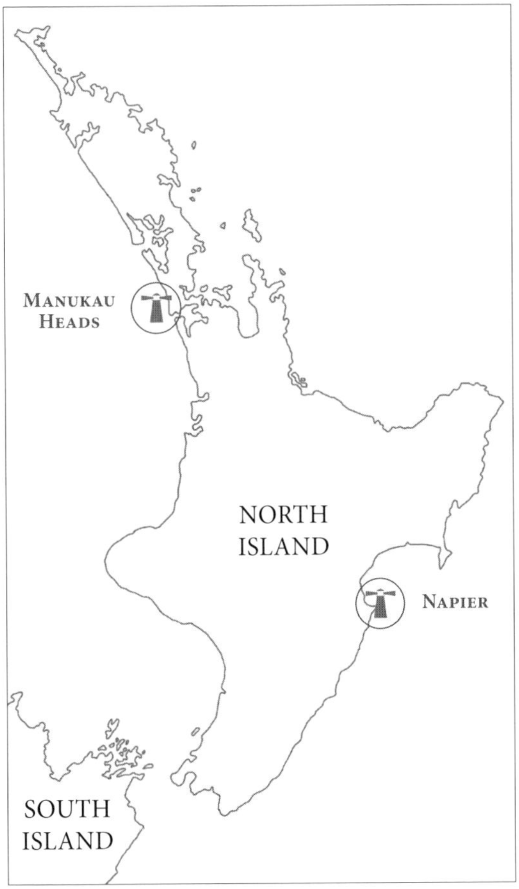

Figure 5.1: Locations of Manukau Heads and Napier Bluff lighthouses © Author

Given Seed's interest in using mineral oil it is appropriate that the first two lighthouses ordered by him used paraffin instead of the usual colza oil, and the revolutionary Doty burner.

Manukau Heads

One of the first acts of the newly elected Auckland Provincial Council was to consider the question of navigation in Manukau Harbour. In his closing address to the first session the superintendent stated that:

> Buoys are being placed in those positions which Commander Drury has pointed out: a flagstaff and semaphore will be shortly erected at Paratutahi, and a Pilot station formed at that place[1]

The erection of the staff 'with telegraphic arms' on the northern side of the entrance to Manukau Harbour (Manukau North Head) was announced in September 1854.[2] The

Figure 5.2: Manukau Harbour showing Heads and Bar (ATL WA-F-3533-F)
© Whites Aviation Collection, Alexander Turnbull Library, Wellington, New Zealand.

telegraph arms were primarily to signal to ships attempting to cross the harbour bar. That this signal was not always clearly understood was shown on 7 February 1863 when the HMS *Orpheus* was wrecked on the bar with the loss of 181 lives. In May the wreck was sold to Messers Coombes and Daldy, who advertised that 'all PARTIES having in their possession any PORTION of the SAID SHIP or any of her appurtenances, or who may hereafter become possessed thereof, will be required to ACCOUNT FOR THE SAME'.[3] Obviously someone profited by the disaster.

Later issues of both the *Auckland Provincial Gazette* and the *New Zealand Government Gazette* contained announcements of beacons, leading marks and buoys in the harbour or at the entrance. The buoys seem to have frequently moved.

The importance of Manukau Harbour to New Zealand's economy was recognised by the Beacons and Lighthouses Committee that reported to the first session of the House of Representatives in 1854.[4] It described the harbour as the

> only refuge for vessels driven into deep bight on a lee shore, well situated for communications with Auckland, southern parts of Colony and Australia,

and pointed to the need for lights to guide ships over the Manukau bar.

The following year the maintenance of buoys in Manukau Harbour was considered by the House of Representatives[5] and again in 1867.[6] In his section of the third report, Balfour indicated that an estimate had been prepared for a temporary sixth order light at Manukau.

It was decided to establish a fifth order light on the head. The order was included in the requisition sent for Farewell Spit and Cape Campbell.

Slight,[7] Dove & Co[8] and Milne & Son[9] were invited to tender for the Manukau lantern, which was designed to be '7ft 3 inches [4.7m] in height from the bottom to the top of astragal'. Morrison, the New Zealand agent, was advised to grant the contract to Dove & Co for £819.[10]

Chances were invited to tender for '130° of a fifth order Dioptric fixed apparatus 37½°. The apparatus is to be fitted with a hemispherical mirror in Edinburgh'.[11] This was recorded in the Chance Brothers' order book on 26 September 1868 along with those for Farewell Spit and Cape Campbell.[12] Milne & Son manufactured the lamps,[13] and Smith supplied stores, tools glasses and oil cisterns.[14]

The complete lantern, apparatus and stores were shipped to New Zealand on the *Memento*, which left St Katharine's Dock, London, in January 1869. In the letter sent to Balfour giving details of the shipment, D & T Stevenson indicated two unusual aspects:[15]

> The lantern has been kept in one piece with the ventilator etc on it so that you will only require to place it on its tower …
>
> Instead of making astragals in the lantern as in your requisition we thought it better to send the panes in one piece bent to the circle; three panes have been sent, two being spare.

These unilateral changes demonstrate that D & T Stevenson were confident that the Marine Department would accept their judgment. The total cost was £170/11/5.[16]

In April 1869 the Marine Department announced the proposed erection of a 'Temporary Light at Manukau Heads... on one of the beacons on the South Head of Manukau Harbour'. It was anticipated that because of the height of the land [500 ft, 152m] the light would be seen at a distance of 30 nautical miles in very clear air. However, Balfour indicated that

> [t]his light is only meant to benefit vessels accustomed to trade on the coast; it is not intended in any way to encourage vessels to attempt to enter Manukau Harbour during the night, nor is it meant to encourage strangers to approach the port.[17]

Perhaps the latter was to encourage use of the Pilot service.

The shipment was sent to Wellington and stored as the proposed tower had not yet been built. As it was still in storage when Balfour died on 29 June 1870, the House of Representatives resolved to send the shipment to Manukau and to instruct Stewart to design the tower and oversee installation. At the time Stewart was overseeing the erection of the Bean Rock and Ponui Passage[18] lights and so the lantern was placed in storage in the Custom House at Onehunga, the Auckland port on Manukau Harbour.

It is not known how far Stewart had proceeded in his design, but he was overtaken by events when in December 1870 the Custom House, along with neighbouring buildings, was burned down and the lighthouse shipment destroyed.

The Marine Department took the opportunity to upgrade the Manukau lighthouse from a temporary fifth order light to a permanent third order fixed light, to illuminate an arc of 180°.[19]

The requisition was sent along with the order for the Napier Bluff lighthouse, which the Marine Department had undertaken to obtain for the Napier Harbour Board. This appears to have been sent directly to D & T Stevenson by the Marine Department and not through the Agent General as was usual.[20]

Although the order had been sent there was still some debate as to the final site for the Manukau lighthouse. In a letter dated 7 May 1873 to Thomas Gillies, the superintendent of Auckland Province, William Reynolds asked that the opinions of the pilot and harbourmaster be obtained. Reynolds also pointed out that the options were the Manukau South Head or Paratutai, the Manukau North Head.[21] O'Rorke, the previous superintendent and now member for Onehunga, had urged South Head, emphasising the closeness to the existing pilot and proposed telegraph station. Captains Johnson and Edwin had recommended Paratutai.

Gilles reply was a telegram[22] agreeing to obtain the opinion of Thomas Wing, the harbourmaster, but also pointing out acidly that the question 'has been twice at least settled and should not be reopened'. In his report Wing recommended the South Head.[23]

The specification for the lantern was a version of the standard specification for a cast-iron and gunmetal lighthouse, as had previously been supplied. The lantern was to be six feet [1.8 m] in height from the bottom to the top of the astragal. The six sole plates were to form a polygon of 12 sides described about a circle of 8 feet 3 inches [2.48 m]. The trimming path was too have an inside diameter of 5 feet 6 inches [1.65 m]. The lantern was to be glazed for 270° and the remaining portion fitted with dormant panes of sheet iron ⅛ inch [3.2 mm] thick.[24]

Milne & Son and Dove & Co were both invited to tender,[25] and the tender for the Manukau lantern (and also the fourth order one for Napier Bluff) was awarded to the latter.

The lantern was despatched from Edinburgh on 12 April 1873 to London for shipment on the *City of Auckland* from the East India Docks.

Chance Brothers were approached to supply the lens along with the pillar and table. The original order, dated 29 October 1872, was for a lens consisting of three panels of 60° and a dioptric mirror of 90° (1½ panels). The table was to be of diameter 3 feet 6 inches [1.07m] and the height from top of table to base of pillar 6 feet 3⅞ inches[26] [1.90 m]. Subsequently the Marine Department decided to widen the dioptric mirror to 120°.[27] Chance Brothers recorded the additional order two days later.[28] The eventual cost was £455.

It is not known what lamps were originally specified, but during the early part of this project, Seed was seeking advice on lamps burning mineral oil rather than vegetable colza. In the eighth report of the Marine Department he informed the House that

> Some valuable information, received from Commander G. P. Heath, Portmaster at Brisbane, and from the Deputy Minister of Marine and Fisheries of Canada, induced the Department, in ordering the light for Manukau and a small light for the Provincial Government of Napier, to suggest to the engineers in Great Britain who superintended the construction of the lantern and apparatus, that lamps to burn kerosene should be adopted; they recommended that Captain Doty's patent burners should be used and these have been ordered.[29]

The order took the form of a telegram dated 8 February 1873 from the Marine Department to the Agent General, ordering D & T Stevenson to 'use Doty burners for Manukau and Napier Lighthouses'.[30] These new burners were being tried out by the Northern Lighthouse Board, and New Zealand was thus employing these burners less than a year after they had first been installed in the Cloch lighthouse on the Clyde.[31] Milne & Son were invited to give an offer for Doty lamps for both lighthouses;[32] their price for the Manukau lamps was £97/16/-[33] although the final invoice was for £98/12/-.

Seed anticipated that these lamps would eventually be used throughout New Zealand, giving a brighter light than the existing colza burners at an annual saving of approximately £407/5/-. He added that:

> a supply [of kerosene ... sufficient for the first year's consumption, has been ordered from England. It is intended to test this oil with the best

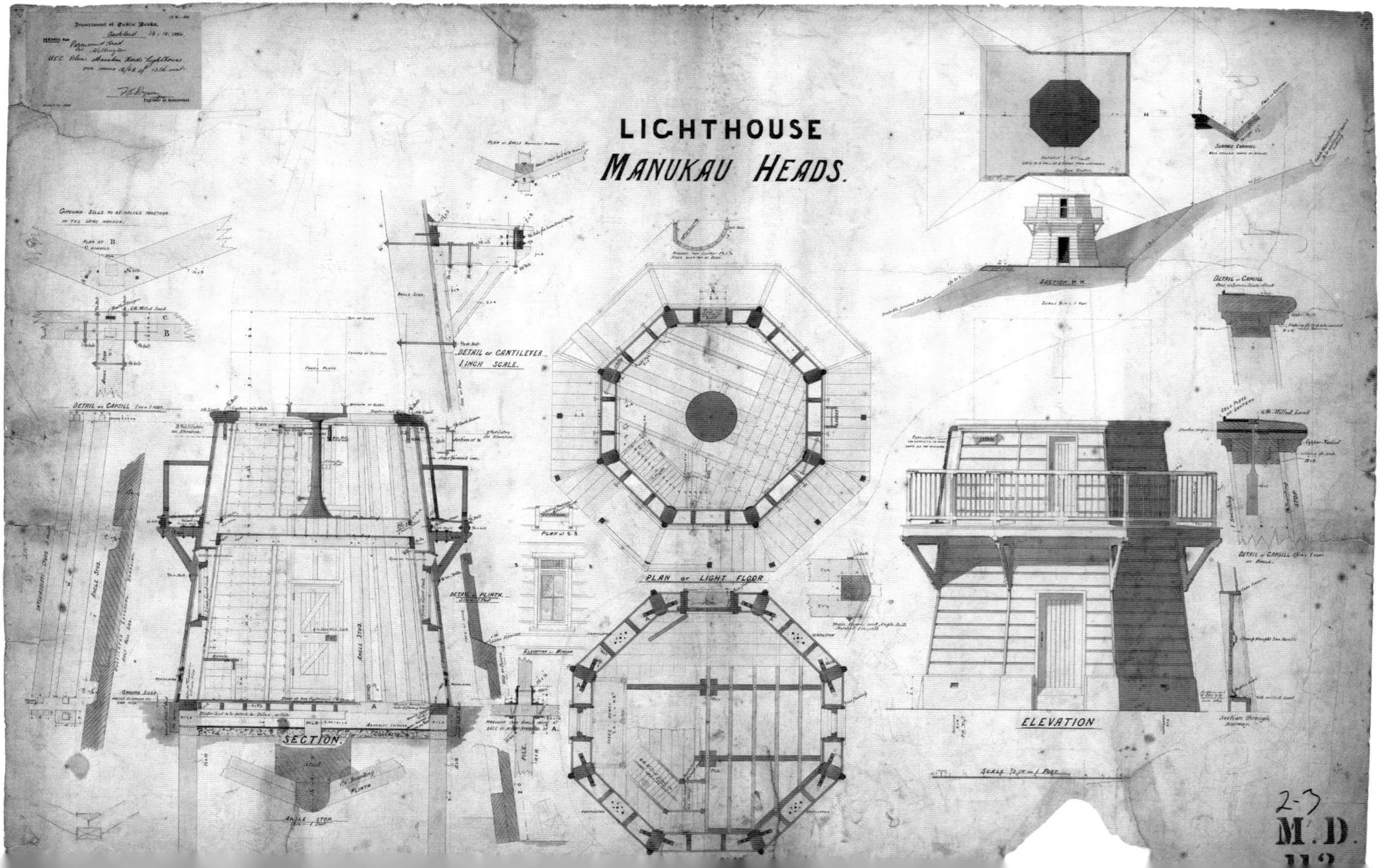
LIGHTHOUSE
MANUKAU HEADS.
DETAIL OF CANTILEVER
1 INCH SCALE.
SECTION H.H.
DETAIL OF CAPSILL
DETAIL OF CAPSILL
PLAN OF LIGHT FLOOR
DETAIL OF PLINTH.
PLAN AT B
ELEVATION OF WINDOW
SECTION.
PLINTH
ANGLE STUD.
ELEVATION
SCALE ½ IN = 1 FOOT.
Section through Doorway.
2-3
M.D.

Figure 5.4: Manukau Heads lighthouse, 1908 (AWNS19080116-15-6) © Sir George Grey Special Collections, Auckland Libraries

sample procurable in the New Zealand market, with a view of arranging hereafter, if possible, to purchase it in the Colony.

This was to be a common feature of Seed's tenure as secretary of the Marine Department – a constant search for alternative supplies of lighthouse oil in an effort to save money. Balfour, Seed, and Chance Brothers all referred to 'kerosene', while D & T Stevenson generally called the fuel 'paraffin'. Paraffin for the Scottish lights was obtained by distilling oil shales.

This also meant that for the first time tanks for paraffin (also referred to as 'oil cisterns') had to be supplied. Specifications were sent out on 13 March 1873 to Smith & McLean of Glasgow[34] and Dove & Co.[35] It was Dove & Co who were given the contract, but six days later they were asked what reduction would be made if the tanks had two coats of paints instead of galvanising.[36] The final cost was £106/14/4.

Youngs Paraffin Oil Company of Glasgow supplied 700 gallons of paraffin in five-gallon tins, at a cost of £76/0/10.[37] Although the original intention had been to ship from Glasgow, Stevensons were instructed to have it sent to London for shipment on the *Ocean Mail*.[38] The Marine Department was later to have problems with shipping paraffin to New Zealand.

Stores were supplied by Smiths,[39] who also supplied joiners' and tinsmiths' tools,[40] and paraffin testers[41] for a total cost of £79/8/-.

The apparatus, lamps and stores were shipped from Granton on 25 November for shipment to New Zealand by the *Ocean Mail* from South West India Docks, London.[42] D & T Stevenson also sent separately a copy of the standard flame and a copy of the instructions which the Northern Lighthouse Board had issued for the guidance of their lightkeepers in the use of the paraffin lamps. They also enclosed instructions for using the paraffin tester.[43]

It was originally assumed that the paraffin was to be sent by rail from Young's refinery in West Lothian to Glasgow, but instead the Agent General instructed that the cases be

[Opposite] Figure 5.3: Construction drawing, Manukau Heads lighthouse (ABPL 8607 W5220/35 MD112-2) © Archives New Zealand, the Department of Internal Affairs, Te Taiwhenua

[Above] Figure 5.5: Abandoned Manukau Heads lighthouse (Footprints(03368-sm) © Gerard Johnson/ Fairfax Media / Auckland Libraries

[Left] Figure 5.6: Replica Manukau lighthouse © Christopher North

sent to Poplar Station so as to be sent along with the rest of the shipment on the *Ocean Mail.*[44]

The usual practice up till that time was for D & T Stevenson to obtain tenders and then submit a summary to the provincial agent or Agent General with a recommendation as to which tender to accept. From the tenders for Manukau and Napier onwards this practice ceased and D & T Stevenson themselves awarded the tenders on behalf of the New Zealand Government.

By this time Blackett had been appointed marine engineer and was responsible for the design of the 20 foot [6 m] tower. This was the first example of the classic New Zealand timber lighthouse of the style described in his ICE paper.

The light was 385 feet a.s.l. and the range 26½ nautic miles.[45] The contracts for the tower and the keeper's dwelling were let in October 1873[46] to Cameron, and the light was exhibited on 1 September 1874.[47]

Incandescent lamps were installed at Manukau in 1916,[48] which would have led to the installation of a revolving dome.

The 1925–1926 annual report stated the intention to automate the light when funds became available.[49] The new acetone-acetylene light was exhibited on 5 March 1929. The fixed light was changed to group flashing, showing two flashes every 30 seconds, with the pattern-flash 1 sec; eclipse 1 sec; flash 1 sec; eclipse 27 secs. The range was reduced to 18 nautical miles.[50]

In 1935 the light was converted to electricity.[51] Due to erosion the original tower had to be removed, and was replaced by a 12-sided poured concrete tower 8 feet high and approx 10 feet diameter.[52]

While the original dome and lens were initially retained, a new fixed dome was designed for Manukau in 1944.[53]

The light was transferred to the Auckland Harbour Board[54] in 1974.

The light was extinguished in 1986 and the lighthouse was abandoned.

Six years later the tower was removed when it was again threatened with erosion. The lantern appears to have been dumped in local farmland, but the lens was placed in secure storage, probably the Harbour Board store. The local community formed a lighthouse trust to build a replica lighthouse on an adjacent site using the rescued lantern, pillar and table, and lens. The design was based on Blackett's original, but unfortunately the architect responsible was not knowledgeable in lighthouses and the proportions are such that the lens is placed too low in the lantern.

Napier Bluff

Hawkes Bay was declared a province on 1 November 1858 and the following year J D Ormond, superintendent of the province, commissioned a report on harbour improvements. However, this concentrated on dredging and drainage with no recommendation of a lighthouse.[55] The council papers for 1868 refer to 'proposed expenditure for a lighthouse at Napier' and 1869 saw a Lighthouse Reserve Act. The Provincial Council was primarily concerned with developing the port, so although the Napier lighthouse was primarily intended to direct shipping into Napier Harbour, it was purchased on the council's behalf by the Marine Department. The Provincial Government reimbursed the General Government and agreed to maintain the light.[56] In practice the lighthouse was operated by the Marine Department and the Provincial Government, and later the Napier Harbour Board reimbursed the Marine Department.

The chosen size was fourth order. The specification[57] describes the lantern as having

> 6 feet internal diameter from glass to glass and 3 feet 6 inches in height of daylight ... gun-metal astragals 1 inch broad and copper roof and ventilator.

The sole plate consisted of four cast-iron pieces forming an eight-sided polygon with an internal diameter of 5 feet 9 inches [1.73 m]. Dove & Co and Milne & Son were invited to tender for the fourth order lantern on 6 November 1872,[58] and the contract was awarded to Dove & Co, six days later.[59] Dove & Co's price was £190.

The specification for the apparatus[60] required the glasswork to be set in gunmetal frames 'the three vertical frames being connected by rings at top and bottom'. The pillar was to be cast iron, and cast in a single piece. The accompanying sketch indicated that the pillar was to be 6 feet [1.8 m] in height.

Milne & Son were invited to submit a price for the spherical mirror.[61]

It seems that Chance Brothers were initially invited to tender for the dioptric apparatus and pillar[62] and subsequently suggested an alteration to the design. Stevenson approved the proposed section but reminded them that the pillar had to be shortened to keep the focal plane at the same level as in the original design.[63]

The day following the reply to Chance Brothers, Milne & Son were also invited to tender for the dioptric apparatus although it is not obvious whether this was for the original design or Chance's modified one.[64] The contract was let to Milne & Son as their price for the dioptric apparatus was lower; £78.10/- for the fourth order dioptric apparatus, and £19.10/- for the spherical mirror.[65]

The telegram that changed the specification for Manukau to Doty burners also included Napier Bluff.[66] For Napier these were two-wick burner Doty lamps made by Milne & Son.[67] Stores were supplied by Smith along with those for Manukau. The cost for Napier Bluff was £21/6/-.[68] As was done for Manukau, the order was later augmented with a supply of joiners' and tinsmiths' tools,[69] and paraffin testing equipment.[70]

The Napier Bluff lantern was finished ahead of that for Manukau, and the Agent General was reminded on 3 April 1873 that '[t]he lantern for Napier Bluff Lighthouse has been packed for some time'.[71] It was despatched from Edinburgh by coastal steamer along with that for Manukau, and shipped to New Zealand on the *Lennox Castle* from East India Docks in April 1873.[72]

The apparatus, lamps, and stores were sent south by Caledonian Railway on 21 May 1873. They were shipped to New Zealand on the *Celestial Queen* from East India Docks in late May 1873.[73]

This tower was also a timber one, probably a Blackett design.

It was 20 feet high [6 m], the light was 160ft [4.8 m] a.s.l. and had a range of 18 nautical miles. It was first exhibited on 5 May 1874. The Notice to Mariners[74] described it thus:

> Fixed Bright Light of fourth order dioptric; will be exhibited from sunset to sunrise from a tower erected on the eastern side of Napier Bluff, Hawke's

> The power of the light will be considerably increased, the present light being replaced by an Aga light with a cluster burner. The light should be visible for a distance of 19 miles.[87]

This new light was not completely automatic. Fraser, the Department's lighthouse expert, informed the Department that it was

> still necessary to have a man in charge to pull down the blinds each morning and put them up in the evening and keep the lens clean, as the lantern is different from the ordinary automatic light which does not require attention'.

It was therefore decided to retain the services of the gaol warder, who until that time had been responsible for the lighthouse. He was paid a salary of £20 p.a. by the Marine Department, which was reimbursed by the Napier Harbour Board.

Its value as a harbour light was markedly reduced by the Richter 7.8 earthquake which shook Hawkes Bay in February 1931. One of the effects of this earthquake was to lift the seabed around Napier about 2.5 m, placing the light further from the sea.

Parts of the Ahuriri lagoon drained completely, and as the area was part of its endowment, the Harbour Board had the area grassed and used for grazing.

Five years later in April 1936 the Merchant Service Guild (representing ships' masters) recommended that the light be moved to the top of the bluff, in the original lighthouse reserve. This proposal was eventually given serious consideration, and in 1940 possible alterations were considered. The minister of marine was asked for formal permission to shift the lighthouse.[88] This permission was required by Section 197 of the Harbour Act 1923.

However, the light had already been doused as a wartime measure and was never relit. It was officially decommissioned in 1945 and demolished in 1948.

Endnotes

1 Acts & Proceedings Auckland Provincial Council, Session 1 p A-65 Address by Superintendent Closing First Session Provincial Council February 1854

2 *Auckland Provincial Gazette*, No. 23 19 September 1854 Notice to Mariners p 132

3 *Daily Southern Cross* Volume XIX, Issue 1820,19 May 1863, Page 1

4 Votes & Proceedings House of Representatives Session 1 1854 p 415(?), Report of Beacon & Lighthouse Committee

5 AJHR 1864 D-No 8, Correspondence Relative to the Maintenance of Buoys in Manukau Harbour

6 AJHR 1867 E – No 36 Third report on the Marine Board Department for year 1867

7 Acc 10706 Vol 36 p 592, letter Stevenson to Slight 26 Feb 1868

8 Acc 10706 Vol 36 p 593, letter Stevenson to Dove & Co 26 Feb 1868

9 Acc 10706 Vol 36 p 594, letter Stevenson to Milne & Son 26 Feb 1868

10 Acc 10706 Vol 36 p 621, letter Stevenson to Morrison 5 Mar 1868

11 Acc 10706 Vol 36 p 644, letter Stevenson to Chances Brothers 10 Mar 1868

12 Chance Archives BS6/12/1/1/2 p 98

13 Acc 10706 vol 36 p 950 letter to John Morrison 4 June 1868

14 Acc 10706 vol 37 p 481 letter to Morrison re New Zealand Lighthouses 7 Jan 1869

15 Acc 10706 vol 37 p 528 letter to Balfour 21 January 1869
16 Acc 10706 vol 38 p 3 Letter to Balfour 5 August 1869
17 *NZG Gazette* Notice to Mariners dated 22 April 1869,(p 201)
18 AJHR 1871G – No 6 Sixth Report of the Marine Department for years 1869–70 and 1870–1
19 AJHR 1872 G – No 30 Seventh Report of the Marine Department for year ended 30 June, 1872
20 Acc 10706 vol 43 p 619 letter to AG London re New Zealand Lighthouses 6 October 1873
21 NZG Archives ACFM/8180/7/2140/73 letter 123 Col Secretary's office from William Reynolds to Superintendent of Auckland (Thomas B Gillies) 7 May 1873
22 NZG Archives ACFM/8180/7/2140/73 Copy telegram 16 May 1873
23 NZG Archives ACFM/8180/7/2140/73 Copy Wing's report, Harbour Dept, Manukau 20 May 1873
24 Acc 10706 vol 113 pp 140–146, Specification of a Lighthouse Lantern for Manukau Heads Lighthouse for the New Zealand Government. October 1872
25 Acc 10706 vol 42 p 373 letter to Dove & Co re New Zealand Lighthouses 15 October 1872, also letter to Milne & Son, same page
26 Chance Archives, BS6/12/1/1/3 p 131 29 October 1872
27 Acc 10706 vol 42 p 986 letter to Chance re Manukau Lighthouse 26 March 1873
28 Chance Archives, BS6/12/1/1/3 p 160 March 28 1873
29 AJHR 1873 H-19A Eight Report of the Marine Department for year ended 30 June 1873
30 Acc 10706 vol 42 p 849 letter to Secretary of Customs Wellington 20 Feb1873
31 Acc 10706 vol 40 pp 869–71 Report to Clyde Lighthouse Trust, 30 April 1872
32 Acc 10706, vol 42 p 927 letter Milne & Son re New Zealand Lighthouses 8 March 1873
33 Acc 10706, vol 42 p 953 letter Milne & Son re New Zealand Lighthouses 14 March 1873
34 Acc 10706, vol 42 p 949 Circular letter to Messrs Smith & McLean, Morrisbank Glasgow dated 13 March 1873
35 Acc 10706, vol 42 p 949 Circular letter Dove & Co 13 March 1873
36 Acc 10706, vol 42 p 965 letter Dove & Co re N Z Lighthouses 19 March 1873
37 Acc 10706 vol 43 p 686 letter to Messrs Smith Morrison & Co London re New Zealand Lighthouses dated 22 October 1873
38 Acc 10706 vol 43 p 689 letter to Youngs Paraffin re New Zealand Lighthouses 24 October 1973
39 Acc 10706, vol 42 p 972 letter Smiths Blair Street re New Zealand Lighthouses 21 March 1973
40 Acc 10706, vol 43 p 64 letter Smiths Blair Street re Circular Manukau Lighthouse 19 April 1873
41 Acc 10706, vol 43 p 88 letter Smiths Blair Street re N Z Lighthouses 23 April 1873
42 Acc 10706 vol 43 p 694 letter to Messrs Smith Morrison & Co London re New Zealand Lighthouses 24 October 1873
43 Acc 10706 vol 43 p 740 letter to Secretary of Customs 29 Oct 1873
44 Acc 10706 vol 43 p 689 letter to Youngs Paraffin re New Zealand Lighthouses 24 October 1873
45 *NZG Gazette* 1874 No.40 Notice to Mariners No 18 of 1874, p 483 29 July 1874
46 AJHR 1874 H-*22* Ninth report of the Marine Department for year ended 30 June 1874
47 *NZG Gazette* 1874 No 40 Notice to Mariners No 18 of 1874 p 483 29 July 1874
48 AJHR 1917, H-15 Marine Department Annual Report for1916–17
49 AJHR 1926, H-15 Marine Department Annual Report for 1925–26
50 *NZG Gazette* 1929 No.8 14 Feb Notice to Mariners No 6 of 1929 p 403 12 February 1929
51 Auckland Harbour Board Archives, Voyager Maritime Museum File SN 149 /1623 copy instructions to foremen and inspectors
52 Auckland Harbour Board Archives memo Electrical Engineer to Chief Engineer AHB dated 26 April 1974 SN149/1623 Voyager MaritimeMuseum, Auckland
53 NZGA ABPL W5221 7457 Box 1 45/11/2 Drawing M.D.8282 MANUKAU HEAD LIGHTHOUSE Cowling for Dome March 1944
54 Auckland Harbour Board Archives SN 149/1623 letter R N Kerr, Director Marine Division, Ministry of Transport to General Manager AHB, 26 February 1974
55 Acts and Proceedings Provincial Council of Hawkes Bay, 1861 Session 3 Report of Select Committee on Harbour Improvements
56 AJHR 1874H-22 Ninth Report of the Marine Department for year ended 30th June 1874
57 Acc 10706 vol 113 p 148 Specification of a Lighthouse Lanterns for Napier Bluff Lighthouses, New Zealand. November 1872

58 Acc 10706 vol 42 p 484, to Milne & Son re NZL, sgd Dick 6 Nov 1872
59 Acc 10706 vol 42 p 503, Dove & Co re Napier Bluff L, 12 Nov 1872
60 Acc 10706 vol 113 p 151,152 Specification of Dioptric Apparatus for Napier Bluff Lighthouse 7 Nov 1872
61 Acc 10706 vol 42 p 486, Milne & Son re Napier Bluff Lighthouse, 7 Nov 1872
62 Acc 10706 vol 42 p 488, Chance re Napier Bluff Lighthouse, 7 Nov 1872
63 Acc 10706 vol 42 p 508, Chance re Napier Bluff Lighthouse, 12 November 1872
64 Acc 10706 vol 42 p 510, Milne & Son re Napier Bluff Lighthouse, 13 November 1872
65 Acc 10706 vol 42 p 553, Milne & Son re Napier Bluff Lighthouse, 21 November 1872
66 Acc 10706 vol 42 p 849 letter to Secretary Customs Wellington 20 Feb 1873
67 Acc 10706, vol 42 p 925 letter Milne & Son re New Zealand Lighthouses 8 March1873
68 Acc 10706, vol 42 p 972 letter Smiths Blair Street re New Zealand Lighthouses 21 March 1873
69 Acc 10706, vol 43 p 64 letter Smiths Blair Street re Circular Manukau Lighthouse 19 April 1873
70 Acc 10706, vol 43 p 88 letter Smiths Blair Street re N Z Lighthouses 23 April 1873
71 Ac 10706 vol 42 p 1002 letter to Agent General London re New Zealand Lighthouses 3 April 1873
72 Ac 10706 vol 43 p 41 letter to Agent General London re New Zealand Lighthouses 11 April 1873
73 Acc 10706 vol 43 p 183 letter to Agent General London re New Zealand Lighthouses 22 May 1874
74 *NZG Gazette* 1874, No 23 Notice to Mariners No 9 of 1874 p 286 24 April 1874
75 Napier Harbour Board Archive, Hawke's Bay Museum, letter Weber to Secretary Napier Harbour Board, 2 April 1877
76 Napier Harbour Board Archive, Hawke's Bay Museum, Minutes Meeting Napier Harbour Board, 3 April 1877
77 Napier Harbour Board Archives letter M77/889 No 446/15 Customs Department, Marine Branch Secretary to Chairman Napier Harbour Board, 17 April 1877
78 *NZG Gazette* 1878 No.8Notice to Mariners, No 1 of 1878, p 79 10 Jan 1878
79 Acc 10706 Vol 88a letter Wilson to Thomas Stevenson 4 January 1878
80 Acc 10706, vol 85 letter Agent General 4 May 1881 & requisition cs 28/81 4 May 1881
81 Acc 10706, vol 85 pg 278 letter from Milne & Son 10 May 1881
82 Acc 10706 vol 52 p 104 letter to E A Smith London re New Zealand Lighthouse 6 June 1881
83 AJHR 1882, H-12 Marine Department Annual Report
84 AJHR 1883 H-15 Marine Department Annual Report
85 NZG Archives ADOE 12261 M1/751 8/41/2 Memo Allport to Minister, Marine 1912/2780 4 May 1915
86 NZG Archives ADOE M1/ 751 8/41/2 Letter Moore agent Boving & Co. to Secretary Marine Department 24 June 1913
87 *NZG Gazette* 1915 No 47 Notice to Mariners No 27 of 1915 p 1072 7 April 1915
88 NZG Archives ADOEM1/751 8/41/2 Memo Campbell Sec MD to Minister, 19 November 1940

6 Cape Foulwind, Tory Channel and Moeraki

The next paraffin burners were ordered for two lighthouses to light the inter-island trade. The Farewell Spit light marked the western end of Cook Strait, and a light was required to guide ships down the west coast of the South island. The lighthouse at Cape Foulwind was designed to do this.

The discovery of gold in Nelson Province had increased shipping into Nelson, much of it coming from Wellington and through the Marlborough Sounds. The sounds were frequently entered through Tory Channel. The need to light this entrance was debated. It was suggested (in the ninth report of the Marine Department) that a light should not be placed as it would encourage inexperienced sailors to attempt the entrance at night.

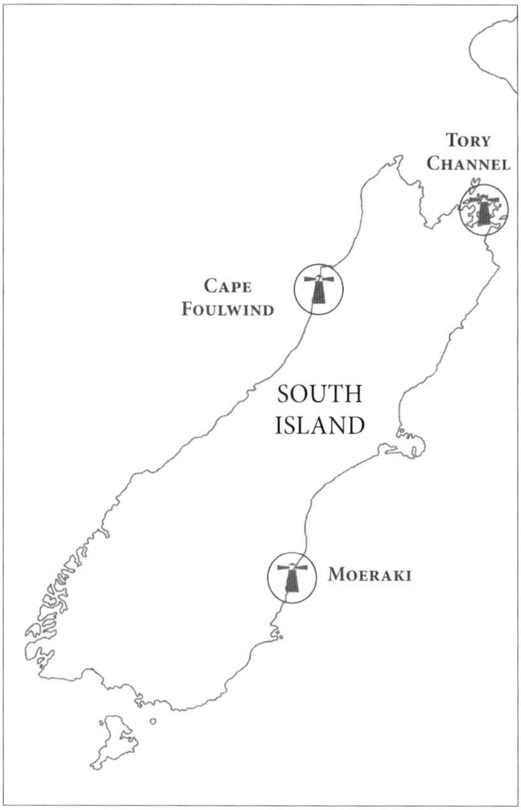

Figure 6.1: Locations of proposed Tory Channel, Cape Foulwind and Moeraki lighthouses © Author

Cape Foulwind

A light at Cape Foulwind was proposed by Edwin and Johnson in their 1873 report.[1] They recommended a second order white light revolving every minute. William Reynolds the commissioner of customs, indicated in his section of the eighth report that the legislature would be asked to grant funds to build seven new lighthouses including Cape Foulwind.[2] In the ninth report Seed reported that the legislature had voted £35,000 for the seven lighthouses.[3]

In November 1873 Johnson and Blackett reported that they

had visited the sites for both Tory Channel and Cape Foulwind lighthouses, and now had sufficient information to allow an order being prepared for lighting apparatus and lantern for each.[4] The orders for the lanterns and apparatus for both lighthouses were received on 2 March 1874 by D & T Stevenson, who prepared the specifications for the second order lanterns.[5]

These were probably identical to the lanterns supplied for Cape Campbell and Farewell Spit six years previously, although these were to prove more expensive. Milne & Son and Dove & Co were invited to tender. The contract for both lanterns was awarded to Dove & Co, who quoted £558 for Cape Foulwind. Manufacture was straightforward, and the lanterns were shipped from Leith by the *Iona* and transferred to the *Soukarz* at the East India Docks. The *Soukarz* sailed in August 1874.[6]

D & T Stevenson suggested an improvement to the apparatus for Cape Foulwind.[7] As the apparatus had to show over 246 degrees and the light was presumed to be full strength up to the limits of this, they felt that a mirror of 114 degrees could not be applied. The extent of the mirror would be only 69 degrees. As this would strengthen only a very small part of the whole, D & T Stevenson recommended leaving the space for the working of the lamps. This advice was taken.

The lens for Cape Foulwind is one of the few where the original specification[8] and drawing[9] are available. The specification was as follows:

> We send you inclosed tracing for a Second Order Holophotal Revolving apparatus required for Cape Foulwind Lighthouse New Zealand for which offers are requested.
>
> The apparatus to be a Second Order Holophotal Revolving Light consisting of 8 lenses, 8 panels of totally reflecting prisms above the lenses and 7 panels below the lenses all of which are to revolve and are to be constructed and fitted in the manner shown in the accompanying tracing.
>
> The prisms are to be adjusted to a frame of 220 feet above the level of the sea and the horizon as seen through the lens is to be 23 mm above the burner. With the burner in this position, the upper prisms are to be adjusted to 29 mm above it and 12 mm behind the axis, and the lower prisms to be 17mm above the burner and 23 mm in front of the axis.
>
> The glasswork armature etc to be in all respects of the same class of workmanship as specified in recent specifications.
>
> The offer to include fitting up of the whole apparatus for trial taking it down and packing it with great care in double cases for transhipment to New Zealand the inner boxes being tin lined (but not soldered up) and delivery of the whole in Edinburgh carriage paid.
>
> All the work to be executed to our entire satisfaction and we are to have the power of rejecting any faulty materials and workmanship which may

be discovered both before and after delivery in Edinburgh and we are to be the sole arbiters in case of any dispute.

The whole work to be finished and delivered in Edinburgh within 20 weeks after the date of the acceptance of the offer. Should the contractor fail to execute the work in the time agreed a reduction of £5 for every week will be made from the contract price for every week that the contract period is exceeded.

The contract price will be paid when the apparatus has been delivered in Edinburgh tried and approved of.

Offers were invited from Chance & Co and Barbier & Fenestre of Paris. The contract was awarded to Barbier & Fenestre at a price of £690.[10] Cape Foulwind was the first lighthouse for which Barbier & Fenestre supplied the apparatus, and they were to provide lenses for most of the lighthouses built after Cape Foulwind and Tory Channel.

The specification[11] for the second order machine etc was similar but not identical to that for Farewell Spit and Cape Campbell. The requirement that the fly be enclosed in a sheet brass box was deleted, and any casting patterns that the contractor had to make had to be 'delivered up at the conclusion of the work' rather than becoming, as required in the 1868 specification, 'the property of the New Zealand Government'. The 1868 requirement was probably the result of an unthinking adherence to a standard Northern Lighthouse Board specification. Such patterns could be retained in the Northern Lighthouse Board store at Granton. It is highly unlikely that the Agent General would wish to build up a collection of industrial *objects d'art*, however beautifully made.

Fortunately the somewhat obscure 1868 paragraph

[t]he cast iron work to be left from the forgings except at the fittings and where otherwise specified is to receive one coat of red lead paint and two coats of dark green paint

was replaced by

[t]he cast and malleable iron work where not otherwise specified to receive one coat of red lead paint and two coats of dark green paint.

The notable difference lay in the instructions for the positioning of the driving weight. The 1868 specification for Farewell Spit stated only that the weight 'is to pass down the centre of the tower'. The specification for Cape Foulwind was more detailed:

The weight rope is to pass from the machine barrel under a grooved pulley not less than 12 inches in diameter whose axle is to be 1⅜ins in diameter and to work in antifriction bushes in the weight similar to those specified for the barrel axle and to have its end secured to the pedestal of the machine case. The weight is to be kept short and the pulley to be sunk in it so as to diminish as slightly as possible the effective fall.

It is not clear why the Cape Foulwind specification is more detailed. Both lighthouses had the same speed of revolution and thus similar falls on the driving weight. Either experience had suggested that more detail was required or there had been an improvement in the design in the interim.

Milne & Son and Dove & Co were invited to tender for the machine, and the contract was let to Dove & Co for £420.[12]

Even though specifications had been drawn up and tenders called, D & T Stevenson suggested

> for consideration whether it might not be desirable to make the apparatus attain its greatest brilliance once every half minute instead of once every minute as in Mr Blackett's requisition provided the characteristics of the neighbouring lights will admit of this being done. We have not sufficient information here to enable us to judge of the characters of neighbouring lights but we would strongly recommend the lights for Cape Foulwind being made with half minute intervals. If we do not hear from you by telegraph on receipt of this letter we shall proceed with the apparatus as requested by Mr Blackett vizt once every minute. The weight will descend down the centre of the tower and your engineer will be good enough to provide for this leaving a space of about 18 inches for the weights.[13]

They also mentioned that due to the increased speed of steam vessels they had been shortening the period of revolving lights especially in narrow seas. The Marine Department agreed to the change[14] and Dove & Co were informed.[15]

The burners were of course Dotys, burning paraffin, and those for Cape Foulwind were manufactured by Milne & Son at a cost of £104/8/-.[16]

The order also called for 30 paraffin cisterns, 15 for each lighthouse. The copy of the specification in the Stevenson Archives is notable for the copyist's anachronistic use of the long 's' (ſ) which by September 1874 had long fallen out of general use.[17] The cisterns were to measure 2 feet10 inches [86.4 cm] high and 1 foot 10½ inches [6.4 cm] diameter with the bottoms of copper cupped ¾ inch [19 mm] deep and filled in with American red pine 1⅛" [28.6 mm] in 'thickneſs'. The specification was otherwise identical to the one prepared for the Six Lighthouses (Chapter 7).

Milne & Son supplied the oil cisterns for both lighthouses,[18] and stores and lamp glasses for both. The total cost of stores and lamp glasses was £148/1/10.[19] Milne & Son also supplied Liverpool lamps for both, with red glasses for Tory Channel and white for Cape Foulwind.[20]

The apparatus, lamps stores etc for Cape Foulwind were despatched by steamer from Granton, to be transferred to the *Border Chief* at South West India Docks.[21] The *Border Chief* sailed in June 1875.

Generally the entire shipment process went smoothly. D & T Stevenson informed the Agent General that the lighthouse materials were ready for packing and shipment. The Agent General would then send instructions as to how the packages [and all items were called 'packages', even when they were crates], were to be addressed and labelled, and which ship they were to be sent to. Generally they were addressed to the secretary of marine if being shipped to Wellington, and to the collector of customs at other ports. The packages were usually labelled with the name (sometimes initials) of the lighthouse, and each contractor was given the numbers to place on their part of the shipment. Detailed lists of contents were provided for each package and sent separately. Most shipments were from London, although occasionally shipments went from Glasgow, particularly when they were orders for paraffin.

The Agent General's letter was usually followed promptly by a note from the shipping agent to the New Zealand Government asking when the shipment could be expected at the London docks. The shipments were either sent from Edinburgh by rail or by local steamer from Leith or Granton which brought the cargo up the Thames to the docks.

In contrast to most shipments, the *Border Chief* arrived in New Zealand with one package missing. This was subsequently found in a London warehouse and despatched to New Zealand with the Centre Island lantern.

Captain Johnson, in his letter of 9 June 1874, reported that the site for the proposed lighthouse had been cleared and he had been able to select the best position for the light. This was on a small schist granite ridge about 15 feet high and 15 feet broad at the top, at an elevation of 190 feet above sea level. This would require a tower 36 feet in height with an illuminated arc extending from S round by W to N 68° E.[22]

Johnston stated in distinctly New Zealand terms that

> the exact position is marked by a large tree, marked by a crow's nest, and bears N. 40° E distant about 6 chains from the original rata marked by Mr Blackett and myself on a former visit.[23]

Details of construction of New Zealand lighthouses frequently refer to 'clearing the bush'; Johnson says it in the report above. It should be kept in mind that 'bush' is the term used in New Zealand not for low shrubby vegetation but for major forest cover, usually indigenous and frequently referred to as rain forest. Consequently clearing the bush for a lighthouse site was a major operation although the felled trees were available for timber on the site. The rimu used in the Cape Foulwind tower was obtained on the site.[24]

By June 1875 the road to the site had been completed and Blackett had begun design and specification for the proposed tower.[25] The tower consisted of a lower open framework and the upper part close-boarded. The contract for the lighthouse tower, dwellings etc on was let to Maher & O'Connor, of Westport, for £2,880/-/-.[26]

The lantern was erected and the apparatus installed under the supervision of John Mills. Mills had been recruited by D & T Stevenson to erect the lanterns and apparatus for the Six Lighthouses ordered in 1875[27] (Chapter 7). On arrival from Scotland he began with

Figure 6.2: Original Cape Foulwind lighthouse (ATL1/1-010094-G)
© F N Jones Collection, Alexander Turnbull Library, Wellington, New Zealand

Cape Foulwind, subsequently erecting the Six Lighthouses.

The Cape Foulwind light was exhibited on 1 Sept 1876.[28]

The Notice to Mariners described it as

> situated on the Cape of that name on the West Coast of the Middle Island of New Zealand. The tower is 53 feet in height from the base to the top of the lantern, and is constructed of timber and painted white; the upper part being close boarded, and the lower part of open framed work.
>
> The light will be a Second Order REVOLVING WHITE LIGHT visible all round the horizon as far as the land will allow. It will attain its greatest brilliance every *thirty seconds.* The light is elevated 190 feet above the sea, and, allowing 15 feet for the height of the eye, will be seen at a distance of 19½ nautic miles in clear weather.

Figure 6.3: Second Cape Foulwind lighthouse (ATL 1/4-102051)
© E R Williams Collection, Alexander Turnbull Library, Wellington, New Zealand

The Doty burner was replaced by an incandescent burner in 1916[29] and the lantern cowl upgraded with a revolving cowl two years later.[30] The Marine Department report on Moeraki light (see below) emphasised that the incandescent light produced more heat than the previous Doty burners and improved ventilation was required.

By June 1923 it had been decided to install a Dalen automatic acetone-acetylene light in Tiri Tiri Matangi lighthouse and it was proposed to do the same at Cape Foulwind.[31] Unfortunately by 1924 the timber structure was beginning to deteriorate. So the decision was made to establish a new lighthouse on an adjacent site equipped with the new lighting system.[32] As the installation on Tiri Tiri Matangi was delayed, Cape Foulwind was the first lighthouse in New Zealand to use the Dalen equipment.[33]

The new tower was white, circular, made of concrete and 30 ft [9 m] high. The new site was 115 feet [35 m] from the former.

The old light was dismantled and the lantern and apparatus placed in storage. The lantern was eventually used in the reinstallation of the lighthouse established on Kaipara North Head.[34]

The new light was exhibited on 21 July 1926.[35] The characteristic remained the same; flashing white each 30 secs: flash 2 seconds, eclipse 28 seconds. The new elevation was 225 ft, and the range 21 miles in an arc of 011° to 242°.

There is little information available on the electrification of Cape Foulwind. This may be because the light had been automated; the Marine Department annual reports tended to concentrate on the manned lights. Electrification was apparently completed in September 1953, as a Notice to Mariners[36] for that year stated that

> [o]n or about 9 September 1953, the light will be altered to show 'flash 1 sec., eclipse 9 secs'.

The change in character was reported in August 1953.

Tory Channel

A light at the entrance to Tory Channel was first suggested in the 1854 report of the Beacon and Lighthouses Committee of the House of Representatives.[37]

Several years later Balfour visited the site, and in his 1865 report[38] suggested establishing a third-order fixed red light on a site on the South Head. He proposed a 'timber framed tower 30 to 35 feet high covered with galvanised iron, and timber dwellings'.

The Marine Department reports of 1866, 1868 and 1870/71 all urged the erection of a lighthouse at Tory Channel. In the latter report, Seed pointed out that Parliament in its previous session had voted a sum for erecting a light at Tory Channel (and Portland Island), but no progress had been made toward erecting these lighthouses as the Government had given directions that they were not to be proceeded with.[39]

Captains Edwin and Johnson in their April 1873 report[40] suggested:

> When the proposed light for Cape Saunders is erected, a Sixth-order white light will be sufficient to mark the entrance of Otago harbour; and the present illuminating apparatus of the Third-order (red) could be removed to Tory Channel entrance, where a light of this order and description is much required, and will complete the lighting of this portion of Cook Strait.

Cape Saunders was not to be lit until 1880.

Meanwhile the need for a light at Tory Channel had been emphasised when on 14 November 1872 the barque *City of Newcastle* was wrecked close to the entrance of Tory Channel with the loss of two crew members.[41] The captain had apparently confused the fixed white light at Pencarrow for the fixed white light at Mana Island.[42] Tory Channel was another of the lighthouses listed by William Reynolds in his section of the eighth report.

The Tory Channel lighthouse was also the subject of a Question in the House in August 1874. The *Marlborough Express*[43] reported that the commissioner of customs [Reynolds] was asked

> why the lighthouse at Tory Channel had not yet been erected; and whether the Government intended to proceed with a work so long contemplated, without delay.

In his reported reply, Reynolds showed a somewhat infirm grasp of New Zealand's geography when he assured the Assembly that

> shortly after the rising of the Assembly last session, action was taken at once, and the 'Luna' was despatched to survey the sites for the lighthouses at both Tory Channel and Cape Campbell [sic]. A report was made and almost immediately a light was ordered from home.

(Johnson and Blackett, who were on the *Luna*, examined the proposed site for the Tory Channel light on the same trip as the site visit to the proposed Cape Foulwind lighthouse.)

Reynolds' appreciation of the niceties of lighthouse procurement was on a par with his geography, as he informed the House that

> there was only one manufacturer of lights required, New Zealand had to take its turn with other countries, and as yet the Government had received no information as to when this light would be shipped.

A review of the correspondence in the Stevenson Archives shows that far from having to 'take its turn', orders from New Zealand for lighthouses were dealt with promptly, with specifications frequently being sent out within a week of receiving the letter from the Agent General. This is one factor which contributed to New Zealand lighthouses having the most up-to-date technology of their time.

The New Zealand Government Archives contains a copy of the specification for Tory Channel light. (This document survived the 1952 fire and is somewhat charred.)

The specification included:

- 135° of 3rd order dioptric apparatus with dioptric glass mirror
- light placed at elevation 145 ft 'above the level of the sea'
- Lantern 3rd order Northern Lighthouse pattern, with gun metal astragals, sole plate and cornice to be cast iron
- Supply spare glass and storm panes
- Blind panes to be supplied 180° as shown in sketch
- Set dome riveting tools, tap and dies for all screws used in construction of the lantern
- Copper frames for red glass and 2 sets spare ruby glass panes

- One set 3 lamps to burn petroleum (or kerosene) oil, to be fitted with best burners invented, made to same gauge as burners for Manukau heads lighthouse. [a reference to Captain Doty burner is crossed out]

- Supply of wicks and lamp glasses

- Stores
  Liverpool lamp
  Oil tanks
  Copper oil measures
  1 oil measure tray
  1 oil funnel
  1 oil feeder
  1 copper oil can
  'and generally all necessary stores for a Petroleum Light'

- Copper wire rope lightning conductor, 190ft with earth plate and platinum point. 3 fixings for timber towers and proper connectors to lantern

- Tracing of sole plate and apparatus stair(?) to be sent earliest opportunity.[44]

The Tory Channel specification showed how clearly Scottish technology had been embraced and how it simplified ordering. The specifier had only to state the size, make a comment about preferred materials and give lengths for the lightning conductor. As has been shown in the description of the Cape Foulwind lighthouse, the two lanterns were ordered and delivered together. The Stevenson Archives do not contain a specification for the Tory Channel lantern, but was probably the standard Northern Lighthouse Board specification.

Dove & Co were also given the contract for Tory Channel lantern, and their tender price was £543 which was only £15 less than their tender for the second order lantern for Cape Foulwind. This indicates the scale of basic costs for manufacturing a lighthouse lantern irrespective of size; the difference primarily represented the extra materials.

As the requisition would have also included a chart of the area to be lighted, D & T Stevenson were able to offer advice on the apparatus i.e.:

> The mirror can be adopted for this apparatus and a section of top panel of prisms furnished which will show over the top of the mirror and give the light required up the channel. The red lamp glass will in this case be preferable to red shades for producing the red light.

The contract for this third order apparatus was let to Barbier & Fenestre for £240,[45] and the contract for lamps to Milne & Son for £98/10/-.[46]

In their letter acknowledging receiving the specification,[47] D & T Stevenson particularly stated that the proposed burners for this light and Cape Foulwind were 'to be on the Doty Principle which are much superior to any other we know of'. It seems that in the final

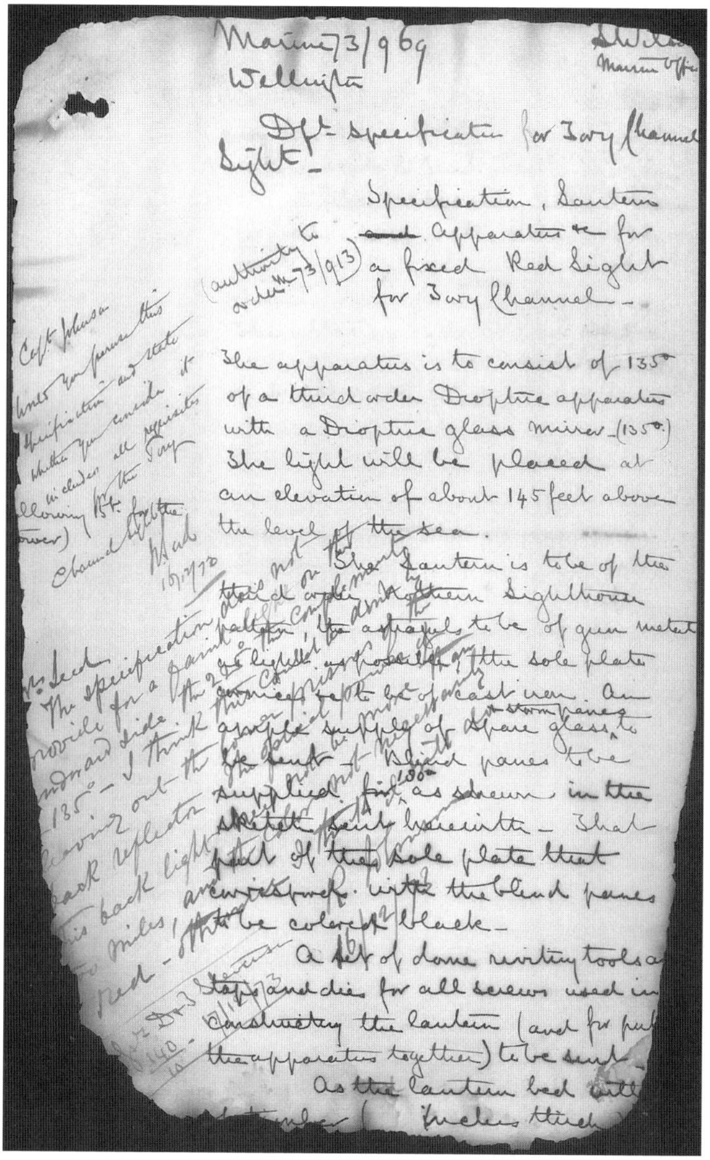
Marine 73/969
Wellington

Dft. specification for Tory Channel Light.

Specification, Lantern Apparatus &c for a fixed Red Light for Tory Channel.

The apparatus is to consist of 135° of a third order Dioptric apparatus with a Dioptric glass mirror (135°). The light will be placed at an elevation of about 145 feet above the level of the sea.

A set of [illegible] tools and dies for all screws used in constructing the lantern (and for put[illegible] the apparatus together) to be sent.

Figure 6.4: First page of the specification for the Tory Channel lighthouse (ADOE 16612 M1/ 749 r 8/39/3)

specification an attempt was made to avoid committing the Marine Department to a given make of burner by crossing out the reference to Captain Doty's burners.

Details of the lantern's despatch are not available, but in mid-September 1874 D & T Stevenson sent to the Marine Department a detailed list, dated August 1874, of the number and contents of the packages shipped for the lantern.[48]

The apparatus, lamps, oil cistern and stores were despatched from Granton by General Steam Navigation steamer on 26 January 1875 and were transhipped to the *Hindostan* at the East India Docks.[49]

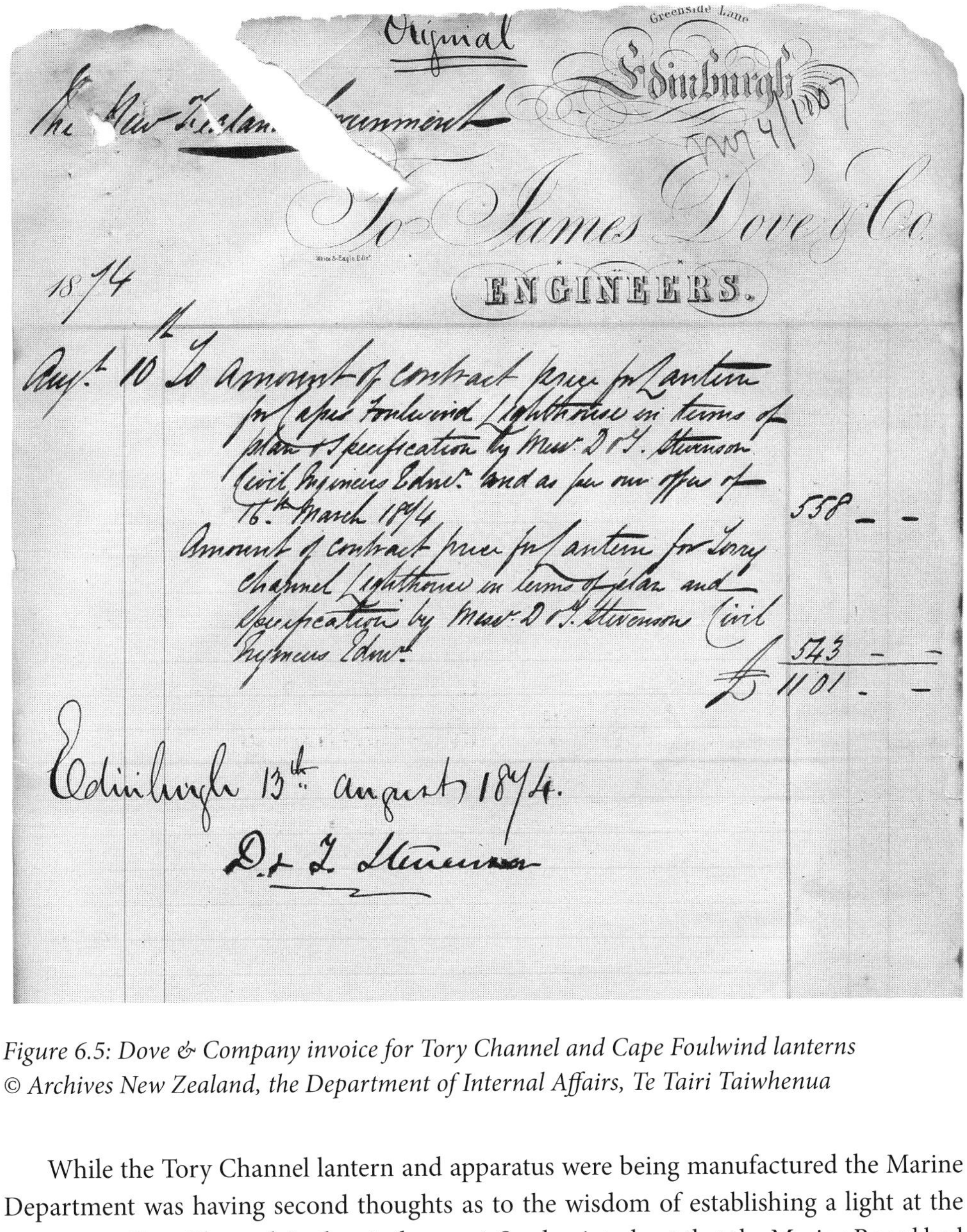

Original

The New Zeal... ...ernment

Greenside Lane

Edinburgh

To James Dove & Co.

ENGINEERS.

1874

Augt. 10 To Amount of contract price for Lantern for Cape Foulwind Lighthouse in terms of plan & specification by Mess. D & T. Stevenson Civil Engineers Edinr. and as per our offer of 16th March 1874 — 558 - -

Amount of contract price for Lantern for Tory Channel Lighthouse in terms of plan and specification by Mess. D & T. Stevenson Civil Engineers Edinr. — 543 - -

£1101 - -

Edinburgh 13th August 1874.

D & T. Stevenson

Figure 6.5: Dove & Company invoice for Tory Channel and Cape Foulwind lanterns © Archives New Zealand, the Department of Internal Affairs, Te Tairi Taiwhenua

While the Tory Channel lantern and apparatus were being manufactured the Marine Department was having second thoughts as to the wisdom of establishing a light at the entrance to Tory Channel. In the ninth report, Seed pointed out that the Marine Board had opted to place lights at Mana Island and Tory Channel as an easier alternative to building a lighthouse on The Brothers, even though the Board regarded 'the Brothers as the most suitable site for a light to facilitate navigation of the Strait'. Not without a touch of drama, he also pointed out that:

> The opinion of shipmasters who trade through the Strait is now, I believe, entirely in favour of a light on the Brothers; and those best acquainted with the entrance to Tory Channel strongly condemn the project of placing a

light there, as they say that owing to the strong tides and narrow entrance, it would not be safe to enter at night, even with the assistance of a light; and that such a light would probably have the effect of luring vessels to destruction, rather than assisting them to avoid the dangers by which the navigation of the locality is beset in dark and stormy weather. They maintain that on no account should the entrance be attempted except in the daylight.[50]

Despite these concerns, Captain Johnson was despatched to Tory Channel the following year to select sites for leading lights to guide ships through the entrance at night.[51] He placed temporary beacons on these sites. Masters of vessels frequenting the channel were asked to comment on the suitability of the course.

Seed's suggestion was that the light ordered for Tory Channel should be used for Cape Egmont, and the Mana light used for one of the proposed new lights. As discussed in Chapter 2, the light on Mana was dismantled and re-erected at Cape Egmont.

In 1875 William Reynolds, made a statement on marine matters to the House of Representatives in addition to the Department's annual report.[52] In this statement he indicated that the Tory Channel apparatus 'will be available for Hokitika, where it is proposed to erect it'. He also stated that the Government proposed ordering at an early date lights for four sites including Moeraki on the Canterbury Coast.

A few months after the tenth report was submitted to the House, the decision was made to use the Tory Channel light at Moeraki, 'where a light was urgently required to guide vessels away from the dangerous reefs lying off that part of the coast'.[53]

In 1881 the Marine Department established a set of leading lights at the entrance to Tory Channel. There is no reference to these in the Stevenson Archive so they were probably ordered directly from Chance Brothers at Smethwick.

Moeraki

The port of Moeraki was established in the 1840s for the export of oil from the nearby whaling stations. With the decline in whaling the port also declined but did not cease operation.

In February 1874 Captain Johnson was asked to visit the coast at Moeraki (as well as Oamaru and areas further south) and to fix a suitable site for a new lighthouse. In his report[54] he referred to the danger afforded to coastal shipping by the Fish Reef off the Moeraki coastline. He concluded that a coastal light was 'seriously needed' and that the light should be erected as soon as possible, and recommended a site near the boat harbour. Johnson suggested that a fourth order light would be sufficient, but it was decided to use the third order apparatus purchased for Tory Channel.

However, the lens had to be modified and Barbier & Fenestre were asked to supply 35° of the central belt and lower prisms only, to be added to the side of the existing apparatus.[55] On 10 April 1875, Milne & Son sent a box containing the apparatus as sent from France, along with wicks, photographs of standard flames and 42 pairs of lightkeepers' spectacles, to London on the East Coast Railway to be shipped to Wellington on the *Camperdown* at the East India Docks.[56]

Figure 6.6: Manifest for Tory Channel lantern
© Archives New Zealand, the Department of Internal Affairs, Te Tari Taiwhenua

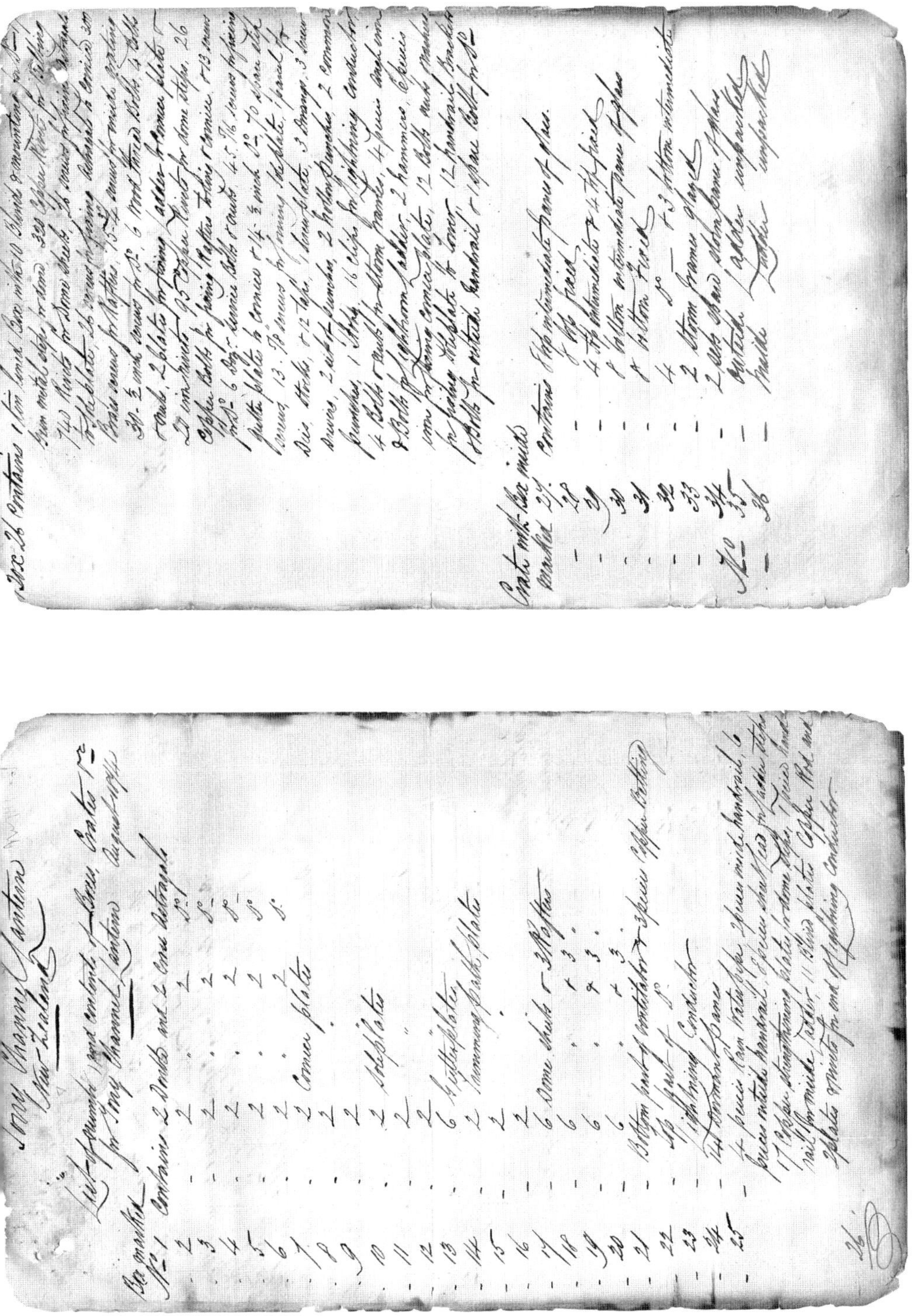

Meanwhile the boxes of burners had been opened and found to be damaged. Seed appears to have taken this calmly, as his letter of 10 January 1876[57] stated:

> I have the honor to inform you that all the burners for Tory Channel Light arrived more or less in a damaged condition, their weight resting on the rack handles having bent and carried away the oil ducts. I should therefore fell [sic] obliged if you would give directions that steps be taken in future to have the burners so packed as to prevent this again taking place.

The letter was copied to Milne & Son, who replied[58] that:

> They were packed so that we considered they could not be injured even by rough usage but from the injury described we will take care to provide against such in the packing of all burners.

It was 1877[59] before plans and specifications for the lighthouse were drawn up and tenders called for. The tower was again a timber Blackett design, 28 ft [8.5 m] high, to be painted white.

The original Tory Channel order had specified red lamp glasses. As it had been decided to establish a white light at Moeraki.[60] Smiths were asked to supply 6 dozen three-wick lamp glasses. These were shipped to New Zealand on the *Hurunui* in September.[61]

The light was eventually exhibited on 22 April 1878.

The Notice to Mariners description was:[62]

> The Tower is 28 feet in height built of wood and painted white. The light will be a Fixed White Light of the Third Order, visible all round as far as the land will allow. The Light will be elevated about 170 feet above the sea, and, allowing 15 feet for the height of the eye, will be seen at a distance of 19 nautic miles in clear weather.

Soon after construction it was realised that the tower had to be further braced against the southern gales.

A similar bracing system was installed on the Brothers light.

The Doty burner was replaced by an incandescent burner which was first lit on 10 May 1917.[63] It worked satisfactorily, giving a more powerful light, for the same consumption of oil. However, as happened at other lighthouses, the heat from light caused the lantern panes to sweat considerably, and Fraser, the lighthouse artificer, recommended installing a revolving cowl.[64] This was done the following year.[65]

In 1926 the nautical adviser suggested that Moeraki be converted to an unwatched light with range 11 miles, and group flashing 3 flashes every 45 seconds, length each flash

[Opposite] Figure 6.7: Moeraki Construction Drawing (M27 65 Drawing MD 40 2-7)
© Archives New Zealand, the Department of Internal Affairs Te Tari Taiwhenua

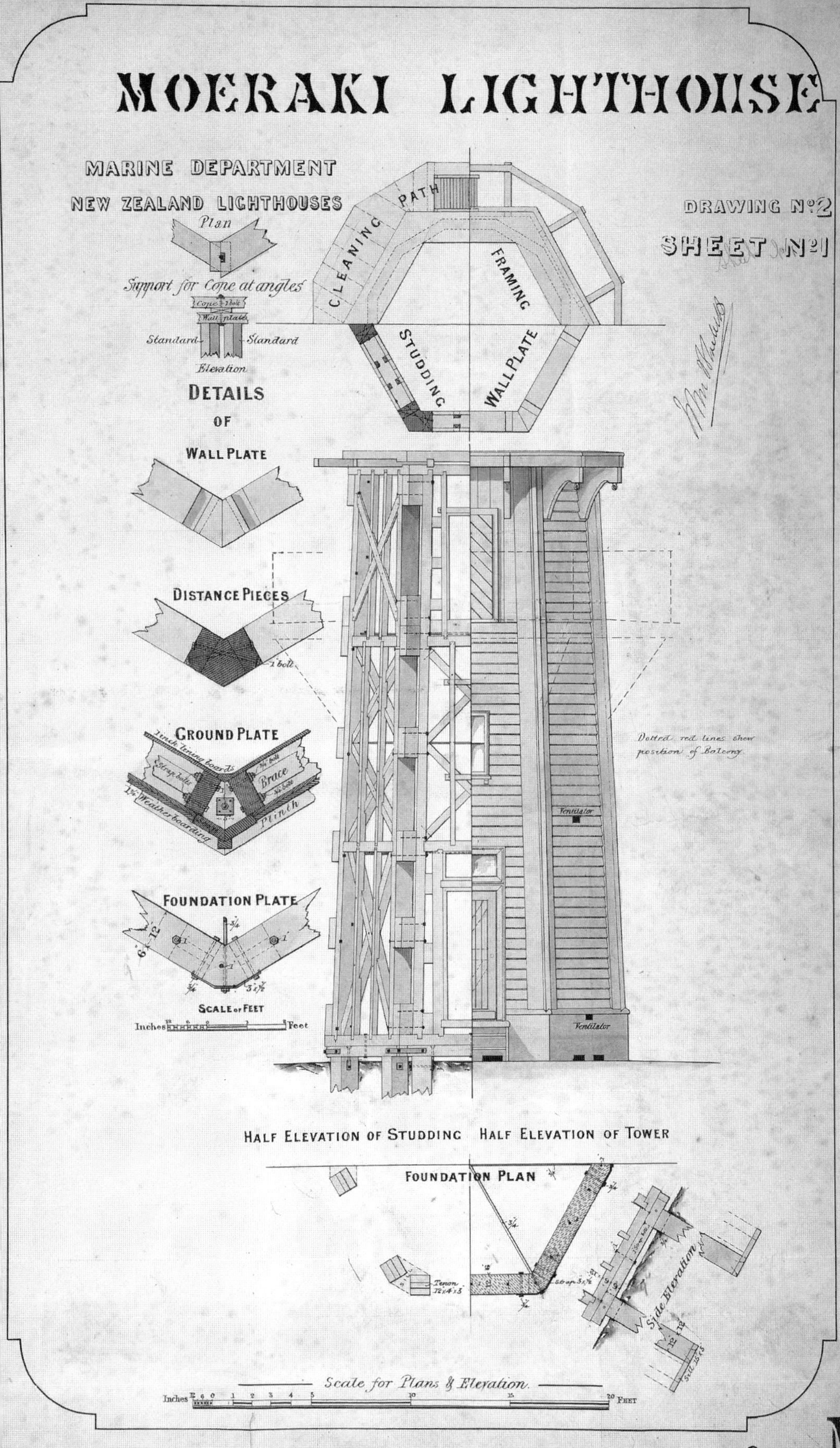

MOERAKI LIGHTHOUSE
MARINE DEPARTMENT
NEW ZEALAND LIGHTHOUSES
DRAWING No. 2
SHEET No. 1
Plan
Support for Cope at angles
Cope
Wall plate
Standard
Standard
Elevation
CLEANING PATH
FRAMING
STUDDING
WALL PLATE
DETAILS
OF
WALL PLATE
DISTANCE PIECES
1 bolt
GROUND PLATE
Brace
Plinth
1½ Weatherboarding
FOUNDATION PLATE
SCALE OF FEET
Inches
Feet
Dotted red lines shew position of Balcony
Ventilator
Ventilator
HALF ELEVATION OF STUDDING
HALF ELEVATION OF TOWER
FOUNDATION PLAN
Tenon
Side Elevation
Scale for Plans & Elevation
Inches
Feet

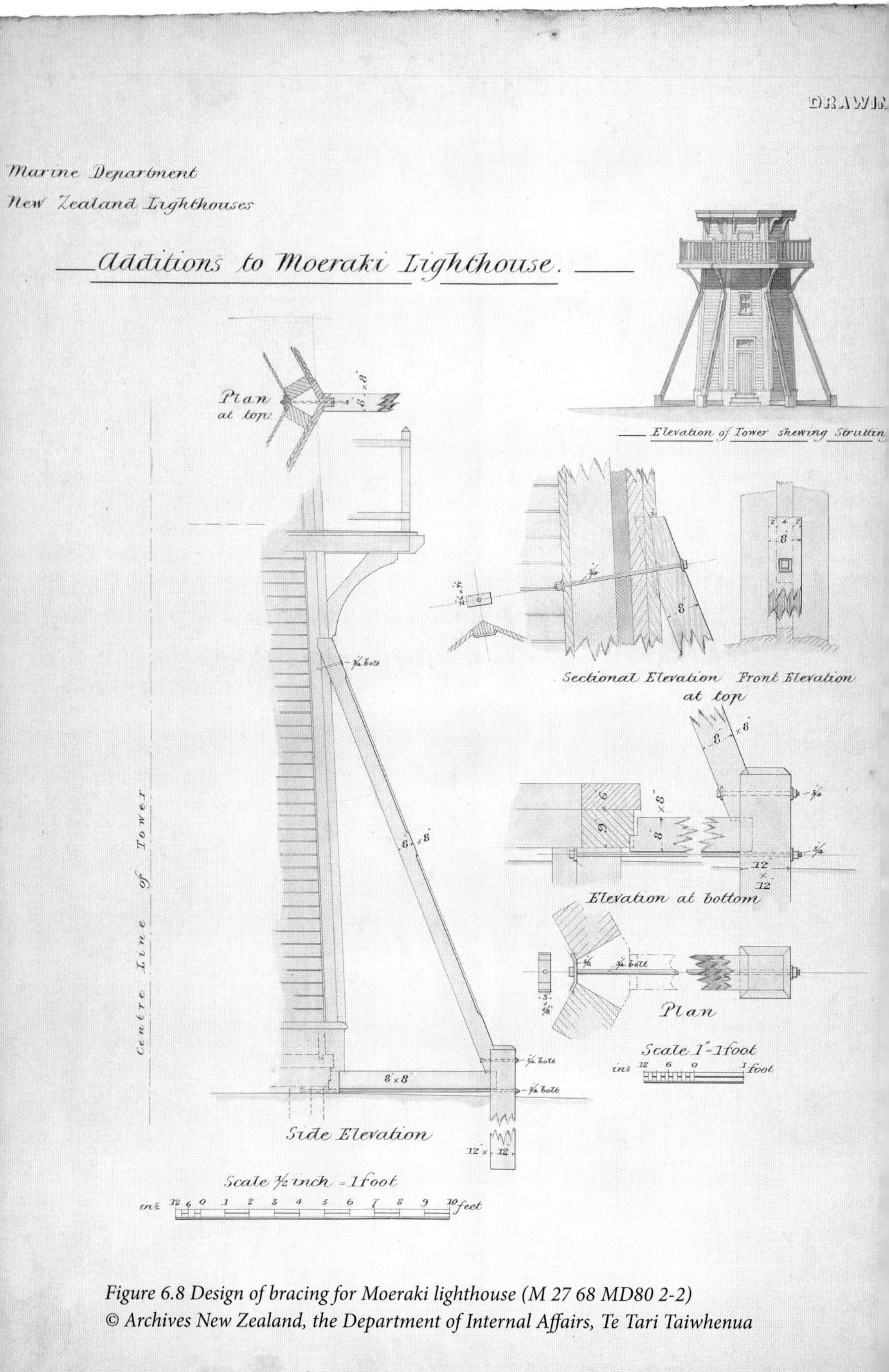

Figure 6.8 Design of bracing for Moeraki lighthouse (M 27 68 MD80 2-2)
© Archives New Zealand, the Department of Internal Affairs, Te Tari Taiwhenua

⅓ sec. The Department indicated in the 1926 report its intention to install the acetone-acetylene lights at three lights including Moeraki if funds became available.[66]

In 1950 the characteristic was changed from fixed to flashing. The annual report for 1941[67] indicated that in addition to building extra accommodation at Moeraki, the Department had constructed power lines for electrification and that the light equipment was on order. However it was almost ten years before electrification was completed. The 1950 report[68] baldly stated that:

> Conversion from fixed to flashing character should be completed in 3 months.

The characteristic recommended by the nautical adviser was; flash 1 sec, eclipse 1 sec, flash 1 sec, eclipse 5 sec.[69]

Endnotes

1 AJHR 1873, H-19 Eighth Report of the Marine Department Report on the Lighthouses on the Coasts of New Zealand

2 AJHR 1873 H-19A Eighth Report of the Marine Department for year ended 30 June 1873

3 AJHR 1874 H-22 Ninth report of the Marine Department for year ended 30 June 1874

4 NZG Archives ADOE 16612 M1/749 8/39/3 Letter R Johnson, J Blackett to Wm Seed re Tory Channel & C.F. 24 Nov 1873

5 Acc 10706 vol 44 p 264 letter to AG London re New Zealand Lighthouses 3 March 1874

6 Ac 10706 vol 44 p 765 letter to Messrs Smith Morrison & Co London re New Zealand Lighthouses 12 August 74

7 Ac10706 vol 44 p 301 letter to Secretary of Customs Wellington re Cape Foulwind Lighthouse 11 March 1874

8 Ac 10706 vol 44 p 352 Circular giving specification for Cape Foulwind lens

9 MS 5885 NLS Drawing for 2nd Order Revolving Holophotal Apparatus for Cape Foulwind, New Zealand March 1874

10 Acc 10706 vol 44 p 395 letter to Barbier & Fenestre re New Zealand Lighthouses 3 April 1874

11 Acc 10706 vol 113 pp 350 –357 Specification of Machine, Machine case, revolving carriage and other works required for fitting up the Apparatus for Cape Foulwind Lighthouse New Zealand 23 March 1874

12 Acc 10706 vol 44 p 437 letter to Dove & Co re New Zealand Lighthouses 14 April 74

13 Acc 10706 vol 44 p 392 letter to William Seed, Secretary Custom (Marine Branch) re New Zealand Lighthouses 3 April 1874

14 Acc 10706 vol 44 p 862 letter to Secretary for Customs (Marine Branch) 22 September 1874

15 Acc 10706 vol 44 p 639 letter to Dove & Co re New Zealand Lighthouses 15 June 1874

16 Acc 10706 vol 44 p 878 letter to Milne & Son re Cape Foulwind Lighthouse 29 Septembet 1874

17 Acc 10706 vol 113 p 612 Specification of Paraffin Cisterns for Torry (sic) Channel and Cape Foulwind Lighthouses, New Zealand

18 Acc 10706 vol 44 p 850 letter to Milne & Son re New Zealand Lighthouses 14 September 74

19 Acc 10706 vol 45 p 15 letter Messrs Milne & Sons re New Zealand Lighthouses 10 November 74

20 Acc 10706 vol 45 p 108 letter Messrs Milne & Son re New Zealand Lighthouses dated

21 Acc 10706 vol 45 p 669 letter to Agent General London re New Zealand Lighthouses 17 June 1875

22 AJHR 1874 H-23 Correspondence Relative to Selection of Site for Projected Coastal Lighthouses No 2 Report Captain Johnson to Commissioner Customs 9 June 1874

23 Metrosideros robusta (Northern Rata), Metrosideros umbellate (Southern Rata)

24 Dacrydium cupressinum

25 AJHR 1875 H-12A Tenth Report of the Marine Department for year ended 30 June 1875

26 *NZG Gazette* 25 September 1875 Tenders p 621 Customs Dept (Marine Branch) Williams H Reynolds

27 Acc 10706 vol 45 p 844 letter to Commissioner of Customs 2 August 1875

28 *NZG Gazette* 1876 No 45 Notice to Mariners 20 of 1876, p 587, 9 August 1876

29 AJHR 1917, H-15 Marine Department Annual Report for 1916–17

30 AJHR 1919, H-15 Marine Department Annual Report for1918–19

31 AJHR 1923, H-15 Marine Department Annual Report for 1922–23

32 AJHR 1924, H-15 Marine Department Annual Report for 1923–24

33 AJHR 1925 H-15 Marine Department Annual Report for1924–25 .

34 NZG Archives M1 box 747 8/35/9 memo Tate (Lighthouse Overseer) to Secretary Marine Department 26 May 1947

35 *NZG Gazette* 1926, No.49 Notice to Mariners No 38 of 1926 p 2177 19 July 1926,

36 NZG Gazette 1953 No 46Notice to Mariners No 63 of 1953 p 1328 6 August 1953,

37 Votes and Proceedings of the House of Representatives, Session 1, 1854

38 AJHR 1865 D No 1 C (Appendix) Report *on the Marine Board Department of New Zealand 1864-5* by the Postmaster General. Report of Mr James M Balfour Wellington

39 AJHR 1871 G - No 6 Sixth report of the Marine Department for year 1869–70 and 1870–1, 12 August 1871

40 AJHR 1873 H-19 Report on the Lighthouses on the Coasts of New Zealand by R Johnson and R A Edwin, dated 3 April 1873 Enclosure in No 2

41 AJHR 1873 H-19AEighth Report of the Marine Department for year ended June 30th 1873 Return of Wrecks

42 AJHR 1874 H-19A Ninth Report of the Marine Department for year ended 30 June 1874

43 Marlborough Express Vol IX, 5 August 1874, p 5 *Tory Channel Lighthouse*

44 NZG Archives ADOE 12661 M1/ 749 8/39/3

45 Acc 10706 vol 44 p 416 letter to Barbier & Fenestre re New Zealand Lighthouses 10 April 1874

46 Acc 10706 vol 44 p 879 letter to Milne & Son re Tory Channel Lighthouse 29 Sept 1874

47 Acc 10706 vol 44 p 305 letter to Secretary of Customs Wellington re Tory Channel Lighthouse 11 March 1874

48 Acc10706 vol 44 pg 862 letter to Secretary Customs Wellington re New Zealand Lighthouses

49 Acc 10706 vol 45 p 191 letter to Agent General London re New Zealand Lighthouses 27 Jan 1875

50 AJHR 1874 H-19A Ninth Report of the Marine Department for year ended 30 June 1874

51 AJHR 1875, H-12ATenth Report of the Marine Department for year ended 30 June 1875

52 AJHR 1875 H-12 Statement on Marine Matters by Commissioner of Customs

53 AJHR 1877 H-29 Twelfth Report of the Marine Department for year ended 30 June 1877

54 AJHR 1874 H-23 Correspondence Relative to Selection of Site for Projected Coastal Lighthouses No 1 & 2

55 Acc 10706 vol 85 letter to Barbier & Fenestre

56 Acc 10706 vol 85 pg 73,letter Messrs Milne & Son 11 April 1876

57 Acc 10706 vol 85 letter from William Seed, Customs Department (Marine Branch) Wellington 10 Jan 1876 Duplicate original via Suez M76/42 No 443/12

58 Acc 10706 vol 85 p 69 letter from Messrs Milne & Son dated 9 March 1876

59 AJHR 1877 H-29 Twelfth Report of the Marine Department for year ended 30 June 1877 Enclosure 1

60 Acc 10706 vol 85, p162 copy specification signed Colonial Secretary Bowen

61 Acc 10706 vol 48 p 635, letter to Agent General, London re New Zealand Lighthouses

62 *NZG Gazette* 1878 No 25 Notice to Mariners No 9 of 1878 p 339, 19 March 1878

63 NZG Archives ADOE 16612 M1 /749 8/39/3 Letter Fraser (Lighthouse Artificer) to Secretary Marine Department 31 May 1917

64 NZG Archives Extract from report by Fraser dated 1 June 1917ADOE 16612 M1 /749 8/39/3

65 AJHR 1919, H-15 Marine Department Annual Report for1918–19

66 AJHR 1926, H-15 Marine Department Annual Report for 1925–26

67 AJHR 1941 H-15 AnnualReport of the Marine Department for the year 1940–41

68 AJHR 1950 H-15 of the Marine Department Annual Report for the year1949–1950

69 NZG Archives ADOE 16621 M1/7318/22/4 pt1 (File titled Centre Island) Memo N W Stuart Nautical Adviser 26 April 1949.

7 1875 'The Six Lighthouses'

In 1875 the New Zealand Government ordered six new lighthouses, and Doty burners to upgrade four other lights. This was the biggest single order ever placed in the course of constructing New Zealand's lighthouses and had a total cost of £16,905/7/6.[1] However, it was in keeping with the times. This was a period of major infrastructure development in New Zealand. The country was settling down after the upheavals of the Land Wars (fought primarily in the North Island) and the government under the premiership of Julius Vogel was expanding New Zealand's communications. Soon after the order was sent to Scotland, Vogel resigned and took up the post of Agent General in London where he was able to oversee the procurement of part of the Six Lighthouses, and the lighthouses for Hokitika, Cape Saunders, and Akaroa Head.

This order included the earliest first order revolving lights procured for New Zealand. The first order light for Cape Saunders/ Nugget was a fixed one.

At this period William Seed, on leave from the Marine Branch, was visiting the United Kingdom when the Six Lighthouses were being ordered. D & T Stevenson were able to consult with Seed directly, leading to some changes in the order.

Although there were no particular innovations in this order, standardisation and the established relationship with D & T Stevenson enabled a big order to be placed with confidence.

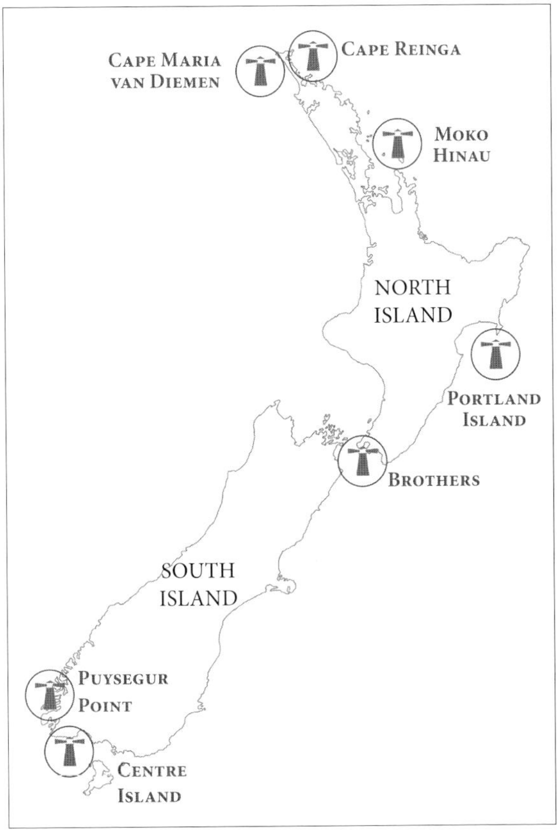

Figure 7.1: Locations of the Six Lighthouses and Cape Reinga © Author

The six lighthouses were:

- The Brothers
- Portland Island
- Centre Island
- Puysegur Point
- Cape Maria van Diemen, (often referred to simply as Cape Maria)
- Moko Hinau.

In correspondence, D & T Stevenson referred to them as 'The Six Lighthouses'.

The Doty burners were ordered to upgrade Nugget Point (referred to as Cape Saunders), Taiaroa Head, Farewell Spit, and Cape Campbell.

Lanterns

The order for the Six Lighthouses was sent directly by the Marine Department in February 1875[2] and not through the Agent General in London.[3] In their acknowledgment of the order (dated 30 April), D & T Stevenson were able to give immediately the heights of the lightrooms; 8 feet for Brothers and Puysegur and 9 feet for the other four lights. As Portland Island was to have a second class lantern and Puysegur a first class lantern these heights were subject to doubt. The letter also indicated that:

> We propose to send as you request first order lanterns for all the lights except Portland Island which will be second class.

This was the combination for which tenders were called. The same letter also indicated that in all cases the driving weights would go down the centre of the tower and would need 28 feet of fall and a space of 12 inches.[4]

Milne & Son and Dove & Co were invited to tender for all six lanterns.[5] This was the first time that *sealed* tenders were specifically required. Milne & Son received a contract for four of the lanterns:[6]

a) One second order lantern for Portland Island Lighthouse for the sum of £509/11/-.

b) One first order lantern for Centre Island Lighthouse for the sum of £905/5/-

c) One first order lantern for Moko Hinau Lighthouse for the sum of £910/4/-

d) One first order lantern for Cape Maria van Diemen Lighthouse for the sum of £905/5.

This left the lanterns for Puysegur Point and The Brothers. Dove & Co had tendered but were asked:[7]

> Please return the Specifications of the lanterns for New Zealand as we must take other offers. Your price being so much higher than the other tenders

Dove & Co apparently made another offer which was still not satisfactory as the next communication said:

> Your offer as amended is much higher than the other tenders and unless you take the work for £950 for each of the two lanterns still required we cannot give you the work. The sum which we have quoted is still much above the other offers. Please say per bearer whether you agree to this.[8]

Dove & Co apparently did, as the next letter accepted 'your offer of this date to furnish two first order lanterns for the New Zealand Government for the sum of £950 cash'.[9]

This bargaining probably ensured that the New Zealand Government got the best price in the circumstances [i.e. Milne & Son not wanting to tender for all six] and a saving of 10/.

However after discussions with Seed, the final agreed order was:

> The Lighthouses of Puysegur, Centre Island, Cape Maria van Diemen and Moko Hinau will have first order lanterns... the lightrooms in each case to be 9 feet from floor to under bed of sole plate. The Lighthouses at Brothers and Portland Island to have second order lanterns the lightrooms for these two lights to be 8 feet high from the floor to the underbed of the sole plate.[10]

It seems that the tender for the second order lantern on The Brothers was transferred to Milne & Son, as in July 1875 Stevenson indicated that Milne & Son should first manufacture the second order lantern for The Brothers.[11] In turn, Dove & Co took over the contract for the Moko Hinau lantern as well as that for Puysegur Point.

The specification for the second order lantern has much in common with that for Cape Foulwind, but there are some slight changes and interesting additions. Most of the straight changes related to materials e.g. both inner and outer skins of the dome are copper instead of copper outer and iron inner, and details of the cornice plate and gutter plate are different. The Brothers/Portland specification includes details of the ladder from lightroom to trimming path. Given the number of lanterns a contractor would expect to be manufacturing a longer period is allowed for completion – 18 months instead of 6. There is also a major change in payments. In the older specification the contractor was entitled to one third of the contract sum when castings were completed and the remainder when the lantern passed inspection and was packed. In the newer specification the contractor became entitled to the full sum when the lantern had been packed and delivered free on board to the export ship. On the positive side, the penalty for late completion was reduced from £10 per week to £5, which became standard for later contracts. In reality Stevenson were loath to enforce this, preferring to gently nag dilatory contractors.

The Stevenson Archive does not contain a copy of the first-order specification; it was probably based on the Northern Lighthouse Board edition but with the same payment schedule and penalty clauses as the second-order specification.

The lantern orders included lightning conductors, and Milne & Son[12] and Dove & Co[13] were provided with the following tower heights to enable the correct length of wire rope to be provided. The heights were:

Brothers 28 ft

Portland Island 28ft

Centre Island 16 ft

Cape Maria 28 ft

Puysegur 58 ft

Moko Hinau 28 ft.

Machines and apparatus

Two days after agreeing the tender for the Brothers and Puysegur lanterns, Dove & Co were sent the specification for the first- and second-order machines[14] which had already been sent to Milne & Son.[15] The original requisition had been for three first order revolving lights and three second order revolving lights. The Marine Department apparently anticipated placing second order apparatus inside first-order lanterns.

Stevensons initially called tenders for the original requisition and the tender for the first order machines was awarded to Milne & Son, and that for the second order to Dove & Co for £548 each.[16] The first order machines were for Centre Island, Moko Hinau, and Cape Maria van Diemen lighthouses, and the second order for Brothers, Puysegur Point, and Portland Island lighthouses.[17]

These first order revolving dioptric lights were the first to be used in New Zealand. Earlier revolving lights such as Cape Foulwind and Cape Campbell had been second order, and the first order revolving light at Dog Island had used different technology. The specification for the machine was probably based on that used for the Northern Lighthouse Board lights.

The second order specification was very similar to that for Cape Foulwind, although it did not include the information which would fix the speed of rotation or the date of completion. The former reflected the uncertainty as to what the final characters would be, and the latter could only be fixed when the tenders were awarded and it was known how many machines each contractor would be manufacturing.

The main differences in the specifications were in the service table and the upper part of the revolving carriage. For Cape Foulwind these were cast in two pieces and bolted together; the newer specification called for them to be cast in a single piece. This probably reflected a change in the Northern Lighthouse Board design. Again, contractors were to be paid when the machine etc was delivered f.o.b the designated export ship.

Chance and Barbier & Fenestre were invited to tender for the three first order and three second order holophotal revolving apparatus sets at the end of April 1875.[18] The contract was let to Barbier & Fenestre for £5,520 for all six lights.[19]

The early part of the specification for the second order apparatus[20] was copied into the Stevenson records by the same hand – with the long 's' rampant – which had recorded the specification for the paraffin cisterns for Tory Channel and Cape Foulwind eight months earlier.

As the Cape Foulwind apparatus specification has not survived, it is not possible to assess whether the second order apparatus for the six lighthouses order marked any major change. Given the time frame it is unlikely.

The apparatus was to consist of

> 8 lenses, 8 panels of totally reflecting holophotal prisms above the lenses of 16 prisms in each, and 7 panels below the lenses of 9 prisms in each all of which are to revolve.

The gap was left in the lower panels to 'permit of acceſs to the lamp'.

The prisms were to be adjusted to a focus of 250 feet above sea level, and the horizon seen through the lens was to be 23 mm above the burner.

The specification for the first order revolving apparatus was not copied into the Stevenson records, but like that for the lantern was probably very similar to the standard Northern Lighthouse Board specification.

At the end of May, Dove & Co was informed that:

> As it is possible that one of the second order machines which you have contracted to supply to the New Zealand Government may not be required will you kindly delay proceeding with it till you receive further instructions,[21]

and Barbier & Fenestre were warned that

> one of the second order apparatus which you have contracted to supply to the New Zealand Government may be converted to a first order fixed apparatus.[22]

After consultations with Seed, the character of the lights was finally determined at the end of June,[23] and Dove & Co were informed that

> it has now been definitely arranged to dispense with one of the machines which you were contracted to supply for New Zealand Government.[24]'

This was the machine for Puysegur Point. It was decided to replace the first order revolving light at Centre Island with a first order fixed light,[25] the first since the one ordered for Cape Saunders and erected at the Nuggets. The revolving light which had been ordered for Centre Island would now be used at Puysegur Point.

The Statement of the Character of Lights also confirmed the change of the Brothers light from first order lantern to second order.[26]

All this extra negotiation (referred to by D & T Stevenson as 'several interviews with Mr Seed') to clarify the order took time and this was reflected in D & T Stevenson's invoice to the Agent General for professional fees. These additional charges were queried by the Agent General, and D & T Stevenson's reply of 10 February 1876 gives an insight into how professional fees were assessed by the consulting engineer.[27]

> We are favoured with your letter of the 8th Instant regarding our account of £54.15.2 on account of 'Lighting of Coasts of New Zealand'.
>
> We beg leave to explain for your information that the Fees in question refer to lengthened discussion and correspondence regarding the scheme for lighting the whole coast of New Zealand, including the present as well as prospective lights, and you will observe that it was necessary to determine this before anything toward designing and constructing the apparatus now under contracts could be proceeded with in order that the requisite characteristic distinctions for the different lights might be adopted.
>
> In our experience with foreign governments as well as with the Board of Northern Lighthouses where any such preliminary work or survey has been done it has been paid for as preliminary charges not coming under the fees charged per percentage which includes the subsequent time in designing work, arranging contracts, superintending execution of work, testing of apparatus and trouble connected with despatching from this country etc. This we may remark is an almost universal principle applicable to all Engineering works. In the case of a Dock, for example or any such work, the Engineering fees for surveying and making Parliamentary plans are paid for as preliminary work and thereafter if the work proceeds, the Engineer prepares the working plans etc and superintends the construction of the work at a percentage on the cost. Similar preliminary charges to those now made were sanctioned in 1862 by the Provincial Government of Otago with reference to lighthouses for Cape Saunders, Taiaroa Head and Dog Island.
>
> We had no agreement in writing on the subject of fees with the Government of New Zealand but there was an understanding that we were to conduct the work on the same principle of charge as we had arranged with other foreign governments and the preliminary fees referred to were charged in accordance with that understanding.
>
> We trust the above statements will satisfactorily explain how the matter stands.

We are etc

One can only guess at the Agent General's reaction at being called the representative of a 'Foreign Government'.

Barbier & Fenestre were invited to tender for auxiliary lights at the Brothers, Portland Island, and Cape Maria van Diemen lighthouses,[28] and were given the contract to supply lens tables for The Brothers and Portland Island, and the auxiliary light condensing apparatus for Cape Maria van Diemen.[29]

The final order for lamps was for four-wick Doty burners for the four first order lights and three-wick for Brothers and Portland Island. The drawing for the lamps shows cross-section of the burner and reservoir with the first order lamps in black and the second order as a red overlay.[30]

It carries the comment:

> The drawing for 4 wick lamp applies to Cape Maria van Diemen and Moko Hinau. It also applies to Centre Island and Puysegur so far as this lamp and fountain but at Centre Island the fountain must be placed outside of the mirror and the descending pipe taken down also at back of mirror. At Puysegur the fountain must be raised above the level of the mirror and the descending pipe lengthened. The drawing for 3 wick lamp applies to 'The Brothers' and Portland Island Lights.

Milne & Son provided four sets of first order lamps at £162 each set[31] and Dove & Co provided two sets of second order lamps at £135 each set.[32]

Although separate specifications were provided for stores[33] (including lamp glasses) and oil cisterns,[34] they were placed as a single tender for the eminently practical reason of packing the stores into the cisterns for despatch. The supplies were to be despatched in six separate shipments in the following order:

Brothers	two months
Puysegur	three months
Centre Island	four months
Moko Hinau	five months
Cape Maria van Diemen	six months
Portland Island	seven months.

The requests for tenders were sent to both Milne & Son and Smiths, and the tender was awarded to Smiths for a price of £315/4/-.[35]

While Seed was in Scotland he was able to interview John Mills, who was appointed to oversee construction of the six lighthouses,[36] and until sailing date he was employed as an inspector in the workshops of Milne & Son to familiarise himself with the lighthouses.[37] For this inspection work he was paid eight shillings a week. He sailed to Wellington on the *Waipa* leaving London on 19 December 1875.[38]

Bryden and Son were asked to provide

> lighthouse signals same as you now furnish to Northern Lighthouses with the necessary tube for a distance between the lightrooms and the keepers' bed rooms of about 100 feet.

Brydens' tender for £20/10/- for each set was accepted on 9 July 1875.[39]

Paraffin

The order for the Six Lighthouses also included initial supplies of paraffin for each light. D & T Stevenson therefore had to obtain paraffin and containers suitable for shipment to New Zealand. The tender called for 1,000 tins to be supplied over a period of 18 weeks[40] i.e.:

> 138 in three weeks from the date of acceptance of offer
>
> 170 in six weeks from the date of acceptance of offer
>
> 215 in nine weeks from the date of acceptance of offer
>
> 170 in twelve weeks from the date of acceptance of offer
>
> 169 in fifteen weeks from the date of acceptance of offer
>
> 138 in eighteen weeks from the date of acceptance of offer

The tins were to be 18" × 9" × 9" (internal diameter) and made of charcoal iron, and would be filled with 5 gallons of paraffin each. Laidlaw and Son of Simon Square[41] and Smiths[42] were invited to tender. It seems that the tender was awarded to Smiths as the accounts do not mention Laidlaw and Son. The cost of the tins seems to have been added to that for stores.

The tins were to be sent to Young's Paraffin Works at Addiesell in West Lothian. Young's were contracted to supply:[43]

Brothers Lighthouse	690 gallons in 3 weeks
Puysegur Lighthouse	845 gallons in 6 weeks
Centre Island Lighthouse	845 gallons in 9 weeks
Moko Hinau Lighthouse	845 gallons in 12 weeks
Cape Maria van Diemen Lighthouse	1075 gallons in 15 weeks
Portland Island Lighthouse	690 gallons in 18 weeks

There seems to have been a deliberate mismatch between the number of tins to be delivered and the number to be filled. This would have ensured that from the second delivery onwards there were spare tins available for later orders. The tins were to be packed four to a box.

An order of this size had the makings of a logistic nightmare, and D & T Stevenson endeavoured to manage the project by specifying the order in which lanterns, machines and apparatus were to be manufactured as well as specifying the delivery of stores, paraffin, and tins.

Barbier & Fenestre were first asked to provide the apparatus in the following order;[44]

Brothers	Second order
Puysegur Point	Second order
Centre Island	First order
Moko Hinau	First order
Cape Maria	First order
Portland Island	Second order

although the request was modified by the comment:

> Perhaps the better way would be to send a second order to Messrs Dove and Co and a first order to Messrs Milne and Son and then keep the order of the apparatus as mentioned above.

The subsequent modifications to the orders for Puysegur Point and Centre Island may well have upset this programme.

The previous day Milne & Son had been informed:

> The following is the order in which the work you are now manufacturing for the New Zealand Government is required:- Lanterns Brother - second order, Centre Island - first order, Cape Maria van Diemen - first order Portland Island - second order Machines etc. Puysegur - 1st order, Moko Hinau - 1st order Cape Maria van Diemen[45]

And in case Milne & Son did not understand, the letter added:

> The lantern for the Brothers to be got ready first and the first machine you have to get ready is that for Puysegur.

The letter to Dove & Co of 10 July instructed:

The lantern for Puysegur ... will be first required. The lantern for Moko Hinau is the fourth required. The machine for the Brothers should be got ready with the least possible delay as the New Zealand Government are most anxious to get this light exhibited to the Mariner.[46]

The request for a price for the lighthouse signals specified that they were to be supplied in the order The Brothers, Puysegur, Centre Island, Moko Hinau, Cape Maria van Diemen, Portland Island, and delivered in London in three, four, five, six, and seven months' time respectively from the date of acceptance of the offer [9 July 1875].

However, procurement did not go smoothly and in February 1876 Milne & Son wrote with elaborate courtesy:

Might we call your attention to the amount of the above (N Z Lighthouses) we have lying packed & ask you to forward if possible the arrangements for shipping as we have a quantity more ready for packing & the full cases take up so much room that they interfere very much with the speedy execution of the works. Portland Island - 2nd class lantern all packed, Puysegur apparatus and first class machine all packed. Centre Island First class fixed apparatus, packing will be ready to send off by Wednesday 9th inst. Moko Hinau will be ready for trial tomorrow (Friday) and can be packed in three days, Cape Maria Machine ready as far as possible but apparatus not to hand yet. Your attention will be esteemed by your ...[47]

Milne & Son couched their communication in terms of workshop space, but as has been commented, the contract price only became payable when the order was delivered f.o.b. to the export ship. Milne & Son had, therefore, a financial incentive to encourage the Agent General's staff to arrange shipping promptly.

The Brothers

The Chief Marine Board considered placing a lighthouse on The Brothers, and in August 1863 Captains Sharp and Johnson visited the islets. Although ideally placed for a lighthouse to guide ships into the narrows between Stephen's Island and Cape Terwhiti, neither of the isles offered easy access. In their report Sharp and Johnson concluded with respect to the northern islet:

1. That the islet only is accessible for the purpose required by the aid of expensive works and in tranquil weather.
2. That the erection of a lighthouse (if possible) on the sight [*sic*] indicated would exceed by several thousand pounds the means placed at the disposal of the Board
3. That from its isolated position and difficulty of communication, as also the total absence of fuel and water, the maintenance of an establishment would be more costly than lighthouses in other situations

4. That the site is doubtless good, being prominently visible passing through the Strait either from the north or south, and if only one light be used, perhaps the best; but if two lights were substituted (one on Mana Island and a second in the vicinity of Wellington Head) they could afford the double advantage of directing vessels to the south entrance of Queen Charlotte Sound (now a great thoroughfare) and in our opinion render the dangerous part of the Strait more easy than though only one light was used.
5. The expense of erecting and maintaining two lights on Mana and Wellington Head would not, from the comparative easy means of access, probably exceed that incurred for the one on the Brothers.[48]

Two years later Balfour visited The Brothers and agreed[49] that although the northern islet would make a good lighthouse site, landing would be difficult and the cost of erection and maintenance prohibitive. He felt that the Mana light was well placed and that the next light to be established should be Tory Channel.

In 1874 the question of placing a light on The Brothers was again raised.[50] In September 1874, after the end of the session which received the eighth report, staff from the Marine Department again visited The Brothers[51] and it was decided to construct the lighthouse there and not on Tory Channel.[52] The report included a sketch by Blackett of the proposed positions of the lighthouse and dwellings.

The Brothers was not one of the possible lights referred to in the eighth report of the year before, for which the legislature was going to be asked to vote funds. As the Tory Channel project was abandoned, it was easy to slip The Brothers into the works programme.

The final agreed characteristic for the light was:

> To be second order Dioptric flashing light with a period of 10 seconds between each flash. To be of the natural colour and shown all around. An auxiliary red light to be shown from a lower window over Cooks Rock. The tower to be not less than 28 feet high. The lantern to be second order.[53]

The position of the proposed lighthouse and the required red zone can be seen in the tracing sent to D & T Stevenson with the original requisition.)

The auxiliary light was 5° arc and was included to indicate the position of Cook's Rock.

The specification for the auxiliary light indicated that the same lights were to be provided for both Brothers and Portland Island lighthouses. The apparatus was to consist of

> [a] circular lens of 700 mm focal distance set in gun metal and fixed in a square malleable iron frame 2" broad and ½" thick … The sole plate is to be of cast iron of ⅝" metal.[54]

Although the lantern for The Brothers was ready by the end of September 1875,[55] it was not until the middle of November that it was despatched from Edinburgh, for shipment on the *Hurunui.*[56] However, a letter from the Agent General in February 1876 and the reply suggests that it was actually shipped on the *Ocean Beauty.*[57] To further complicate matters, the letter dated 29 February to the secretary of the Marine Department stated 'Lantern for Brothers Lighthouses shipped per *Chaudiere* to Hawkes Bay';[58] the writer had probably confused the Brothers lighthouse with the other second order light for Portland Island.

The apparatus, machine, signals, lamps and stores were sent to London by General Steam Navigation steamer from Granton on 18 December 1875 for shipping by the *Penshaw.* On the same day 35 boxes of paraffin were shipped from Grangemouth, to also be shipped on the *Penshaw.* The *Penshaw* shipment also included the additional Doty burners ordered for Nugget, Taiaroa Head, Farewell Spit and Cape Campbell.[59]

The eleventh report stated that the lantern and apparatus for The Brothers (as well as Puysegur Point, Centre Island and Moko Hinau) had arrived in New Zealand.[60]

In his 1874 memo Blackett had suggested that some site works be done before the arrival of the lantern. He suggested:

> 1. The improvement of the landing-place at the southern end, and the erection of a derrick with lifting gear at the same place
> 2. The cutting of a track from here to the site of the dwelling houses, and from thence to the northern end of the island, where there exist facilities for constructing and alternative landing-place for goods and material, &c
> 3. The erection of a derrick and lifting gear, &c., at this landing-place
> 4. [numbered 5 in the printed version] The necessary levelling and other preparation of the foundation of the light tower, dwellings, and stores.

An attempt was made to do some preliminary work in March 1876 when a work party was despatched. This party erected a house (sent ready-framed) and a small store, cut tracks, and partly prepared landing places. The sites for the tower and dwellings were partly excavated. However, after two months it was realised that a reliable steamer service to support the construction could not be provided, and the work party was withdrawn on 4 May.

Later that year the newly acquired Marine Department steamer *Stella* arrived in New Zealand, and a second work party was landed on 18 November 1876. The work was almost complete when Blackett wrote his report for the commissioner of customs, dated 30 June 1877.[61]

The light was exhibited on 24 September 1877,[62] and the Notice to Mariners stated that:

> The tower is 28 feet in height, built of wood and painted white. The light will be a SECOND ORDER FLASHING WHITE LIGHT, visible all round,

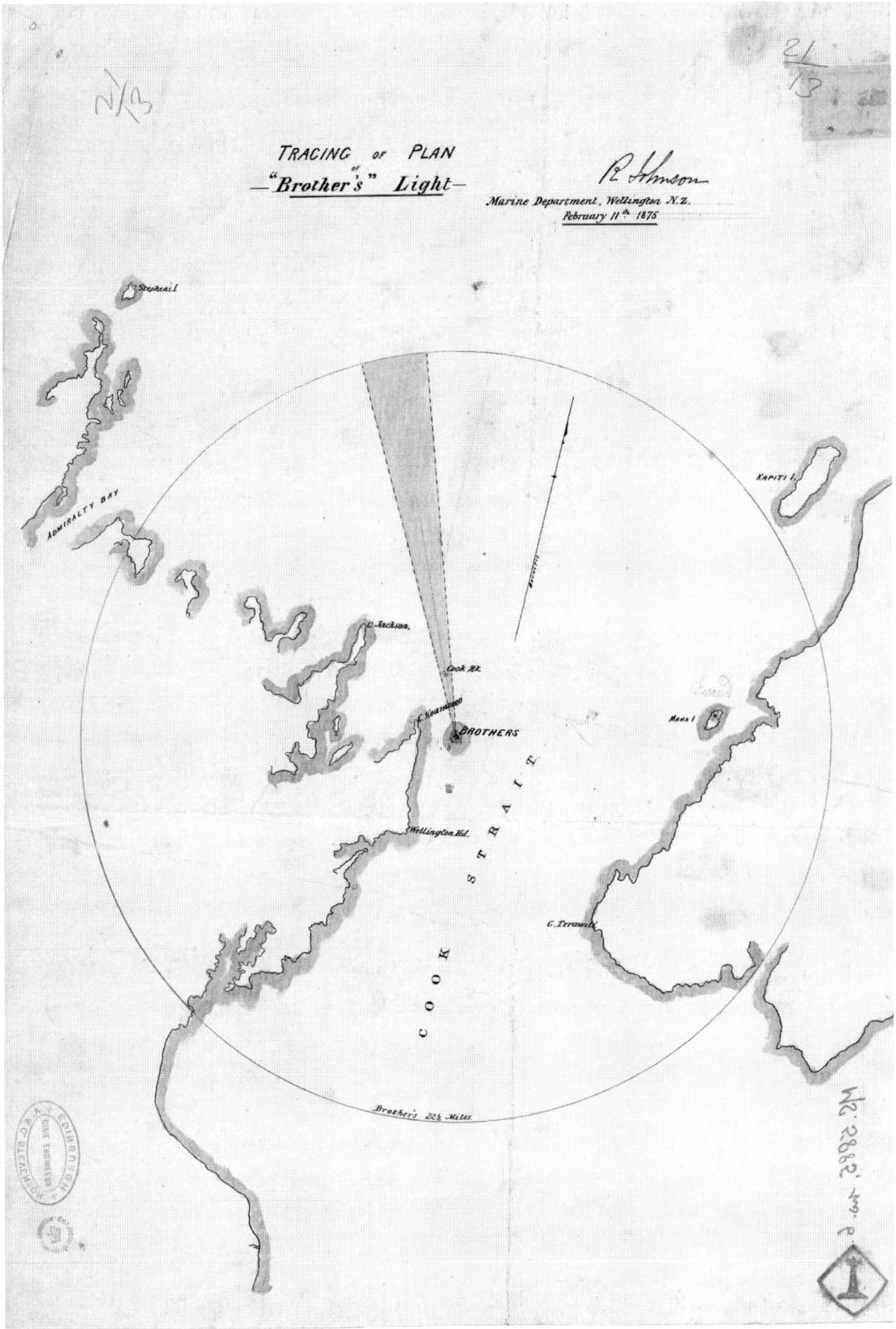

Figure 7.2: Site for The Brothers lighthouse and arc over Cook's Rock (Ms 5885)
© National Library of Scotland

> SHOWING A FLASH EVERY TEN SECONDS. The light will be elevated about 258 feet above the sea, and, allowing 15 feet for the height of the eye, will be seen at a distance of 22 nautical miles in clear weather, and at lesser distances according to the state of the atmosphere.
>
> From the lower part of the tower a FIXED RED LIGHT, having an arc of 5°, will be shown in the direction of Cook's Rock.

The Notice to Mariners also announced the discontinuing of the Mana Island light.

An interesting feature of the construction of the tower was its ballasting. Pieces of rock were placed between the lining of the wall and the outer weatherboards to give it greater stability in the Cook Strait gales.

The tower was later reinforced with struts, as was done at Moeraki.

Wilson, in a letter to Thomas Stevenson of January 1878, pointed out that the driving weight was too heavy and it was necessary to drill a hole in it to lighten it.[63] The Brothers is the nearest New Zealand has to a rock lighthouse, although the site is sufficiently big for keepers' accommodation and ancillary buildings to have been constructed separately from the lighthouse. It was manned by keepers only on a four-week rotation: three weeks on the lighthouse and one week in the Marine Department store in Wellington.

In 1882 ruby lamp glasses for the auxiliary light were ordered, along with glasses for Dog Island, Portland Island and Cape Maria van Diemen.[64] Interestingly the specification required that half of the order for the red glasses was to be sent by mail steamer, and the remainder of the order by sailing ship; presumably freight rates were cheaper on sailing ships. Certainly lighthouse materials were almost always sent by steamer. The glasses were supplied by Smiths and a later order, for skins for washers for pressure lamps, was filled by Milne & Son.[65] Two boxes of red glassware were sent from Edinburgh by rail on 7 February 1882, for shipment on the *Australia* at Royal Albert Docks.[66] However, they appear to have been shipped on the *Hallamshire*[67] so the boxes may have arrived too late for the *Australia*. The remainder of the order was sent south by Caledonian Railway on 4 March 1882 for shipment on the *Helen Denny* at the East India Docks.[68]

An incandescent burner was installed by early 1909.[69]

When Captain Bollons of the Marine Department steamer *Hinemoa* inspected The Brothers in May 1918, he found that rings and rollers of the machine were badly worn and in need of replacement.[70]

W A Fraser, the then lighthouse artificer, suggested rather unfortunately that the Marine Department 'get rings made in Colony' *[71] and advised the ordering of new roller paths and rollers (bevelled ball bearing type) at once; as the existing system was old-fashioned with level-topped paths and plain rollers. He provided a specification and suggested that the replacements be ordered through Messers D & C Stevenson, 84 George St Edinburgh.[72]

It was 28 months before the specification and requisition were sent to the New Zealand High Commissioner for transmission to Edinburgh.[73] In their reply, D & C Stevenson commented on the present satisfactory life the 'machine having been supplied to our firm's

* New Zealand had ceased to be a colony 11 years earlier.

Cape Reinga lighthouse looming out of the fog

Cape Brett lighthouse

Tiri Tiri Matangi lighthouse

Cape Egmont lighthouse and Mount Taranaki

East Cape lighthouse with luxuriant flax

Window washing on Castlepoint lighthouse

Cape Palliser lighthouse

Baring Head lighthouse

The inter-island ferry passes the Brothers lighthouse

Farewell Spit lighthouse after rain

The non-Scottish Cape Saunders light

Nugget Point lighthouse and the rocks from which it Is named

Kahurangi lighthouse

Cuvier Island lighthouse

Dog Island lighthouse: A tall tower on a flat island

Centre Island lighthouse

[Left] Figure 7.3: The original Brothers lighthouse (ABPL W5221 8848 box 19 [3]) © Archives New Zealand, the Department of Internal Affairs, Te Tari Taiwhenua

[Below] Figure 7.4: The Brothers lighthouse with struts © Archives New Zealand, the Department of Internal Affairs, Te Tari Taiwhenua

design almost 50 years ago'. They had to change the design to take standard bearings as 'Milne & Son's patent' bearings of the specification were no longer available. In addition, they also specified spares (one roller, an additional set of bearings with duplicate set balls) and a supply of grease. They recommended acceptance of the tender from and Steven and Struthers of Glasgow, for £83/10/- fob London, £82/10/ fob Glasgow.[74] The replacement rollers were shipped from Glasgow on the SS *Maimoa* in late April 1922[75] and arrived in

New Zealand about 31 July 1922.[76] Repairs began on 24 August 1923 and the light was relit on 1 September.[77]

The lighthouse continued to use incandescent lamps until electrification. In 1950 preliminary work began to electrify the Brothers light and establish a radio beacon.[78] This was completed by March 1956.[79]

Portland Island

A light on Portland Island was suggested as early as 1861 when Captain Johnson sketched for the Marine Board his plan for lighting the coasts. His original proposal was for a second order fixed white light.

When he submitted his 1870 Memorandum on System of Lighthouses, he suggested that a second order revolving white light would be better.[80] This recommendation was repeated in Edwin and Johnson's report of April 1873, with the added recommendations that the light should revolve every two minutes and show a red sector over Bull Rock.[81] This was repeated in the following year's report to the House of Representatives.[82] Portland Island was one of the seven lighthouses for which the legislature was asked to vote funds by Reynolds, the commissioner of customs.[83]

The final agreed characteristic for Portland Island was:

> to be a second order dioptric revolving light with a period of half a minute to be of natural colour and to show all around. An auxiliary red light to be shown from a lower window over Bull Rock. The tower to be 28 feet high. The lantern to be second order.[84]

The auxiliary light was identical to that for the Brothers, above.

The lantern was completed and packed in time to be mentioned in Milne & Son's letter of 3 February 1878 requesting shipping information.

Instructions finally arrived, on 15 February 1878, to send the lantern to East India Docks for shipment on the *Chaudiere* to Napier.[85] The packages for the lanterns and the signals were sent from Granton on 19 February on the General Steam Navigation Company steamer.[86]

On 17 March Barbier & Fenestre reported that they had despatched the apparatus for Portland Island to Edinburgh,[87] and Dove & Co reported the machine and apparatus as ready for inspection on 28 April 1876.[88] The packages containing the apparatus, machine, lamps and store were sent by steamer from Granton on 24 May for shipment to Napier on the *Waimea* from South West India Docks.[89]

It had been intended to send the 34-case shipment of paraffin for the new lighthouse with the apparatus, but the cargo was refused as the *Waimea* was an emigrant ship. The paraffin had already been despatched from Grangemouth, but the shipping agent was able to intercept it and arrange for it to be loaded onto the *Waikato* at the South West India Docks for carriage to Wellington.[90]

In June 1876 Blackett was able to report that a work party had carried out preliminary works: a track up the cliff on the northern end of the island and along the top to the

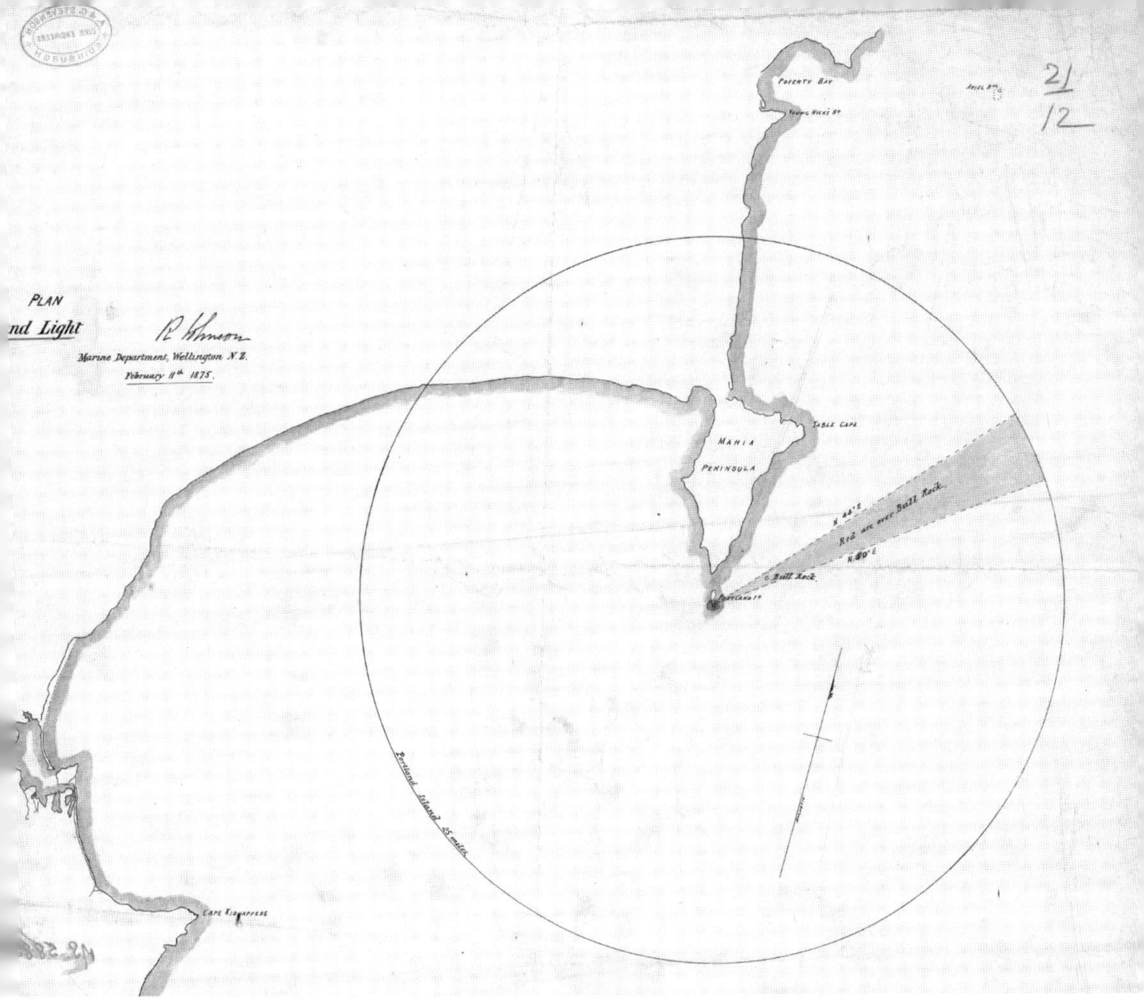

Figure 7.5: Tracing of the plan of Portland Island light, signed Johnson 11 Feb 1875(Ms 5885) © National Library of Scotland

lighthouse site on the southern end; a store for tools and goods, and fencing of the lighthouse reserve. The ironbark timber (from Australia) for the lighthouse was also unloaded.[91] At the time of Blackett's report, tenders for construction of the lighthouse had been advertised and were being considered.

The tender was awarded to H M Shepherd on 16 October 1876: £1,946, with a contract period of ten weeks.[92] Construction took longer than anticipated, and the light was finally exhibited on 10 February 1878. The Notice to Mariners described the new lighthouse thus:

> The tower is 28 feet in height, is built of timber, and painted white. The light will be a second order Revolving White Light, visible all around the horizon as far as the land will allow. It will attain its greatest brilliancy every 30 seconds.[93]

Figure 7.6: Portland Island lighthouse (PH-ALB98-p75n2) © Auckland War Memorial Museum – Tāmaki Paenga Hira

The range of the light was 24 nautical miles in clear weather, and

> from the lower part of the tower was a fixed red light, having an arc of about six degrees, [which] will be shown in direction of Bull Rock, which bears NE from lighthouse, distant 4 miles.

Figure 7.7: Portland Island lighthouse ashore (ABPL W5221 8848 19/2)
© Archives New Zealand, the Department of Internal Affairs, Te Tari Taiwhenua

In 1882 lamp glasses were supplied for the auxiliary light along with those for The Brothers and Cape Maria van Diemen.

The next main purchase was in early 1911 when incandescent lamps were ordered for eight lighthouses including Portland Island,[94] but new lamps were not installed at Portland Island until May 1918. Fraser, the lighthouse artificer, installed the new apparatus and a revolving cowl with a tripod for holding the lightning conductor.[95] Fraser returned in March 1931 to replace the Chance incandescent burner with a Smith-Ford burner.[96]

The 1949 annual report stated that estimates had been approved, and that specifications were being prepared for complete electrification and installation of standard radio beacons at 6 lighthouses one of which was Portland Island.[97] However, Portland Island was the last New Zealand lighthouse to be electrified. A new steel tower was constructed, and the new light was exhibited on 20 February 1957.[98]

By 1960 the abandoned wooden tower had deteriorated badly. The lantern and lens were moved to the nearby town of Wairoa and placed on a new wooden tower. Like that at Manukau Heads, the new tower is not an exact replica. Although shorter than the original it is in the Blackett spirit.

Centre Island

As early as 1861 Centre Island in Foveaux Strait had been suggested as the site of a lighthouse.[99] Edwin and Johnson in their 1873 report recommended a first order light revolving once in every two minutes, on either Rugged or Centre Island. Reynolds included a lighthouse at the 'Western Entrance to Foveaux Strait' among the seven lighthouses listed in the eighth report 1873.[100]

In late 1874, Johnson visited both Centre and Rugged Islands, and took advice from the harbourmasters of Port Chalmers and Bluff, the port for Invercargill on the Foveaux Strait. Johnson then suggested suitable sites on both islands[101] and for Centre Island recommended 'a second order dioptric light showing over an arc from N.N.E. round by E. and S. to N.W. by W.' In his supplementary report of January 1875[102] he recommended that a light be placed on Centre Island rather than Rugged Island due to the ease of landing and the availability of good granite for construction.

Although Johnson had recommended a second order light, the final specification was for the light to be

> first order dioptric fixed lights strengthened by a dioptric mirror. The light to be of the natural colour over and arc of 197 deg towards Stewart Island and to show red from about ½ a mile Southward of Fish Reef around to Pakia Point also about ½ a mile Southward of Escape Reef round to a point marked "boat landing" on the chart. Between the red zones the Northward the light will not be visible. Sixteen feet will be sufficient height of this tower.[103]

Consequently the lantern had two sets of red panes and 21 whole and 6 half dormant panes.

The specification for the Centre Island fixed apparatus, along with a tracing showing the arcs of the apparatus and mirror, were sent to Barbier & Fenestre on 24 June 1875.[104]

They were informed that:

> The mirror is to be 790 mm focal distance and according to the section which you sent to us on 7th March 1874, but if you could without much trouble make the top of it to clear the ray going to the lower edge of the lower prism of top panel it would be an advantage.

Stevenson sent their acceptance of Barbier & Fenestre's tender of £1250 on 3 July.[105]

The specification was longer than that for Cape Saunders, and the apparatus consisted of the lens and its framing as well as the pillar and table for the lamp.[106] The apparatus was to consist of

> 197° of first order fixed apparatus, 163° of a dioptric spherical mirror and additional panels of top and bottom prisms to the extent of 90° on one side

> of the apparatus and 22½° on the other … The top panels are to consist of 18 prisms and the lower 8, the cylindric refracting belt is to have diagonal joinings inclined 25° from the vertical and made up at the sides to the vertical by triangular pieces of the proper size and form.

The framework was to be of gunmetal. The spherical mirror was to have 11 double reflecting prisms. The apparatus was to be set

> to a height of 260 feet above the level of the sea and the horizon as seen through the central belt is to be 25 mm above the burner making the top of the burner 21 mm below the focal plan of the apparatus. With the burner in this position the upper prisms are to be adjusted to 30 mm above it and 9 mm behind the axis and the lower prisms to be 10 mm above the burner and 38 mm in front of the axis. The focus of the dioptric mirror is to be placed 1¼ inches above the central refracting belt.

The pillar stand was to be 9 feet in height and cast in two pieces, with a socket connection 10 inches long. The table was also to be cast in two pieces bolted together. This was in contrast with the second order specification, where the table was cast in a single piece.

Milne & Son were asked to manufacture the Centre Island lantern immediately after that for The Brothers,[107] and the lantern was ready for packing on 8 December 1875.[108] However, there was a delay in shipping and on 29 December D & T Stevenson wrote to the Agent General that

> [t]he contractors for the lanterns for Centre Island, Cape Maria van Diemen & Portland Island are pressing to get them shipped.[109]

This seems to have worked, as on 31 December D & T Stevenson were able to instruct the contractors to send the packages for the lantern, two ladders, and the lighthouse signals to Glasgow by Caledonian Railway for shipment on the *Pomona* to Dunedin. The lantern for Cape Maria van Diemen was sent to London at the same time.[110]

Milne & Son had also received the apparatus for Centre Island from Barbier & Fenestre for packing, and were asked by Stevenson to provide a steadying frame.[111] The tested apparatus was among the lighthouse materials cluttering up the workshop when Milne & Son wrote their letter of 3 February 1876 (see above). The apparatus was sent to Glasgow on 14 February for shipment on the *Horsa* along with that for Puysegur Point.[112]

Due to delay in providing labelling details, the 43 cases of paraffin oil for Centre Island were not despatched from Grangemouth until 25 April 1875.[113] This was for shipment on the *West York* from South West India Docks.[114] On 29 July 1876 William Seed wrote that lantern and apparatus had arrived safely in the colony.[115]

The tender for erection of tower, three dwellings, and three stores, was awarded on 6 January 1877 to Mr William Boyd for £2,087 1s 9d, in the same announcement as that for

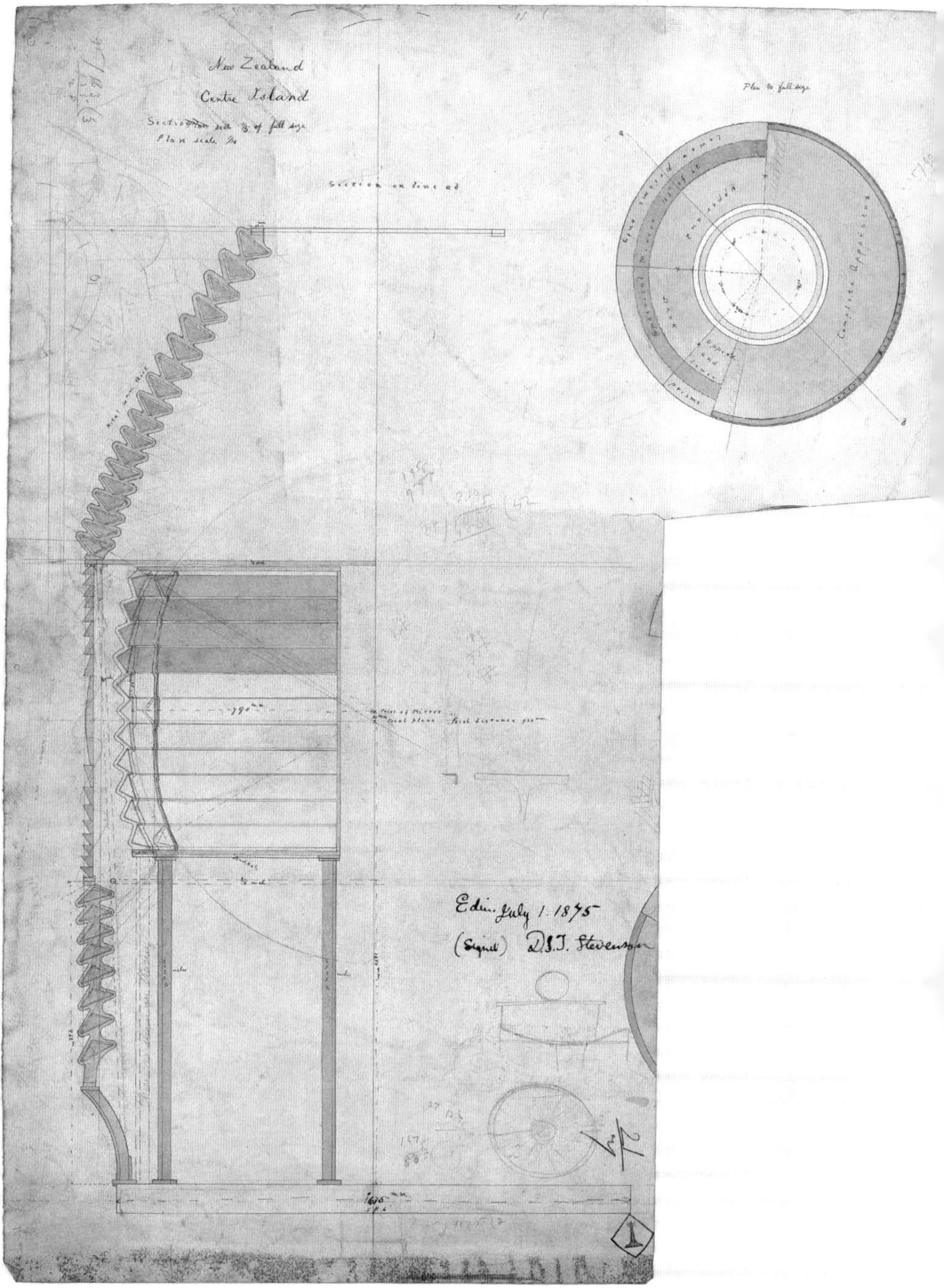

Figure 7.8: Plan and section optical apparatus for Centre Island lighthouse (Ms5885)
© National Library of Scotland

Figure 7.9: Centre Island lighthouse (ABPL 8848 W5221/19/26) © Archives New Zealand, the Department of Internal Affairs, Te Tari Taiwhenua

Puysegur Point.[116] The completion date was set at 6 November 1877.[117] It was completed to contract, with John Mills supervising the setting up of the lantern and apparatus. The light was exhibited on 16 September 1878.

The Notice to Mariners[118] described the light as:

> The Tower is 20 feet in height, built of timber and painted white. The light will be first order fixed showing white from bearing North 78° West, round by the north over an arc of 176° to a bearing of South 82° East,. The light will show red from this latter bearing to South 54° East, and red between the bearings North 78° West and South 4° East, and between this latter bearing and the bearing of South 54° East the light will not be shown.

The expected range was 22.5 nautical miles in clear weather.

The installation of incandescent lamps was announced in the 1909 annual report[119] and confirmed in the 1910 annual report.[120]

In April 1949 it was suggested that the light should be electrified[121] and changed from fixed to group flashing with the pattern flash 2sec, eclipse 1sec, flash 2sec, eclipse 8sec.[122]

Electrification of the Centre Island light was finally completed in 1955.[123]

Puysegur Point

A lighthouse at Puysegur Point had not been on the list of lights recommended by Edwin and Johnson in their 1873 report, although they did suggest a second or third order light at Dusky Bay, to the north of Puysegur Point. Puysegur Point was visited by Johnson in

1874. He described the available site as 135 feet high, with an illuminating arc of 180°. The summit was flat, but covered with dense bush.[124]

He recommended that a light be placed there as opposed to a nearby site at Windsor Point for the following reasons:

> 1. It lies nearer to the quarter whence vessels make the land, and therefore would be sighted sooner;
> 2. Having a less arc to cover, the rays could be reflected from the back to strengthen the sea arc, and reflected so as to show more in the direction of vessels making the land than would a light on Windsor Point;
> 3. The land adjoining Puysegur Point is lower than at Windsor Point, therefore fog is not likely to roll so low down;
> 4. Puysegur Point is only one mile and a quarter from Otago Retreat, and a road could be easily formed;
> 5. A light on [Puysegur Point] would be of great service to any vessels bound for Preservation Inlet, while one at Windsor Point would be utterly useless for this purpose;
> 6. A light on Puysegur Point would show more strongly over Bellamu Reef, and therefore more certainly discerned in foggy weather than one at Windsor Point. The proper light for this position would be a holophotal revolving white light of the first order dioptric; and as the atmosphere is very humid, I recommend that the buildings be constructed either of iron or stone; the latter is abundant here.

Johnson's reference to a very humid atmosphere was something of an understatement. Puysegur Point lies on the south-west corner of the Fiordland National Park, which has an average annual rainfall of 7,000 mm. He neglected to mention another factor of the site – voracious sandflies. These are biting insects, not unlike Highland midges only bigger and more hungry.

The final agreed specification was

> Puysegur to be first order Dioptric flashing light with a period of 10 seconds between each flash. The light to be of the natural colour and strengthened by Dioptric mirror. To be first[125]

The tracing provided by the Marine Department showed that the expected radius was 20½ miles and light was only required for a limited arc.

Consequently the lantern was designed with dormant panes but the apparatus was strengthened with a dioptric mirror.

Barbier& Fenestre were asked to provide the extra 135° dioptric mirror.[126] Their tender for £204 was accepted on 3 July 1875.[127]

By 13 November 1875 D & T Stevenson could inform the Agent General that the 'lantern for Puysegur lighthouse will be ready to be sent off in the course of a fortnight'.[128] There seems to have been some confusion in the office of the Agent General, as four days later D & T Stevenson wrote:

> We have received your letter of 16th Instant. We have no instructions to the ports to which Puysegur Lantern is to be sent. Perhaps all the lanterns and apparatus should be sent to Wellington leaving the authorities there to arrange the carriage of the package to their respective destinations.[129]

Soon afterwards D & T Stevenson contacted Patrick Henderson and Company of Glasgow, although it is not obvious whether this was at the request of the Agent General

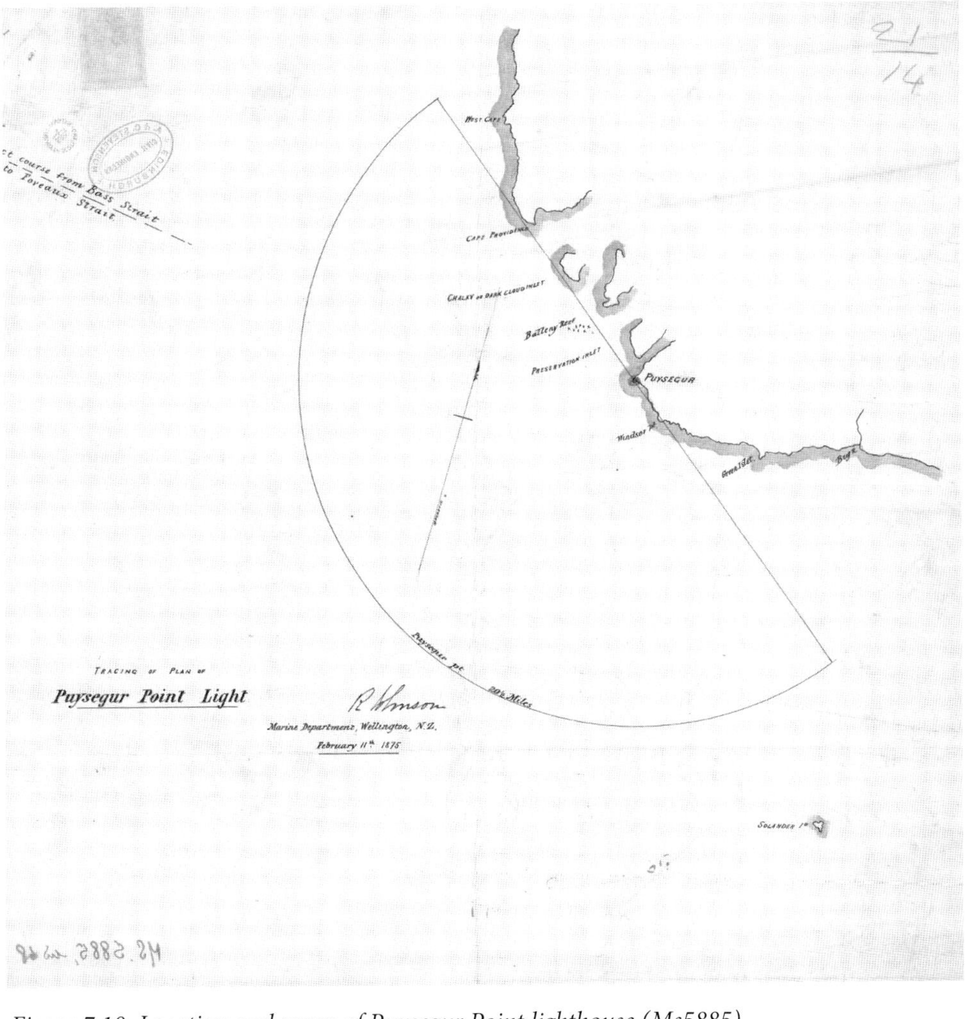

Figure 7.10: Location and range of Puysegur Point lighthouse (Ms5885)
© National Library of Scotland

or was D & T Stevenson's own initiative.[130] Puysegur Point is within Southland Province, which originally had been part of Otago. Henderson and Company had a close connection with Otago Province, particularly in carrying immigrants to the new settlement under contract to the Otago Provincial Council.

D & T Stevenson had anticipated sending the Puysegur lighthouse materials to Bluff. However, while Henderson and Company had a ship for Bluff, they would only land passengers there and not cargo. On 3 December the Agent General instructed Stevenson (by telegram) to despatch the lantern, addressed to the 'Minister of Public Works, Otago', to Henderson for shipment on the *Wellington* to Otago, which meant Port Chalmers.[131] It was despatched by 7 December.[132]

On 12 January D & T Stevenson contacted Youngs and enquired:

> Please let us know how many tins of paraffin you have got ready for the New Zealand Lighthouses. Those for Puysegur, Centre Island and Moko Hinau will shortly be required. On your letting us know how many tins have been prepared & filled with paraffin we shall send an inspector to examine them.[133]

These comments suggest that D & T Stevenson's careful planning to stagger deliveries at three-weekly intervals was coming slightly adrift. Fortunately, the following day Youngs replied that they had 86 cases containing four tins each 'now ready and filled with New Zealand Lighthouse oil the required length of time, ready for inspection'.[134] That is, they had 344 tins containing 1720 gallons which was sufficient for Puysegur Point and Centre Island Lighthouses, with six left over for Moko Hinau. In the end the six extra tins were split between the Puysegur Point and Centre Island orders, thus giving even multiples of four for packing.

By 14 January 1876, the lamps, apparatus, and machine were ready for packing,[135] but 3 February found them still in Milne & Son's workshop, as has been previously commented. However, the agent general had been negotiating with Henderson and Co, and D & T Stevenson were directed to see that the apparatus, lamps, machine, stores and signals were sent to Glasgow to be shipped on the *Horsa* along with those for Centre Island.[136] They were sent on 14 February.[137] D & T Stevenson were directed not to send the paraffin ordered for the two lighthouses, as the ship refused to take it;[138] the *Horsa* may have been carrying immigrants, and immigrant ships were forbidden to carry lighthouse fuel.[139] The 43 cases of paraffin for Puysegur were finally despatched to London on the Carron Company steamer on about 19 February 1876, to be shipped to Wellington by the *Pym* from the East India Docks.[140]

A work party was landed on the site in February 1875 to clear and form a road from the landing place to the proposed lighthouse site, to clear the site for the lighthouse, dwellings and stores, and to erect a small wooden store. The party also landed heavy timbers, mainly Australian iron bark, which had been procured by the government because of their special size and species. The party were withdrawn in November 1875 with all but a short section

Figure 7.11 Puysegur Point lighthouse station (ABPL 8848 W5221/19 [26]) PU31ph

of the road completed and having worked through the winter.[141] The Marine Department were however, sufficiently confident to issue a Preliminary Notice of the proposed lighthouse in October 1877.[142]

The contract to erect tower, three dwellings, and three stores was awarded on 6 January 1877 to Mr W Birss for £2,984 2s, with the work to be completed by 6 January 1878.[143] Blackett expressed the opinion that progress was not altogether satisfactory, but was confident that it would be completed in the contract period.[144] He reported on 30 June 1878 that the contract works had been satisfactorily completed, and the tower was ready to receive the lantern and apparatus.[145]

The new light was exhibited on 1 March 1879, and the Notice to Mariners[146] described it as follows:

> The Tower is 40 feet in height, built of timber and painted white. The light will be a first order flashing white light, visible all around as far as the land will allow, and showing a flash every 10 seconds.

It was described as having a range in clear weather of 19⅓ nautical miles. The lighthouse was described as being 'on the point of the same name, at the south-western extremity of the Middle Island, New Zealand'. (What is now called the South Island or Te Wai Pounamou continued to be called 'Middle Island' until about 1900.)

By 1895 the rollers had become badly worn, and in June of that year the Agent General transmitted a requisition for rollers for Puysegur and Moko Hinau lighthouses.[147] As they had made the original equipment, Milne & Son were approached[148] to give a price for

> 12 steel bearing rollers for Puysegur and 12 for Moko Hinau Lighthouse. These rollers must be of the first quality of Sandersons mild cast steel, perfectly free from flaws of every kind and properly forged and carefully turned to one size and the edges being turned to a flat curve as shown. Each roller to be bushed with gun-metal for spindles ⅝inch in diameter. The price to include packing properly in a tin lined case and delivery free on board export ship in London. Each roller to be well coated with black varnish to prevent rust on the voyage, care being taken that no finger marks are left before or after varnishing.

The price quoted by Milne & Son was not what D & T Stevenson had expected, and in a letter dated 9 July they asked 'Is there no mistake about this offer?[149] It is not clear whether the price was higher than expected, in which case Stevenson would be expected to 'negotiate' on behalf of their client as they had done for the Puysegur lantern, or was lower and may have been a clerical error. Hence D & T Stevenson, as befits a firm of professional civil engineers, gave Milne & Son the opportunity to correct the mistake. Milne & Son's final offer was £55/8/-, which Stevenson recommended for acceptance 'as it is reasonable'.[150]

Although the offer was accepted on 17 July 1895, there was delay in manufacturing the rollers, and two months later Stevenson were writing:

> The Agent General is complaining that the steel rollers are not yet ready for shipment. Please say when they will be ready for inspection[151]

Five days later (19 September) D & T Stevenson were able to report that the rollers had been inspected and passed and could be 'sent forward for whenever you desire it for shipment'.[152]

On 23 September, Milne & Son were given instructions on labelling the package, and were instructed to send it to London for shipping on the SS *Louie* at the Royal Albert Docks.[153] They were despatched by the 28th.[154] Once in New Zealand, the rollers appear to have been placed in the Marine Department store and left there, resulting in departmental confusion as the 1903 Marine Department report stated: 'New rollers *are being procured* from England for Akaroa Head and Puysegur Point Lighthouses'.[155] This might reflect that the order for Puysegur Point should have been for rings as well as rollers and so nothing could be done until new rings were obtained.

Two years later clarity prevailed and the 1905[156] Marine Department report stated:

> New rings have been obtained from Scotland for Puysegur Point Lighthouse to replace those which have been in use since the light was first exhibited

Figure 7.12: Puysegur Point lighthouse on fire (ABPL 8848 W221/19 [26] PU04ph)
© Archives New Zealand, the Department of Internal Affairs, Te Tari Taiwhenua

Figure 7.13: Reconstructed Puysegur Point lighthouse (ABPL 8848 W221/19 [26] PU91ph)
© Archives New Zealand, the Department of Internal Affairs,Te Tari Taiwhenua

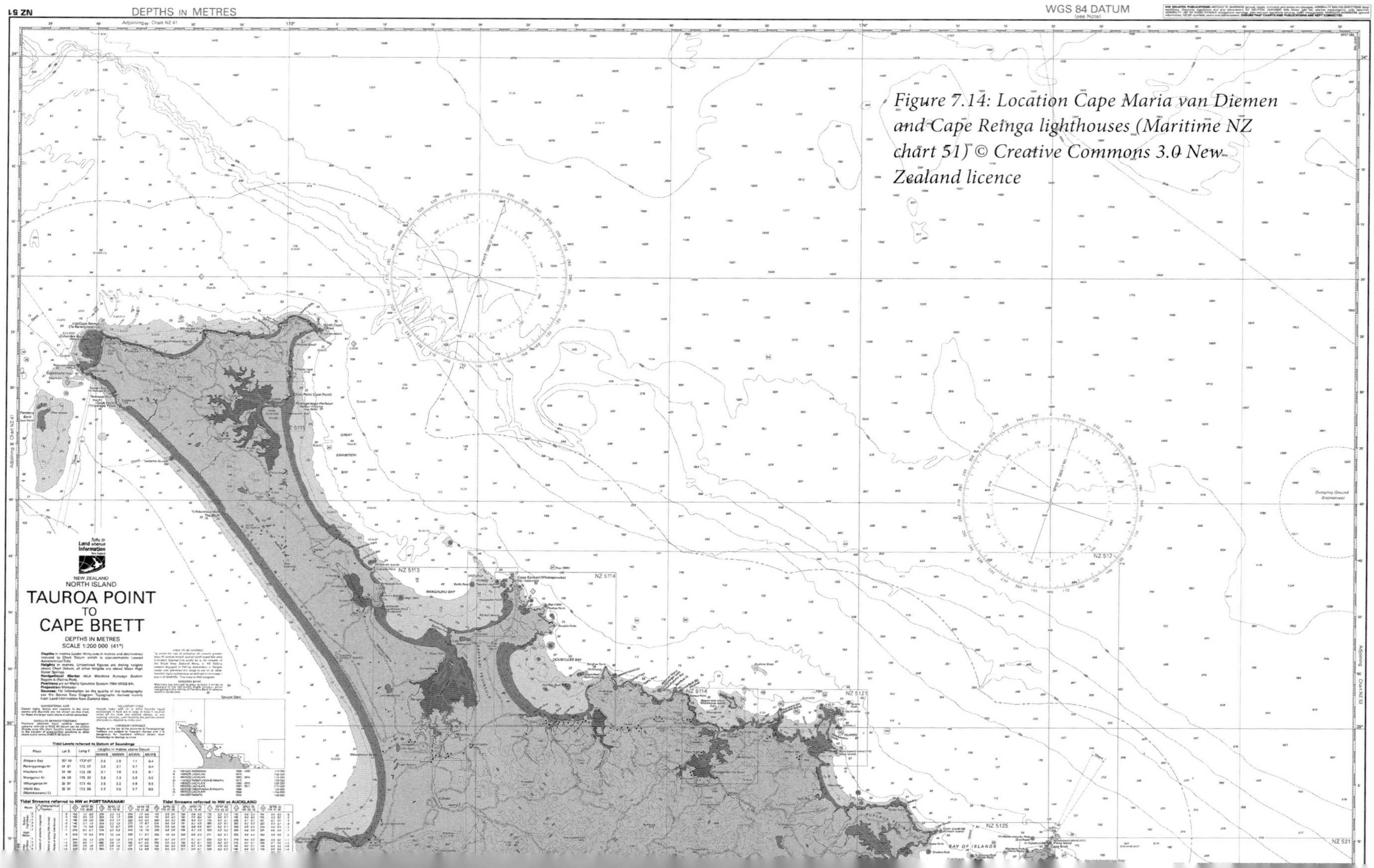

Figure 7.14: Location Cape Maria van Diemen and Cape Reinga lighthouses (Maritime NZ chart 51) © Creative Commons 3.0 New Zealand licence

on 1st March 1879 and they are now being fitted by Mr Scott, Lighthouse Artificer.

It had initially been planned to start installation of the rings and rollers on 26 April,[157] but they did not begin until 3 May, when the flashing light was replaced by a fixed white light.[158] The repairs were completed by 26 May 1905, two days before the issuing of the Marine Department report.[159]

The 1909 annual report indicated that incandescent lamps were being installed at Puysegur Point.[160]

In 1939 the nautical advisor first suggested electrification of the Puysegur Point light.[161] The decision was brought hastily forward in 1942 when the tower was destroyed in a fire set by a local eccentric (probably mentally ill) who had apparently found the flashing light disturbing his sleep and decided to do something about it.

A second fire was also set in the principal keeper's cottage, but this was extinguished.

The reconstructed lighthouse was a mixture of old and new. The new lighthouse was 5.5 m high with a concrete base painted white. The second order lantern which had been removed from Godley Head in 1941 was reconditioned and erected, and the lens which had been replaced at Cape Foulwind was transferred to Puysegur. Melbourne Marine Engineering Company of Lyttelton supplied the new iron sections, constructed the lighthouse, and installed the diesel generator.

The lighthouse resumed operation on 14 October 1943.

Cape Maria van Diemen (Motuopao) and Cape Reinga

Cape Maria van Diemen in the far north of the North Island was named by the Dutch explorer Abel Tasman after the wife of the governor of the Dutch East Indies (now Indonesia). Although the lighthouse was called Cape Maria van Diemen it was in fact on the island of Motuopao, 200 m off the tip of the cape.

Supplies were sent to the lighthouse station by a cable system from the mainland.

Edwin and Johnson had recommended that a lighthouse be placed on Cape Maria van Diemen in their 1873 report. They wrote:

> English and Australian ships bound to Auckland first make land at about Cape Maria van Diemen; and as an extensive traffic exists and strong currents are experiences in the vicinity of the Cape, we have no hesitation in recommending that a First-order light, flashing every ten seconds, be placed in this important position.[162]

Cape Maria van Diemen was placed on Reynolds' list of seven lighthouses in his transmission letter attached to the eighth report.[163]

Johnson revisited the site in 1874 as well as Cape Reinga and the Three Kings Islands. He found the island 'well suited for a light [and] large enough to keep a cow or two'. On technical matters he wrote:

Figure 7.15: Proposed Lighthouse Site on Moutuopao Island © Archive New Zealand, the Department of Internal Affairs, Te Tari Taiwhenua

> The light from here would show from N.25°E. round to N., W., S., and S.56°E., or over an arc of 261°, and right in the track of vessels making for Auckland or the east side of New Zealand from Europe or the southern Australian ports. I therefore consider this position an important one, and would recommend the erection here of a first-order revolving light on the holophotal system, showing white, except over the Columbia Reef, where it should be red.[164]

He dismissed Cape Reinga as a site for a lighthouse, as its height of 456 feet was 'far too great for a lighthouse'.

The tracing sent to D & T Stevenson with the original specification showed the red arc.

The subsequent lantern design marked the Columbia Reef by placing 13 full and 3 half red shades in the lantern. There were also 12 dormant panes.

The red panes were dispensed with and replaced by the auxiliary light, probably after discussions with William Seed. Hence, the final agreed specification was:

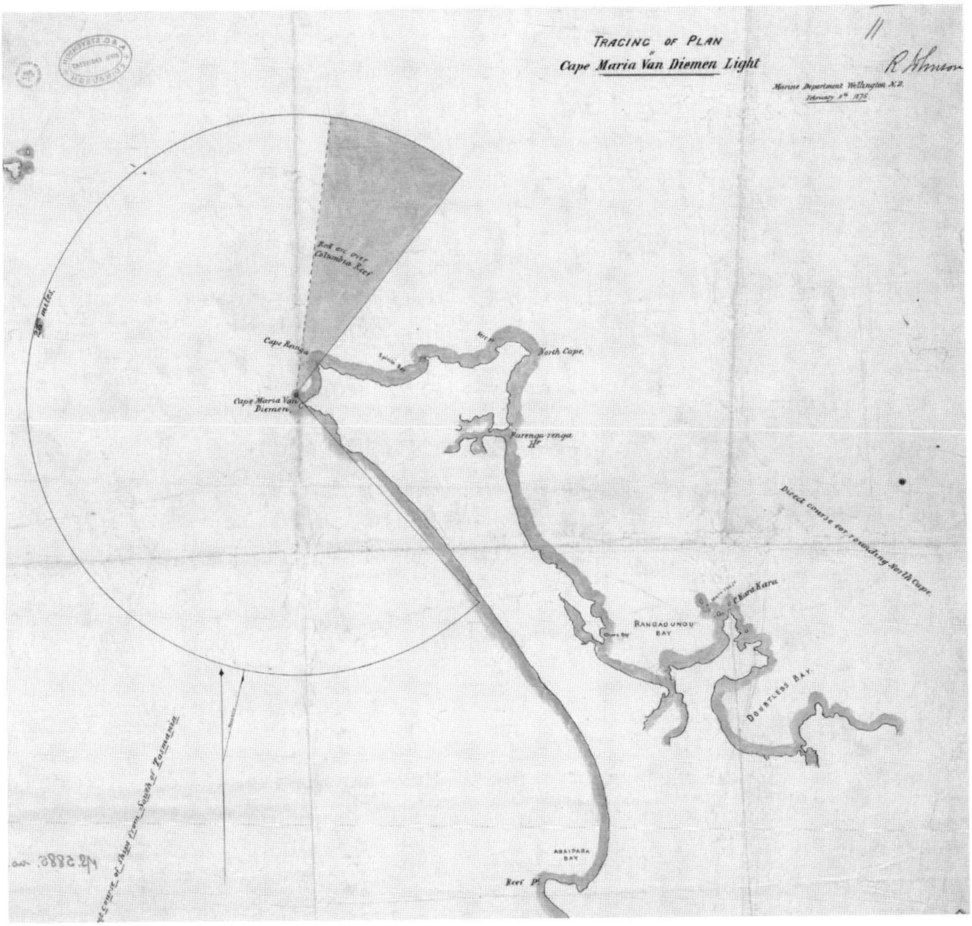

Figure 7.16: Proposed arcs for Cape Maria van Diemen lighthouse (Ms 5885)
© National Library of Scotland

a first order revolving dioptric light with a period of one minute of the natural colour. An auxiliary red light to be shown from a lower window over Columbia reef, the main lights showing white all around as far as the land will allow. No dioptric mirror to be used here. The tower to be not less than 28 feet in height.[165]

The auxiliary light for Cape Maria van Diemen was [a] condensing light. The specification described the apparatus as a

condensing light for 30° and is to consist of a 180° of a 5th order fixed apparatus same section as the 5 supplied for Japan viz 187.5 mm focal distance, two reflecting prisms and ten totally reflecting prisms. These prisms may be reduced in number and increased in size if the Contractor should prefer and it can be done cheaper but the extreme width of the apparatus will not be increased.[166]

The final design is shown below.

Barbier & Fenestre despatched this apparatus to Edinburgh on 13 March 1876, along with the apparatus for Portland Island.[167]

Smiths were asked to add to their order for lamps for Cape Maria van Diemen, 72 Argand ruby glasses and 1 gross of Argand paraffin wicks.[168] Milne & Son were asked to provide the Doty lamps for the light.[169]

By 25 December 1875 the lantern was almost completely packed, and on the last day of 1875 Bryden and Milne & Son were instructed to send the packages containing the signals[170] and the lantern[171] to London to be shipped on the *La Escocesa*, which sailed for Auckland from the West India Docks in late January 1876. The cargo arrived in Auckland in April, and the collector of customs was asked to store it in the basement of the Custom House. The experience with the Manukau light had not been forgotten, and the collector was asked to insure the packages against fire.[172]

At the time of Milne & Son's 3 February letter the machine was 'ready as far as possible, but the apparatus was not to hand'. Given the crowded conditions in the workshop the delay in the apparatus may have been convenient. After some delay the apparatus, lamps and stores were sent from Granton on 16 and 17 June 1876 to be loaded on the *Orari* at the South West India Docks for shipment to Auckland.[173] The shipment arrived in mid-September 1876 and joined the lantern in storage.[174] A year later both shipments were still in storage and the insurance was renewed.[175]

Young's had supplied the paraffin for The Brothers, Puysegur Point, Centre Island and Moko Hinau in multiples of four tins, as these made up whole boxes. Consequently by the time the paraffin was to be supplied to Cape Maria van Diemen there were only 86 cases (344 tins) left. Young's suggested that they be divided up, 52 to Cape Maria van Diemen and 34 to Portland Island.[176] Stevenson agreed to this.[177] The paraffin oil for Cape Maria van Diemen was eventually sent by steamer from Grangemouth in mid-July and was shipped on the *Tamaya* from London on 25 July 1876.[178]

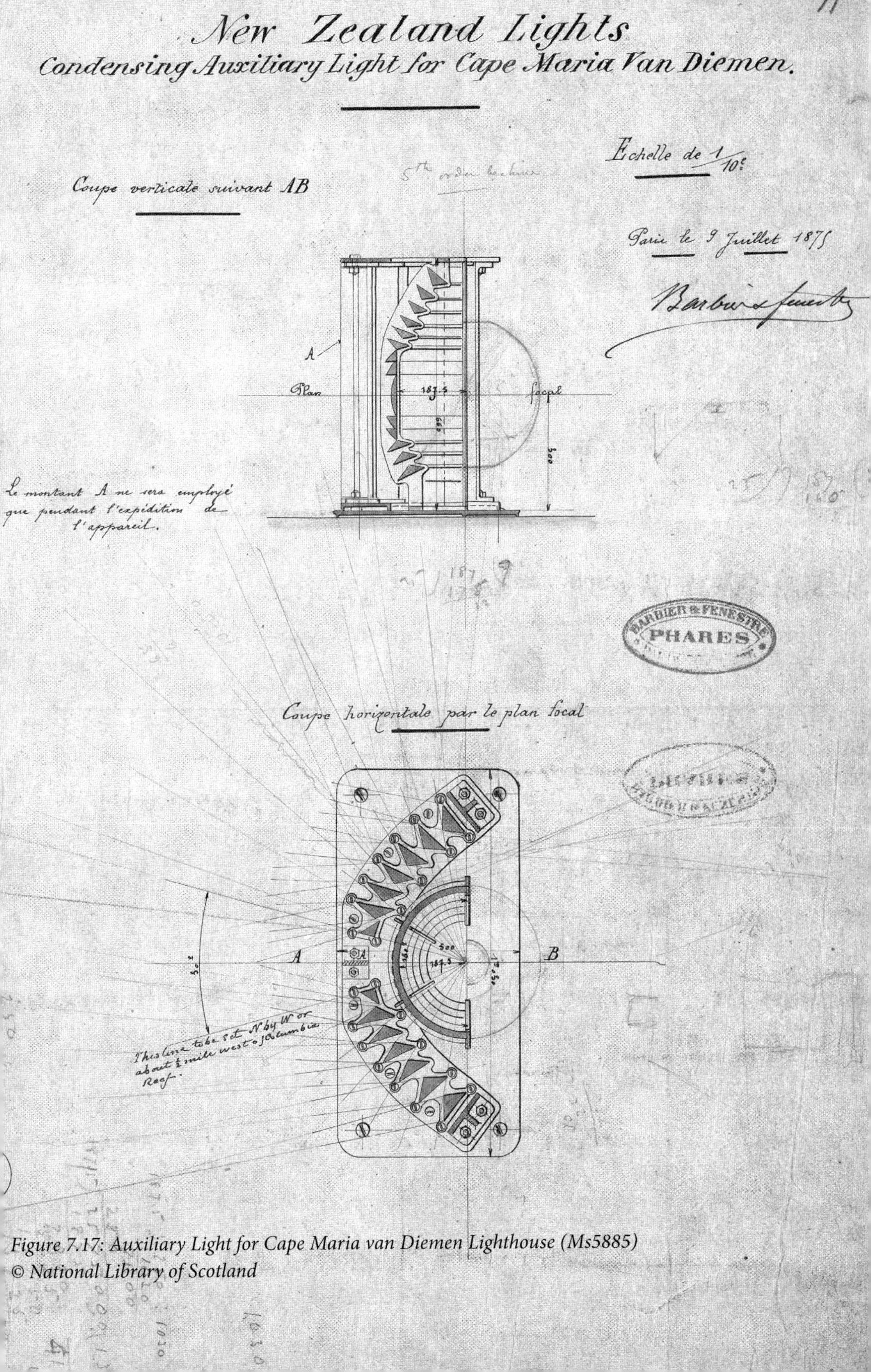

Figure 7.17: Auxiliary Light for Cape Maria van Diemen Lighthouse (Ms5885)
© National Library of Scotland

A work party was landed on Motuopao Island in late August 1877 and Blackett anticipated that the buildings would be ready for erection of the lantern and apparatus in August 1878.[179] Although a preliminary Notice to Mariners had been issued in September 1878 announcing the anticipated exhibition in January 1879, the completed lighthouse was not exhibited until 24 March 1879.[180] The delay occurred because the wicks (3 gross each concentric No 2, No 3, 2 gross Argand wicks for auxiliary light) had not been sent with the apparatus and had to be specially requested.[181] Although the light could be exhibited, this was not the end of the problem as the 'wick for the auxiliary light was sent in short lengths [instead of long lengths] and consequently will not do'.[182] Smiths were instructed to send 24 *yards* of the wick in question.[183]

The Notice to Mariners described the lighthouse thus:

> The tower is 20 feet in height, built of timber and painted white. The light will be a first order revolving white light, visible all round as far as the land will allow, attaining its greatest brilliancy once every minute ... [range] 24½ nm in clear weather. From the lower part of the tower a fixed red light, having an arc of about 30° in azimuth will be shown in direction of the Columbia Reef.

Smiths supplied ruby Argand lamp glass for the auxiliary light in 1882, along with those for Dog Island and the auxiliary lights at Portland Island and The Brothers.[184]

The Doty burners were replaced by incandescent burners in by early 1909[185] and, like other lighthouses with them, had to have a revolving cowl fitted in 1915.[186] By 1910 the rings and rollers had like those at Puysegur Point become badly worn. The new ones were installed in January 1911.[187]

In 1938 an electrification survey was carried out[188] but as the lighthouse was on a small island, the decision was made to dismantle the Cape Maria van Diemen lighthouse and build a new lighthouse on Cape Reinga. In early 1941, the light was extinguished and everything salvageable moved to Cape Reinga, including the lantern and apparatus.[189]

Figure 7.18 Cape Maria van Diemen lighthouse (PH-NEG-1095 ID 35629) © Auckland War Memorial Museum – Tāmaki Paenga Hira

Figure 7.19 Cape Reinga lighthouse(ATL 1/2-092135-F)
© Alexander Turnbull Library, Wellington, New Zealand

Cape Reinga

The new Cape Reinga lighthouse was constructed of reinforced concrete to the same design as Baring Head.

The Cape Maria van Diemen lantern and lens were installed and diesel generators supplied for the electric light. Cape Reinga was automated in 1987.

Moko Hinau (Burgess Island)

In the nineteenth century, one of New Zealand's main communication lines to Europe was mail sent through San Francisco. Edwin and Johnson in their 1873 report commented:

> Vessels from North America and the South Seas bound to Auckland make the land about the Great Barrier [island], and as there is an increasing trade between Auckland, Fiji, and other South Sea Islands, we propose that a First-order light, revolving once every minute, showing alternately red and white, be placed on the Moko Hinau Islands. The light will also be of value to the coasting trade, as it commands the whole entrance of the Hauraki Gulf from the northward.[190]

Johnson visited the islands again in 1874 and recommended that the lighthouse should be placed on the main island (Burgess Island) and commented that the site 'is right on the track of the San Francisco steamers as well as the South Sea traders'.[191]

Moko Hinau was also on the list of lighthouses for which Reynolds anticipated asking the legislature to vote funds..[192] The final agreed design had two major changes from Edwin and Johnson's original. It was

> [t]o be first order Dioptric flashing light with a period of 10 seconds between each flash. To be of the natural colour and shown all around. The tower to be 20 feet high.[193]

The lantern was the other one tendered for by Dove & Co, and was complete and ready for inspection by 19 January 1876.[194] The lantern and signals were sent by steamer from Granton on 2 February 1876 for shipment to Auckland by the *Bebington* from the East India Docks.[195] They had arrived in Auckland by 30 June[196] and were placed in store, and stayed there for six years. Moko Hinau was the last of the Six Lighthouses to be constructed and exhibited.

Barbier & Fenestre had by this time sent the lens to Milne & Son, and the Moko Hinau apparatus was meant to be trialled on 4 February. However, in their letter of 5 February 1876 Milne & Son pointed out that:

> According to your instructions we had the above machine and apparatus ready and lighted for inspection on Friday afternoon but it was not inspected. Could you kindly arrange to have it inspected at an early date as we can do nothing to 'Cape Maria' till it is passed and taken down.[197]

The inspection must have been carried out, as on 15 February 1876 Milne & Son reported that 'Moko Hinau apparatus will be all packed in 8 days', but could not resist saying '[w]e could not get it packed until we got some clearance'.[198]

The packages were eventually sent by General Steam Navigation steamer from Granton on 26 February 1876, and were shipped to Auckland on the *Inverallan* from East India Docks.[199] Instructions on shipping the paraffin oil did not come until about 10 weeks later. Young's despatched the oil from Grangemouth sometime between 13 and 17 May. It was sent by steamer to London for shipment by the *Loch Fergus* from the South West India Docks.[200]

In the 1878 Marine Department report Blackett commented that

> It was hoped that the land required for the site for this lighthouse would have been acquired before this; but as no reasonable terms can be arranged with the owner, it will be necessary to take it under 'The Public Works Act, 1876' arrangements for doing which are now in hand.[201]

The land was taken under the Act, and the proclamation published in the *Gazette* referred to 'Island Pokohinau', as Pokohinau is the alternate spelling of the Māori name for Burgess Island.[202] In his own 1879 report, Blackett commented:

> The land necessary for this was taken under 'The Public Works Act, 1876' and preparations were made to being the work, when instructions were received that the work should be suspended until further orders.[203]

About the same time, the Auckland Harbour Board requested that the lighthouse be established on Moro Tiri, one of the Hen and Chickens Islands, and the marine engineer was instructed to report on the site.[204] This probably led to the suspension of work on Moko Hinau.

In the 1880 Marine Department annual report, R Johnson (now the acting secretary) wrote 'I would strongly urge upon the Government to have it erected at once, as it is much required'.[205] The following year Seed contented himself with:

> The apparatus and lantern for the light at Mokohinau are still in store at Auckland, no instructions having as yet been given to proceed with the erection of this light.[206]

Eventually instructions were issued in November 1881[207] and a work party from the Public Works department under the supervision of David Scott, the lighthouse artificer, was despatched to build the lighthouse in February 1882.[208]

Figure 7.20: Moko Hinau Lighthouse (AWNS_19020925_5_4)
© Sir George Grey Special Collection Auckland Libraries, New Zealand

In contrast to the other five lighthouses which were wooden, it had originally been intended to use local stone for the tower. As this was unsuitable the design was changed to a concrete tower with an ashlar masonry lightroom and an iron balcony.[209]

The lighthouse's completion was announced in the 1883 Marine Department report. The light was exhibited on 18 May 1883. The Notice to Mariners stated that:

> The tower is 46 feet in height from base to top of Cowl and painted White. The Light will be a First Order flashing white light, visible all around seaward, and will show a flash every 10 seconds.

The range would be 26 nautical miles in clear weather.[210]

The exhibition of Moko Hinau was accompanied by an alteration in the lighthouse at Tiri Tiri Matangi:

> On the same night that the Moko Hinou [*sic*] Light is exhibited a section of the Tiri Tiri Light will be altered to show *red over an arc of about* 23° from over Flat Rock on to Kawau Island, the *rest of the light remaining white* as at present. The Red Light will shoe between the bearings from seaward of S. ¼ E. and S.S.E. ¼ E.

By 1895 the rollers and rings were wearing out and 12 new rollers were ordered along with the 12 for Puysegur Point. The repairs were carried out between Monday 19 January 1903 and Sunday 1 February 1903.[211]

In 1915 a revolving cowl was ordered in anticipation of installing incandescent lamps at Moko Hinau,[212] but neither it nor the lamps were installed for another three years.[213] The lamp installation was less than satisfactory as the following year the Marine Department reported that since the incandescent burner had been installed the light had been showing two distinct flashes between the main flash when at a distance of from two to eight miles. This was put down to refracted light from the lens.[214] In 1938 some thought was given to replacing the Chance Brothers lamp with a Smith-Ford burner,[215] but this was not carried out.[216]

Instead the department decided to electrify the lighthouse, with initial surveys in 1938[217] and the conversion completed by mid-1941.[218]

Endnotes

1 Acc 10706 vol 47 p 244 letter to Secretary of Customs (Marine Branch) Wgtn N Z re New Zealand Lighthouses 18 July 1876

2 AJHR 1875, H-12. Statement on Marine Matters,

3 Acc 10706 Vol 45 p 482, letter Stevenson to Commissioner of Customs Wellington, 30 April 1875

4 Acc10706 vol 45 p 482 letter to Commissioner of Customs (Marine Branch) Wgtn N Z re New Zealand Lighthouses

5 Acc 10706 vol 45 p 454 letters to Dove & Co, and Milne & Son re New Zealand Lighthouses 24 April 1875

6 Acc 10706 vol 45 p 478, letter to Milne & Son & Son re New Zealand Lighthouses 30 April 1875

7 Acc 10706 vol 45 p 489 letter to Dove & Co re New Zealand Lighthouses 1 May 1875

8 Acc 10706 vol 45 p 484 letter to Dove & Co re New Zealand Lighthouses 1 May 1875

9 Acc 10706 vol 45 p 484 letter to Dove & Co re New Zealand Lighthouses 1 May 1875 (i.e. two letters on same page)
10 Acc 10706 vol 45 p 714 letter to Commissioner of Customs (Marine Branch) Wgtn N Z re New Zealand Lighthouses 29 June 1875
11 Acc 10706 vol 45 p 706 letter to Milne & Son re New Zealand Lighthouses 9 July 1875
12 Acc 10706 vol 45 p 911 letter to Milne & Son re New Zealand Lighthouses 30 August 1875
13 Acc 10706 vol 45 p 911 letter to Dove & Co re New Zealand Lighthouses 30 August 1875 (two letters same page)
14 Acc 10706 vol 45 p 488 letter to Dove & Co re New Zealand Lighthouses 3 May 1875
15 Acc 10706 vol 45 p 488 letter to Milne & Son re New Zealand Lighthouses 30 April 1875
16 Acc 10706 vol 45 p 488 letter to Dove & Co re New Zealand Lighthouses
17 Acc 10706 vol 45 p 515, letter to Barbier & Fenestre 10 May 1875
18 Acc 10706 vol 45 p 474 letter to Chance & Co re New Zealand Lighthouses 30 April 1875 and p 475 letter to Barbier & Fenestre, 30 April 1875
19 Acc 10706 vol 45 p 496 letter to Barbier & Fenestre re New Zealand Lighthouses 7 May 1875
20 Acc 10706 vol113 pp 616–618 Specification of Second Order Dioptric Holophotal Revolving Light for New Zealand 30 April 187
21 Acc10706 vol 45 p 488 letter to Dove & Co re New Zealand Lighthouses 28 May 1875
22 Acc 10706 vol 45 p 581 letter to Barbier & Fenestre re New Zealand Lighthouses 28 May 1875
23 Acc 10706 vol 45 p 692 re New Zealand Lighthouses 23 June 1875
24 Acc 10706 vol 45 p 700 letter to Dove & Co re New Zealand Lighthouses 24 June 1875
25 Acc 10706 vol 45 p 698 letter to Barbier & Fenestre re New Zealand Lighthouses 24 June 1875
26 Acc 10706 vol 45 p 692 New Zealand Lighthouses *Statement of Character of Lights*
27 Acc 10706 vol 46 p 550 letter to Agent General, 10 February 1876
28 Acc 10706 vol 45 p 733 letter to Barbier & Fenestre re New Zealand Lighthouses 3 July 1875
29 Acc 10706 vol 45 p 758 letter to Barbier & Fenestre re New Zealand Lighthouses 12 July 1875
30 Ms 5885 NLS 19 New Zealand Lights 4 & 3 wick paraf(f)in lamps for 1st & 2nd order revolving lights May 1875
31 Acc 10706 vol 45 p 761 letter to Milne & Son re New Zealand Lighthouses 12 July 1875
32 Acc 10706 vol 45 p 762 letter to Dove & Co re New Zealand Lighthouses 12 July 1875
33 Acc 10706 vol 113 pp 510–1 Specification of Stores and Lamp glasses required for six lighthouses for New Zealand July 1875
34 Acc 10706 vol 113 pp 511–2 Specification of Paraffin cisterns for Six Lighthouses for New Zealand July 1875
35 Acc 10706 vol 45 p 814 letter to Smiths Blair Street re New Zealand Lighthouses 23 July 1875
36 Acc 10706 vol 46 p 236 letter to Agent General 4 Dec 1875
37 Acc 10706 vol 45 p 706 letters to Milne & Son, and Dove & Co re New Zealand Lighthouses 26 June 1875
38 Acc 10706 vol 46 p 157 letter to Secretary of Customs (Marine Branch) Wellington N Z re New Zealand Lighthouses 22 September 1875
39 Acc 10706 vol 45 p 752 letter to Messrs W Boyden & Son George Street re New Zealand Lighthouses 9 July 1875
40 Acc 10706 vol 46 p 23 Specification of Paraffin Tins for the New Zealand Government
41 Acc 10706 vol 46 p 19 Letter to Laidlaw and Son
42 Acc 10706 vol 46 p 20 Letter to Smiths, Blair Street
43 Acc 10706 vol 46 p 24 Specification for Paraffin for New Zealand Lighthouses
44 Acc 10706 vol 45 p 515 letter to Barbier & Fenestre 10 May 1875
45 Acc 10706 vol 45 p 706 letter to Milne & Son re New Zealand Lighthouses 9 July 1875
46 Acc 10706 vol 45 p 756 letter to Dove & Co re New Zealand Lighthouses 10 July 1875
47 Acc 10706 vol 85 p 31 letter Milne & Son to Stevenson, 3 Feb 1876 re New Zealand Lighthouses
48 AJHR 1874 H-23 Correspondence Relative to Selection of Site for Projected Coastal Lighthouses No 5 Report of visit to islet Brothers in Cook Strait, Sharp, and Johnson August 1863
49 AJHR 1865 D No 1 C (Appendix) Report on the Marine Board Department of New Zealand 1864–5. Report of Mr James M Balfour Wellington
50 AJHR 1874 H-22 Ninth Report of the Marine Department for year ended 30 June 1874
51 AJHR 1874 H-12A TenthReport of the Marine Department for the year ended 30 June 1875 Appendix A
52 AJHR 1875 H-12. Statement on Marine Matters by
53 Acc 10706 vol 45, p 692 re New Zealand Lighthouses 23 June 1875

54 Acc 10706 vol 113 p 610 Specification for Auxiliary Lights for Portland Island, The Brothers and cape Maria Van Diemen
55 Acc 10706 vol 46, p 1 letter to Milne & Son re New Zealand Lighthouses 30 September 1875
56 Acc 10706 vol 46, p 156 letter to Agent General London re New Zealand Lighthouses 13 Nov 1875
57 Acc 10706 vol 46, p 226 letter to Agent General London re New Zealand Lighthouses s 21 December 1875
58 Acc 10706 vol 46, p 678 letter to Secretary of Customs (Marine Branch) Wellington N Z re New Zealand Lighthouses 29 February 1876
59 Acc 10706 vol 46, p 279 letter to Agent General London re New Zealand Lighthouses 13 December 1875
60 AJHR 1876 H-26 Eleventh Report of the Marine Department for year ended 30 June 1876
61 AJHR 1877 H-29 Twelfth Report of the Marine Department for year ended 30 June 1877. Enclosure 1 Mr Blackett to Commissioner of Customs
62 *NZG Gazette* 1877 No 77 Notice to Mariners No 31 of 1877, p918 31 August 1877
63 Acc 10705 vol 88a letter LHB Wilson to Thomas Stevenson 4 January 1878
64 Acc 10706 vol 52, p 770 letter to Messrs Smiths George St re New Zealand Lighthouses 19 January 1882
65 Acc 10706 vol 52, p 794 letter to Milne & Son re New Zealand Lighthouses 30 January 1882
66 Acc 10706 vol 52, p 773 letter to E A Smith London re New Zealand Lighthouses 7 February 1882
67 Acc 10706 vol 52, p 871 letter to Agent General London re New Zealand Lighthouses 23 February 1883
68 Acc 10706 vol 52, p 904 letter to E A Smith London re New Zealand Lighthouses 4 March 1882
69 AJHR 1909 H-15 Marine Department Annual Report for 1908–09
70 NZG Archives M1/716/8/12/5; copy letter Bollons on Hinemoa to Secretary Marine 3 May 1918
71 NZG Archives ADOE 16612 M1/716 8/12/5; letter Fraser to Secretary Marine 17 July 1918
72 NZG Archives ADOE 16612 M1/716 8/12/5; report Fraser 22 November 1918
73 NZG Archives ADOE 16612 M1/716 8/12/5; memo to High Commissioner, 24 March 1921
74 NZG Archives ADOE 16612 M1/716 8/12/5; R B 40/88, letter D&C Stevenson to NZ High Commissioner 29 August 1921
75 NZG Archives Invoice from Steven& Struthers M1/716/8/12/5; 24 April 1922
76 NZG Archives M1/716/8/12/5; letter to Fraser dated 6 July 1922
77 *NZG Gazette* 1923 No 63 Notice to Mariners No 40 of 1923, p 2216 15 August 1923
78 AJHR 1951 H-15 Marine Department Report for the year ended 31 March 1951
79 AJHR 1956 H-15 Marine Department Report for year ended 31 March 1956
80 AJHR 1871G - No 6 6th (incor'ing 5th) Report of the Marine Board Department for year 1869–70 and 1870–1 Memo from R Johnson Relative to Additional Lighthouses on the Coast of New Zealand
81 AJHR 1873, H-19. Report on the Lighthouses on the Coasts of New Zealand
82 AJHR 1874 H-23 Correspondence Relative to Selection of Site for Projected Coastal Lighthouses No 3 Report Captain Johnson to Commissioner Customs
83 AJHR 1873 H-19A Eighth report of the Marine Department for year ended 30 June 1873
84 Acc 10706 vol 45, p 692letter Agent General to Stvenson re New Zealand Lighthouses 23 June 1875
85 Acc 10706 vol 85 p 52 letter Agent General to Stevenson, 15 Feb 1876
86 Acc 10706 vol 85 p 57 letter Milne & Son to Stevenson, 19 Feb 1876 [
87 Acc 10706 vol 85 p 71 letter Barbiere & Fenestre to Stevenson, 17 March (Mars) 1876 in French
88 Acc 10706 vol 85 p 84 letter James Dove & Co to Stevenson, 28 April 1876
89 Acc 10706 vol 46 p 996, letter to Smith Morrison re New Zealand Lighthouses 24 May 1876
90 Acc 10706 vol 85 p 96 letter Smith Morrison to Stevenson, 26 May 1876
91 AJHR 1876 H-26 Eleventh Report of the Marine Department for year ended 30 June 1876. Appendix A, Report by Marine Engineer 15 June 1876
92 AJHR 1877 H-29 Twelfth Report of the Marine Department for year ended 30 June 1877. Enclosure 1
93 *NZG Gazette* 1878 No 8 Notice to Mariners, No 1 of 1878, p 79 10 Jan 1878
94 AJHR 1911 H-15 Marine Department Annual Report for1910–1911
95 NZG Archives ADOE 16612 M1/753 8/47/3 letter Fraser – Secretary Marine Department dated 6 July 1918
96 NZC Archives ADOE 16612 M1/753 8/47/3 letter Fraser to Secretary Marine Department dated 24 March
97 AJHR 1949, H-15 Marine Department Annual Report fot the year1948–49 report 9
98 NZG Archives AADX W3148 889 /36 63/143 p2 Letter Turner marine engineer to secretary for marine 14 February 1957, also announced as Notice to Mariners No. 32 of 1957

99 AJHR 1862 2, D-15 Papers on Lighthouses in Foveaux Straits

100 AJHR H-19 Report by R Johnson and R A Edwin on Lighthouses on the Coasts of New Zealand, Enclosure in No 2

101 AJHR 1874 H-23 Correspondence Relative to Selection of Site for Projected Coastal Lighthouses No 2 Report Captain Johnson to Commissioner Customs

102 AJHR 1875 H-12A. Tenth Report of the Marine Department, Appendix A, No 3

103 Acc 10706 vol 45 p 692 re New Zealand Lighthouses 23 June 1875

104 Acc 10706 vol 45 p 698 letter to Barbier & Fenestre re New Zealand Lighthouses 24 June 1875

105 Acc 10706 vol 45 p 737 letter to Barbier & Fenestre re New Zealand Lighthouses d 3 July 1875

106 Acc 10706 vol 113 pp 618–622 Specification of First Order Dioptric Fixed light Apparatus for Centre Island, Foveaux Strait

107 Acc 10706 vol 45, p 706 letter to Milne & Son re New Zealand Lighthouses 9 July 1875

108 Acc 10706 vol 46, p 263 letter to AG London re New Zealand Lighthouses 8 December 1875

109 Acc 10706 vol 46, p 344 letter to Agent General London re New Zealand Lighthouses 29 December 1875

110 Acc 10706 vol 46 p 353 letter to Secretary of Customs (Marine Branch) Wellington N Z re New Zealand Lighthouses 31 December 1875

111 Acc 10706 vol 46, p 441 letter to Milne & Son re New Zealand Lighthouses 25 January 1876

112 Acc 10706, vol 85 p 49 letter Milne & Son to Stevenson 15 Feb 76

113 Acc 10706 vol 85 p 82 letter from John Fyfe (Youngs paraffin) dated 24 April 1876

114 Acc 10706, vol 85 p 76 letter Agent General to Stevenson 20 April 1876

115 Acc 10706 vol 85 p 115 letter William Seed to Stevenson, 29 July 1876, No 401/13

116 *NZG Gazette* 1877 No. 7 Tenders p 61 13 January 1877

117 AJHR 1877 H-29 Twelfth Report of the Marine Department for year ended 30 June 1877. Report from Mr Blackett included in above

118 *NZG Gazette* 1878 No. 84. Notice to Mariners No 22 of 1878 p 1234 22 August 1878

119 AJHR 1909, H-15 Marine Department Annual Report for 1908–09

120 AJHR 1910, H-15 Marine Department Annual Report for 1909–10

121 NZG Archives ADOE 16612 M1/753 /8/22/4: memo Marine Engineer to Secretary Marine Department 28/4/49

122 NZG Archives ADOE 16612 M1/753 /8/22/4: copy memo Nautical Adviser.26/4/49

123 AJHR 1956 3, H-15 Marine Department Report for year ended 31 March 1956

124 AJHR 1874 H-23 Correspondence Relative to Selection of Site for Projected Coastal Lighthouses No 2 Report Captain Johnson to Commissioner Customs 9 June 1874

125 Acc 10706 vol 45,p 692 re New Zealand Lighthouses 23 June 1875

126 Acc 10706 vol 45 p 698 letter to Barbier & Fenestre re New Zealand Lighthouses 24 June 1875

127 Acc 10706 vol 45 p 738 letter to Barbier & Fenestre re New Zealand Lighthouses 3 July 1875

128 Acc 10706 vol 46 p 156 letter to Agent General London re New Zealand Lighthouses 13 November 1875

129 Acc 10706 vol 46 p 164 letter to Agent General London re New Zealand Lighthouses 17 November 1875

130 Acc 10706 vol 46 p 188 letter to Agent General London re New Zealand Lighthouses 25 November 1875

131 Acc 10706 vol 46 p 232 letter to Dove & Co re New Zealand Lighthouses 3 December 1875

132 Acc 10706 vol 46 p 256 letter to William Seed re New Zealand Lighthouses 7 December 1875

133 Acc 10706 vol 46p 394 letter John Fyfe (Youngs paraffin) re New Zealand Lighthouses 12 January 1876

134 Acc 10706 vol 85 letter Youngs to Stevenson 13 Jan 1876

135 Acc 10706 vol 46, p 408 letter to AG London re New Zealand Lighthouses 14 January 1876

136 Acc 10706, vol 85 p 33 letter AG to Stevenson 3 February 1876

137 Acc 10706, vol 85p 49 letter Milne & Son to D & T Stevenson 15 February 1876

138 Acc 10706, vol 85 p 33 letter Agent General to D & T Stevenson 3 February 1876

139 Acc 10706 vol 85 p 96 letter Smith Morrison to D & T Stevenson, 26 May 1876

140 Acc 10706 vol 85 p 56 letter from John Fyfe (Youngs paraffin) 21 February 1876

141 AJHR 1876 H-26. Eleventh Report of Marine Department for year ended 30 June 1876, Appendix A

142 *NZG Gazette,* 1877 No. 89. PRELIMINARY NOTICE p 1041 22 October 1877

143 *NZG Gazette,* 1877 No.7 Tenders p 61 13 January 1877

144 AJHR 1877 H-29 Twelfth Report of the Marine Department for year ended 30 June 1877 Enclosure 1

145 AJHR 1878 H-12. Annual Report of the Marine Department Enclosure 1

146 *NZG Gazette* 1879, No.11 Notice to Mariners No 3 of 1879, p118 24 January 1879
147 Acc 10706 vol 63 p 968 letter to Agent General 27 June 1895
148 Acc 10706 vol 63 p 996 letter to Milne & Son 5 July 1895
149 Acc 10706 vol 63 p 1005 letter to Milne & Son 9 July 1895
150 Acc 10706 vol 64 p 5 letter to Agent General 15 July 1895
151 Acc 10706 vol 64 p 212 letter to Milne & Son 14 September 1895
152 Acc 10706 vol 64 p 225 letter to Agent General 19 September 1985
153 Acc 10706 vol 64 p 231 letter to Milne & Son 23 September 1895
154 Acc 10706 vol 64 p 243 letter to EA Smith (shipping agent) 28 September 1895
155 AJHR 1903 H-15. of the Marine Department Annual Report for 1902-033
156 AJHR 1905 H-15. Marine Department, Annual Report for 1904-05 15 May 1905
157 *NZG Gazette* 1905 No. 29 Notice to Mariners, No 23 of 1905 Marine Dept, Wellington, 27 March 1905
158 *NZG Gazette* 1905 No. 41Notice to Mariners, No 31 of 1905 p 1063 , 27 April 1905
159 *NZG Gazette* 1905No 52 Notice to Mariners, No 38 of 1905 p 1342 , 26 May 1905
160 AJHR 1909 H-15. Marine Department Annual Report for 1908-09
161 NZG Archives ADOE 16612 M1/756 8/48/4 pt 1 Memo Nautical Advisor (Whitford) to Secretary Marine dated 23 August 1939
162 AJHR 1873 Report by R Johnson and R A Edwin on Lighthouses on the Coasts of New Zealand, Enclosure in No 2 dated 3 April 1873
163 AJHR 1873 H-19A Eighth report of the Marine Department for year ended 30 June 1873
164 AJHR 1874 H-23 Correspondence Relative to Selection of Site for Projected Coastal Lighthouses No 3 Report Captain Johnson to Commissioner Customs 10 June 1874
165 Acc 10706 vol 45,p 692 re New Zealand Lighthouses 23 June 1875
166 Acc 10706 vol 113 pp 610–611 Specification of Auxiliary Lights for Portland Island, The Brothers and Cape Maria van Diemen
167 Acc 10706 vol 85 p 70 letter from Barbier & Fenestre 13 March 1876[in French]
168 Acc 10706 vol 46, p 570 letter to Smiths Blair Street re New Zealand Lighthouses 15 February 1876
169 Acc 10706 vol 46, p 773 letter to Milne & Son re New Zealand Lighthouses 29 March 1876
170 Acc 10706 vol 46 p 349 letter to Messrs W Bryden & Son George Street re New Zealand Lighthouses 31 December 1875
171 Acc 10706 vol 46, p 350 letter to Milne & Son re New Zealand Lighthouses 31 December 1875
172 NZG Archives BBAO 5544 17a 1876/312 Memo Secretary Marine Branch, Wellington to Collector H M Customs Auckland 22 April 1876
173 Acc 10706 vol 47, p 96 letter to Smith Morrison London re New Zealand Lighthouses 17 June 1876
174 NZG Archives BBAO 5544 179a 1876/830 letter to Collector Customs Auckland, from Seed, Sec Customs, Customs dept (Marine Branch). 18 September 1876
175 NZG Archives BBAO/5544/19a/1877/329 Memo William Seed, Sec and Inspector of Customs
176 Acc 10706 vol 85,p 82 letter from John Fyfe (Youngs paraffin) [24 April 1876
177 Acc 10706 vol 46 p 879 letter John Fyfe (Youngs paraffin) re New Zealand Lighthouses 25 April 1876.
178 Acc 10706 vol 47 p 217 letter to Smith Morrison London re New Zealand Lighthouses 11 July 1876
179 AJHR 1878 H-12 Annual Report of the Marine Department Enclosure 1,
180 *NZG Gazette* 1879 No.28 Notice to Mariners No 4 of 1879 p 329 6 March 1879, p
181 Acc 10706 vol 85 copy S175/78 of Requisition 3 January 1879
182 Acc 10706 vol 85, p 241 letter Agent General to Stevenson, 14 May 1879,
183 Acc 10706 vol 50 p 451 letter to Agent General London re New Zealand Lighthouses 27 May 1879
184 Acc 10706 vol 52 p 770 letter to Messrs Smiths George St re New Zealand Lighthouses 19 Jan 1882
185 AJHR 1909, H-15 Marine Department Annual Report for1908–1909
186 AJHR 1915, H-15 Marine Department Annual Report for 1914–1915
187 *NZ Gazette* 1911No.1 Notice to Mariners, 1 of 1911, p 29 9 January 1911
188 AJHR 1938 H-15 Marine Department Annual Report for the year1937–38
189 AJHR 1941, H-15 Marine Department Annual Report for the year1940–41
190 AJHR 1873 H-19 Report by R Johnson and R A Edwin on Lighthouses on the Coasts of New Zealand, Enclosure in No 2

191 AJHR 1874 H - 23 Correspondence Relative to Selection of Site for Projected Coastal Lighthouses No 3 Report Captain Johnson to Commissioner Customs
192 AJHR 1873 H-19A 8th Report of the Marine Department for year ended 30 June 1873
193 AJHR 1873 H-19A 8th Report of the Marine Department for year ended 30 June 1873
194 Acc 10706, vol 85, letter James Dove & Co to Stevenson, 19 Jan 1876
195 Acc 10706, vol 46 p 490, letter to Agent General London 3 February 1876
196 AJHR 1876 H - 26 Eleventh Report of the Marine Department for year ended 30 June 1876
197 Acc 10706 p 36 vol 85, letter Milne & Son to Stevenson 5 Feb 1876
198 Acc 10706, vol 85 p 49 letter Milne & Son to Stevenson dated 15 Feb 76
199 Acc 10706 vol 46, p 648 letter to Milne & Son re New Zealand Lighthouses 25 February 1876
200 Acc 10706 vol 46, p 979 letter to Agent General London re New Zealand Lighthouses 17 May 1876
201 AJHR 1878 H-12 Annual Report of the Marine Department Enclosure 1,
202 *NZG Gazette* 1879 p 242, PROCLAMATION by Governor dated ?
203 AJHR 1879 H-10 Annual Report of the Marine Department, The Marine Engineer to the Secretary, Marine Department .
204 AJHR 1879 H-10 Annual Report of the Marine Department
205 AJHR 1880 H-13 Marine Department Annual Report
206 AJHR 1881 H-27 Marine Department Annual Report
207 AJHR 1882 H-12 Marine Department Annual Report
208 AJHR 1882 H-12 t Marine Department Annual Report, Report by the Marine Engineer to the Secretary, Marine Department
209 AJHR 1883 H-15 Marine Department report Annual Report
210 *NZG Gazette* 1883 No.49 Notice to Mariners No 17 of 1883, p 757 , 26 May 1883
211 *NZG Gazette 1902* No103 Notice to Mariners 88 of 1902, p 2713 , 17 Dec 1902
212 AJHR 1915 H-15 Marine Department Annual Report for 1914–15. Report by Marine Engineer
213 AJHR 1918 H-15 Marine Department Annual Report for 1917–1918
214 *NZG Gazette* 1919 No88 . Notice to Mariners No 38 of 1919 p 2405 14 July 1919,
215 NZG Archives ADOE 16612 M1/733 8/25/2 Letter Fraser to Sec MD 6 Jan 1938
216 NZG Archives ADOE 16612 M1/733 8/25/2 memo Campbell Sec MD to Fraser 19 Jan 1938
217 AJHR 1938 H-15 Marine Department Annual Report for the year 1937–38
218 AJHR 1941 H-15 Marine Department Annual Report 1940–41

8 Timaru and Hokitika: Incandescents and Gas Burners

The lighthouses at Timaru Harbour and Hokitika were not high order landfall lights, but were still built to Scottish designs and used Scottish technology, as did the second Timaru light at Tuhawaiki (Jack's) Point to the south. They are of interest, as they marked changes in New Zealand lighthouse illumination: Hokitika was the first light to use gas burners with fuel drawn from the town supply, and the second Timaru lighthouse was the first to use the incandescent burners which gradually replaced Doty paraffin burners.

Timaru Harbour

The port of Timaru is on the South Canterbury coast. It was not a natural harbour, and had begun its maritime career as a space for hauling up *waka*, the large canoes used for voyaging and trading around the New Zealand coast.

A reef of lava acts as a submerged breakwater, and in the early days of colonial settlement ships calling at Timaru were forced to anchor in the roadstead thus created. Cargo and passengers were taken off and put on board by small boats, initially whaleboats and then surfboats organised by professional landing companies. The roadstead was exposed to the south-east, and many ships were wrecked or grounded in Timaru before harbour improvements were made. Balfour was among the civil engineers consulted and in his report of November 1865, he recommended a jetty and breakwater on the southern side of the roadstead. Ironically, he was drowned when the surfboat

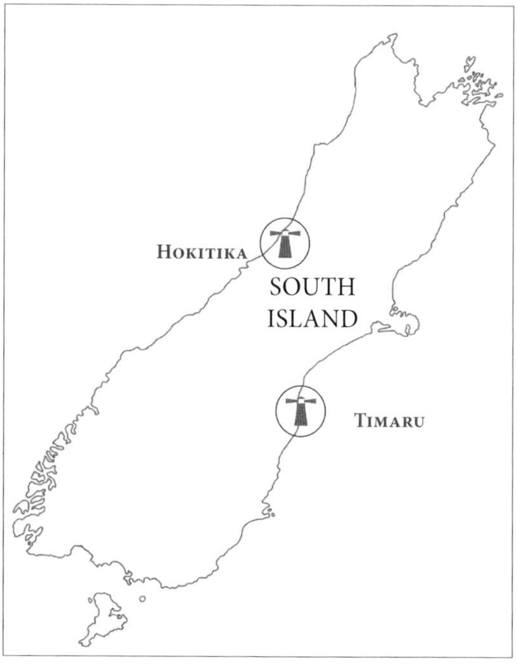

Figure 8.1: Location of Timaru and Hokitika lighthouses © Author

Figure 8.2 Timaru township 1875 showing location of lighthouse (964/46.01 1875)
© South Canterbury Museum

ferrying him to a steamer in the Timaru roadstead overturned. In 1870 an attempt was made to provide some shelter by constructing a groyne, but unfortunately it was soon demolished by rough seas.[1] The 1875 map of Timaru showed the site of the proposed lighthouse with still no harbour structures.

Johnson visited Timaru as part of his 1874 South Island site visits and recommended a harbour light. He suggested white fixed light of fifth order dioptric, with holophotal mirrors showing over a 180° arc from N.6°W to S.9°E., with a tower of about 25 feet.[2]

At this time, administration of Timaru Harbour was the responsibility of the Canterbury Provincial Government. Therefore the Marine Department arranged to obtain the lighthouse lantern etc. from Scotland on their behalf, as had been done for Napier Bluff. Before the lighthouse was completed the Timaru Harbour Board had been established and the completed lighthouse was transferred to them.

The New Zealand Government sent the requisition and order for the new lighthouse on 24 October 1874. Hence this order was sent only a few months before the major one for the Six Lighthouses described in Chapter 7. D & T Stevenson acknowledged the order on 17 December 1874[3] and that day invited tenders from Milne & Son and Dove & Co for the lantern.[4] Milne & Son seem to have had a larger lantern in stock, but the offer to supply this was declined.[5] The contract went to Dove & Co for £183.[6]

The specification describes the lantern as

> 6 feet internal diameter from glass to glass and 3 feet 6 inches in height of day light. The lantern to have gun metal astragals 1 inch broad and copper roof and ventilator, the roof to be single sheet copper No 14 Birmingham wire gauge ... and 9 glass and 7 dormant panes to give a lighted arc of 180°.

The sole plate was

> cast iron in 4 pieces forming a regular polygon of eights sides and measuring 5 feet 9 inches in internal diameter.

Again the lantern was to be cast using supplied patterns.[7]

Barbier & Fenestre were asked to tender for the fifth-order apparatus and dioptric mirror on 26 December 1874,[8] and their offer of £76 dated 31 December was officially accepted on 4 January 1875.[9] Barbier & Fenestre suggested a modification to the design – a change in the hingeing of the mirror that resulted in a change in the height of the apparatus. Stevensons approved the change and indicated that the critical factor was not height but the distance of the bottom of the pillar to the focal plane which had to be 7¾ inches.

Milne & Son were invited to provide the lamps (two fountains with burners complete) and a stand. The stand was to be identical to that supplied for the fifth order Japanese lighthouse apparatus.[10] (When the tenders were sent out for the lighthouse at Hokitika, contractors were asked to supply lanterns and apparatus identical to that supplied for Timaru, including the backup Doty burner.)

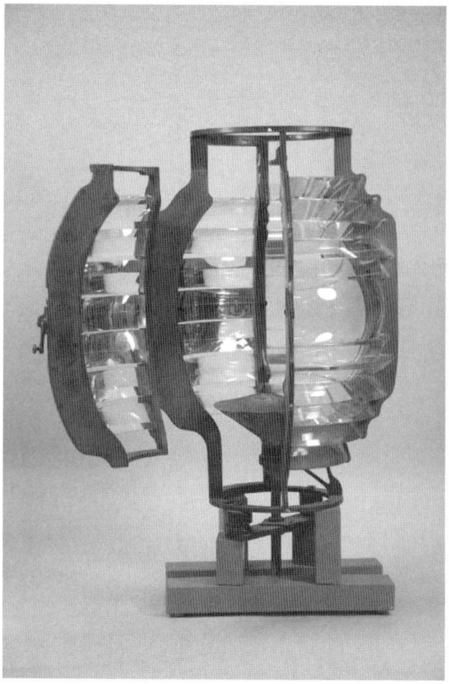

Figure 8.3: Lens, Timaru Harbour lighthouse © South Canterbury Museum

Figure 8.4 Design Timaru Harbour Lighthouse – Exterior (2004/070.37 Sheet # 1) © South Canterbury Museum

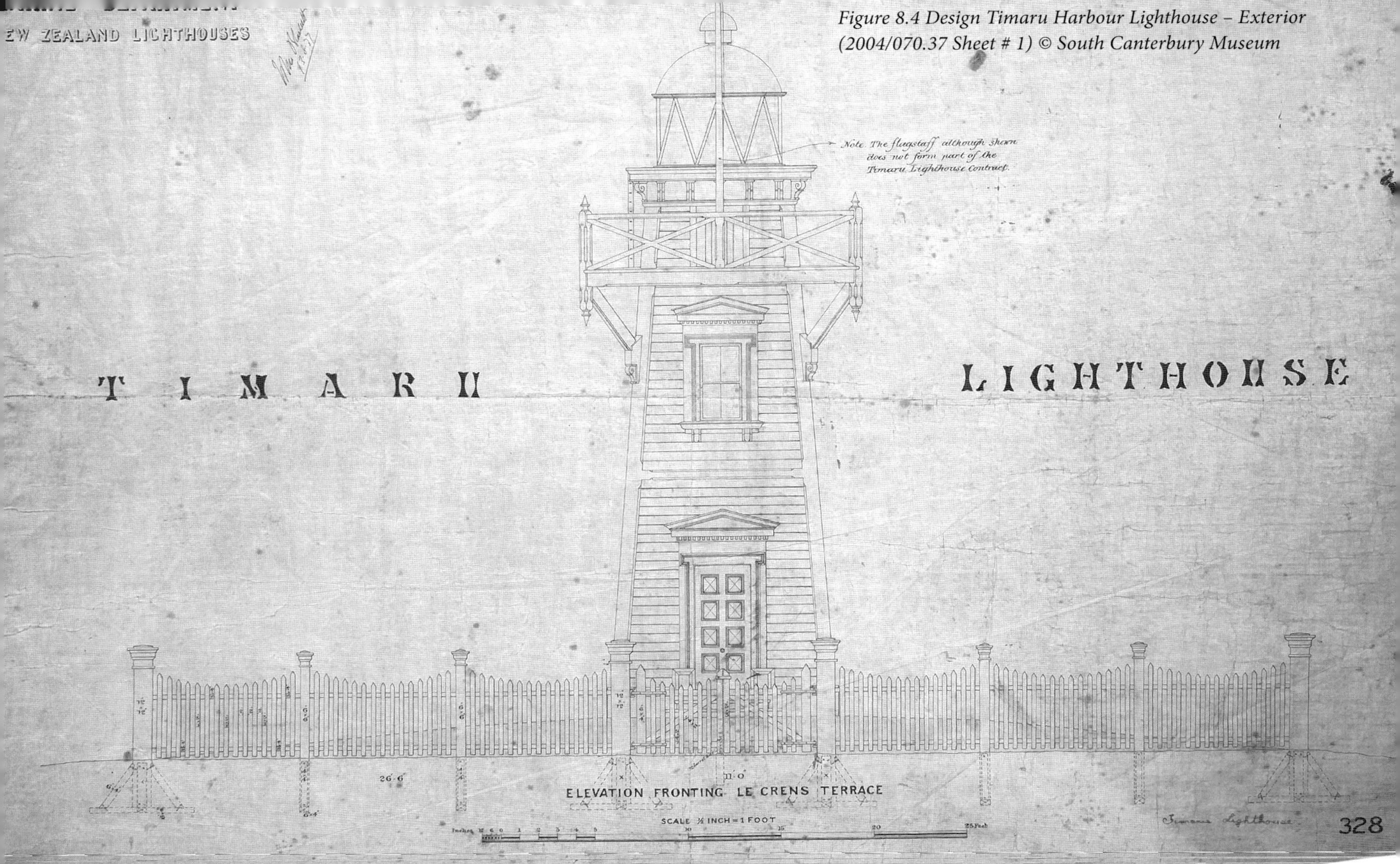

Milne & Son and Smiths were both invited to tender for the stores.[11,12] The contract was given to Smiths, whose price was £18/6/4.[13]

The lantern was sent by General Steam Navigation Company steamer from Granton on 3 April 1875 for transfer to the *Star of China* at the South West India Docks.[14] Lens, apparatus and stores were sent from Granton on 19 May 1875 and shipped to New Zealand on the *Blairgowrie* from East India Docks.[15] Both shipments were addressed to 'Secretary of Customs, Canterbury'.

The Marine Department undertook the erection of the lighthouse, as it would do for its own lights. However, construction was not straightforward. Initially plans and specifications for a timber tower were drawn up, and tenders called for its construction.[16]

In December 1876 Seed explained in a letter to the mayor of Timaru[17] that owing to the passing of a municipal regulation prohibiting the erection of timber structures within the town, the advertisement had to be withdrawn [*]. New designs were then prepared for a stone tower, but some delay occurred in calling for tenders for its erection owing to the uncertainty that existed as to who was to be at the expense of maintaining the light when it was completed.

> An Act, however, having been passed last session for forming a Harbour Board at Timaru, the Government not doubting that the body when constituted will at once cheerfully undertake to provide the cost of its maintenance, determined at once to call for tenders for the stone tower, so as to save further delay in proceeding with the erection of this light, which is considered necessary for vessels, approaching the anchorage at night.

The resulting tenders for the stone tower and an additional dwelling house were too high, so the contract was withdrawn and despite the by-law the tower design was changed back to a wooden one, along with the house.[18]

In designing this tower Blackett allowed some flourishes not found on other wooden lighthouses, no doubt because it was to be placed in a residential area. An elaborate front door, complemented by neo-palladian window surrounds was placed on top of the standard framing .

The contract was finally let to G T Waters for a cost of £573 2s 5d,[19] but construction overran the contract period and the light was finally exhibited on 1 July 1878.[20] The Notice to Mariners[21] gave the following description:

[*] Early Timaru had no fire brigade and a limited water supply, and was at the mercy of prevailing dry north-westerly winds. A fire on Monday 7 December 1868, later known as 'The Great Fire of Timaru', destroyed 39 buildings in the town's commercial district. Subsequently the borough council passed an ordinance requiring all new buildings in the area to be built of brick or stone with slate, tile, or metal roofs. The by-law was published in the *Canterbury Provincial Gazette* of 13 July 1869. This was seven years before the design of the Timaru lighthouse. In his letter, Seed was sliding over the Marine Department's misjudgement in not checking provincial regulations. The passing of the Act which abolished the provincial governments the previous year may have contributed to the lack of foresight. (See Andersen J. *Jubilee History of South Canterbury*, Whitcombe and Tombs, 1916.)

Figure 8.5 Design Timaru Harbour Lighthouse –Framing (2004/070.37 Sheet # 2) © South Canterbury Museum

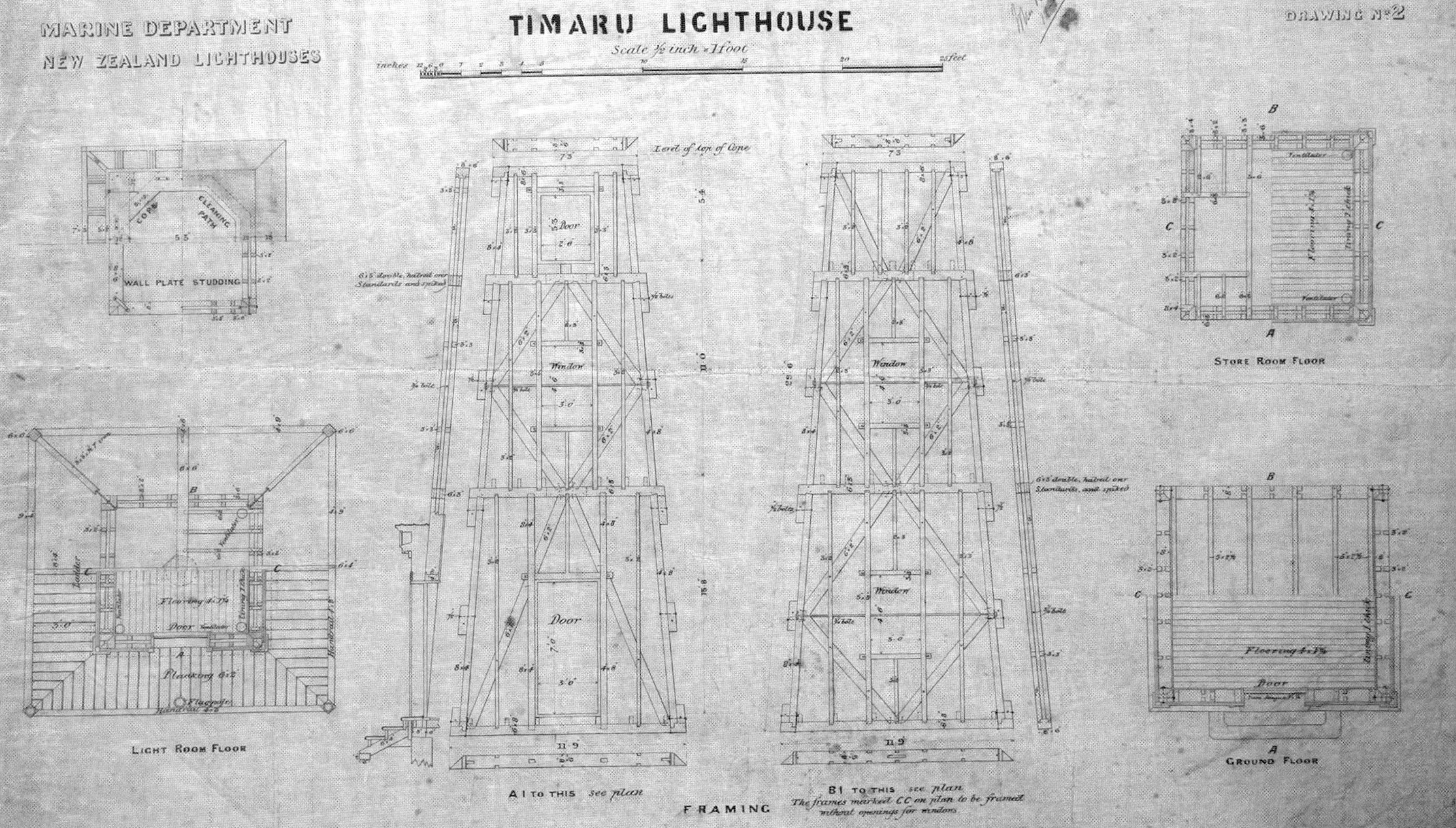

> The tower is 30 feet in height, is built of timber and painted stone colour. The light will be a Fifth Order Fixed White Light, and will be visible from N.20° W. round by W. to S. 20°E. The light is elevated about 85 feet above the sea, and allowing 15 feet for the height of the observer's eye, will be seen at a distance of about 14½ nautical miles in clear weather.

Figure 8.6: Timaru lighthouse on original site, 110 Le Crens Tce (7671) © South Canterbury Museum

In 1882, a 45° green sector was added to the light.[22] This green light was to show between west and south-west as viewed from the sea, and was exhibited from 1 May 1882. The green sector was intended 'as a guide to vessels approaching Timaru during the night or bringing up in the roadstead'.[23] There was a later change to this sector. A Notice to Mariners of May 1891 stated that the green sector was 22° from SSW to S as viewed seawards. This was to show over the wreck of the *Lyttelton*.[24]

In late 1883 it was decided to change the light to gas burners as in Hokitika and Napier, and the Marine Department undertook to act on behalf of the Timaru Harbour Board and obtain the burners and associated stores. Marine Department Requisition Marine 456/38 which included the burners was sent to Stevensons by the Agent General on 28 January 1884. The relevant part of the requisition called for

> Three gas burners suitable for the fourth order light supplied for Timaru; to be of good power as the light has now been shaded green on certain bearings. The burners should be Suggro make and one gross of white glasses to suit the burners to be also supplied.[25]

The specification referred to the 'fourth-order light supplied for Timaru'. Presumably Stevensons were able to ensure that the lamps sent out were suitable for a fifth order lantern. If the specification had gone to anyone other than Stevensons an interesting situation would have developed.

The burners and the other stores were sent by rail from Edinburgh in mid-May for shipment on the *Victory* from East India Docks.

When the lighthouse at Tuhawaiki Point south of Timaru Harbour was exhibited, the white light of the harbour lighthouse was changed to red to avoid confusion.[26] The Timaru Harbour Board electrified the light in 1924.[27] It remained in operation until 1970, when it was decommissioned. The lighthouse was removed from its site on Le Crens Terrace in 1980 and re-erected on another site. In November 2010 it was again relocated, this time to a site on the cliff north of the harbour.

Timaru: Tuhawaiki (Jacks) Point

In January 1874 the Canterbury Provincial Council passed a resolution requesting the superintendent[28]

> to place on the Supplementary Estimates a sum not exceeding £1,000 for the purpose of erecting a Harbour Light at Patiti Point, Timaru.

In his address to the next (41st) session, Rolleston, the then superintendent, made no reference to the proposed coastal lighthouse

By late 1902 it had been decided that a coastal light was needed, but apparently Patiti Point was still the assumed site as the 1903 annual report indicated that plans and specifications had been drawn up for the lighthouse and the keeper's house at that location. The Marine Department planned to re-use the fourth order lantern and apparatus, including the Doty burner which had been removed from Matiu/Somes when it was upgraded to a second order light in 1899.[29]

Figure 8.7: Tuhawaiki Point lighthouse, (1999/112.15) © South Canterbury Museum

Sometime after the report was transmitted it was decided to take up Johnson's suggestion of a light at Tuhawaiki Point south of Timaru,[3*] and the whole Matiu/Somes lighthouse tower, with lantern and apparatus, was utilised.

This was exhibited on 1 July 1904. The Notice to Mariners described the light thus:[30]

> a dioptric fixed white light of the fourth order, is about 94 ft above the level of the sea at high water, and is visible to seaward over an arc of 168 degrees of the horizon from a distance of about 14 miles from the deck of a vessel 15ft above the sea.
>
> The tower is an octagonal cast iron structure, and is painted white. It is 29 ft in height from the surface of the ground at the base to the top of the lantern

The Marine Department first suggested introducing incandescent burners into New Zealand's lighthouses in the Marine Department report for 1903 where it indicated that 'The Trinity House… has adopted the Matthew's incandescent burner on the Kitson system in two of its lighthouses', and that one of these burners had been ordered for possible adoption in New Zealand.

The Chance Brothers Archives contain the order (Mar 282/125 P B 74/37) which is headed with the reference 'Jacks Point, Old Soames Island. This is dated 21 September 1903 and was for 'Incandescent Petroleum Installation for a 4th order fixed apparatus' and included two 'Matthews' burners of 50 mm diameter complete.[31]

There appear to have been problems with this order. The 1904 annual report announces only the establishment of the Tuhawaiki Point lighthouse, and the annual report for the following year stated that an incandescent burner for Tuhawaiki Point had been ordered from England.

The burner was probably delivered in late 1905 or early 1906, as that year's annual report stated that:

> A Matthew's incandescent burner has been procured for this light and also the necessary apparatus to make it an occulating light. It is necessary that this should be done as it is found that at present ship's lights and other lights in the neighbourhood were liable to be mistaken for it. The work of erecting the apparatus and new burner is now being carried out by Mr Scott, the department's lighthouse artificer.[32]

The upgraded light was exhibited on 23 August 1906, with an occulating pattern of light for 7 seconds and a dark interval of 3 seconds.[33]

This marked the start of the use of incandescent burners in New Zealand.

[3*] Tuhawihi Point is named after Hone Tuwawihi, a paramount chief of the Ngai Tahu *iwi* (tribe) of the South Island of New Zealand, who was drowned when his craft was lost on the reef offshore of the point. He was one of the signatories of the Treaty of Waitangi when it was brought to the South Island. He also used the European name Jack, hence the alternative title 'Jack's Point'.

The 50 mm diameter was the largest incandescent burner manufactured, and would have been too big for the fourth order light. Hence when the light proved satisfactory, the Marine Department ordered the smaller 35 mm incandescent burners manufactured by Chance Brothers. The order (Marine 2.12 of PB108/50 NZG) dated 11 December 1907 was for three 35 mm Chances incandescent vapour burners and necessary equipment.[34]

The next two lighthouses to receive incandescent burners were Pencarrow and Stephen's Island.[35] From then onwards all new lighthouses except Baring Head and Cape Reinga were equipped with incandescent burners. Existing lighthouses were upgraded from Doty burners to incandescent if they were not automated with acetone-acetylene apparatus.

In 1929 it was decided to install a new 800 mm lens at Tuhawaiki Point and to automate it with an acetone-acetylene automatic flashing apparatus,[36] with the light being maintained by the Timaru Harbour Board. The upgraded light was exhibited on 29 May 1930, and the new pattern showed one flash every 3 seconds, i.e. flash 1 second, eclipse 2 seconds.[37]

Hokitika

Hokitika, situated where the river of the same name flows into the Tasman Sea, was an important port during the West Coast gold rush. The 1865 report of the president of the Marine Board commented that:

> At Hokitika several vessels have been stranded, a result which might have been anticipated from the dangerous character of the rivers on the West Coast.

One particular hazard was the series of moving sandbanks just off shore from the river mouth commonly referred to as the Hokitika bar.

(Given the West Coast's historic lifestyle, non-engineering comments on the Hokitika bar often imply remarks about the area's 'flexibility' with respect to licensing hours.)

One of the bar's 1865 victims was the *New Zealand*, wrecked in August. She was built by Tod and McGregor of Partick (who built *St Kilda*, the steamer purchased for the Otago Provincial Government), and was a twin-hulled iron paddle steamer designed primarily for inshore cruising on the Caledonian Canal and the Firth of Clyde. Launched in1856 as the *Alliance*, she was rebuilt in 1863 for running the blockade imposed on the southern Confederate States of America. Eric Graham describes her as

> arguably, the most unsuited hull design for open-sea passages and the slowest runner to leave the Clyde for the blockade.

On her first attempt to enter the Savannah River she ran aground and was boarded by a party from the USS *South Carolina*. She was subsequently sold at the Boston Prize Court and her name changed to *New Zealand*. After further modification, including the addition of sails, she was despatched to New Zealand via the Cape of Good Hope. Her safe arrival in Lyttelton was in contrast to the earlier part of her blockading career, when she limped damaged into Nassau in the Bahamas after the trip from the Clyde via Cork. She served as

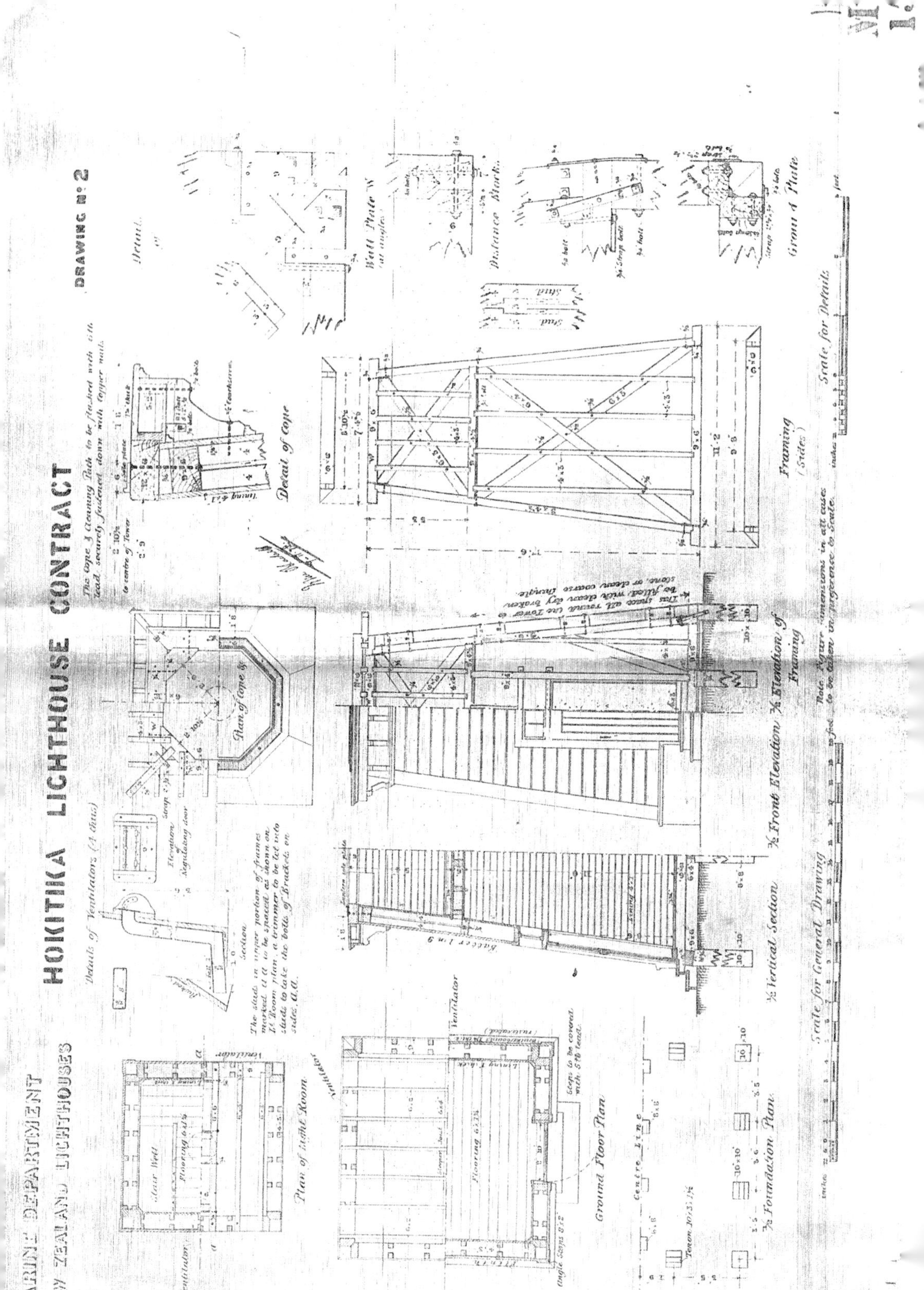

Figure 8.8: Hokitika lighthouse structure © Hokitika Museum

a passenger steamer on the Lyttelton–Dunedin run before being sent to the West Coast.[38] In this case, the Scottish contribution was to *demonstrate the need* for a lighthouse at Hokitika.

In their 1873 report, Edwin and Johnson indicated that Hokitika already had a fixed light, which the 1879 Notice to Mariners (see below) described as a 'White Light … shown from the Flagstaff'. It seems that there was some demand for an upgrade, as the eleventh report[39] stated that

> [a] site has been selected for the lighthouse, and drawings and specifications have been prepared. In a short time tenders will be called for construction.

The table headed 'Amount Expended on New Lighthouses' in the report indicated that £1,279/4/7 had been spent on Hokitika up to 30 June 1876.

The tower was designed on the understanding that the lantern and apparatus ordered for Tory Channel would be used at Hokitika. As discussed in Chapter 6, the Tory Channel lantern and apparatus were used for the Moeraki light as this was considered to be of greater importance Hence Blackett's 1877 comment that 'Nothing more has been done towards the erection of [Hokitika] lighthouse'.[40]

Later that year the decision was made to go ahead with the light for Hokitika. The fact that the minister of marine was Richard Seddon, member for Westland, may have been a contributing factor.

The specification for the lighthouse contract[41] showed that Blackett designed a timber tower 17 feet 6 inches high with trussed frames of ironbark (from Australia) and kauri. Most of the tower was kauri, with totara used for the piles. The outside was clad with weatherboard, and the inside lined with tongue and groove board. The space between weatherboards and lining was to be filled with 'clean dry broken stone, or clean coarse shingle' to ballast the tower. The specification clearly stated that the 'erection of the lantern will not form part of the contract'. The contract was let to M W Jack.

In contrast to the other New Zealand lighthouses, which by now all used Doty paraffin burners, it was decided to obtain a gas burner as the lighthouse could be supplied from the town's municipal supply. This was the first use of gas burners in New Zealand lighthouses. The requisition was sent to D & T Stevenson on 10 January 1878 and regrettably was described by the Agent General as being for 'a harbour light for Ho Kitika'.

The requisition stated:
This light is to be a fixed white light of fifth order

- One hundred and eighty degrees of apparatus together with a spherical mirror is to be supplied.
- The lantern is to be 6 ft gun metal lantern as supplied for Timaru. Dormant Panes to be supplied for 180°.
- The lighthouse will be placed on ground elevated 100 feet above the level of the sea, and will be 16 feet high, and built of wood.

- The light is to burn gas and approved burners for this purpose are to be supplied; but a paraffin lamp, two burners and a small supply of stores are to be supplied in order that paraffin may be burnt in case of any accident to the town gas works, by which the light will be supplied. The gas burners to be so fitted that they can be quickly removed and the paraffin burner & lamp substituted in case of necessity.
- No tanks or tools need be sent, Lightning Conductor fitted with necessary fittings for Timber Tower, earth plate etc to be supplied, also one approved suitable gas meter.[42]

The document in the National Library of Scotland includes a marginal note 'apparatus same as Timaru, Barbier & Fenestre £76'.

Dove & Co were invited to furnish a lantern identical to the one they had supplied for Timaru at the same price – £183.[43] Similarly Barbier & Fenestre were invited to furnish 'a fifth order Dioptric fixed apparatus for 180° and a spherical mirror for 180° with pillar and table similar to what you supplied in 1875 for Timaru Lighthouse at the same price'.[44] Both firms agreed. After receiving the specification Barbier & Fenestre pointed out that the lens was designed with seven prisms in the upper section. Their usual practice was not to include a seventh prism in a fixed light apparatus because it often became heated and blackened by the lamp. They requested permission to dispense with the seventh prism.[45] This was given.[46]

Barbier & Fenestre were also asked to furnish two fountains and burners complete, and one stand same as they supplied in 1875 for the fifth order apparatus in Japan.[47] This was the emergency paraffin burner. It was also the only occasion when they were asked to supply burners for the New Zealand lighthouses. Their price was £9.[48]

The Suggro gas burners with their accompanying meter were provided by Dove & Co for the sum of £6/13/4.[49]

By 15 May the lantern was completed and ready for inspection.[50] The lantern was packed and ready for shipment by 13 June 1878.[51] It was despatched by steamer from Granton on 19 June 1878 for transhipping to the *St Leonards* at the East India Docks. Also in the shipment were general

Figure 8.9: Hokitika lighthouse © Hokitika Museum

lighthouse stores and the new fountains for Nugget Point lighthouse.[52] The apparatus, burners and stores (supplied by Smiths) were shipped from Granton on 21 August for the *Otaki* at the East India Docks.[53]

The marine engineer reported on 30 June 1879 that the 'tower is now about ready for the reception of the lantern and apparatus',[54] and the light was exhibited on 22 September 1879.[55]

The Notice to Mariners included the following:

> The tower is 18 feet in height, is built of timber and painted white. The light will be Fifth Order Fixed White Light, and will be visible all around seaward. The light is elevated 122 feet above the level of the sea, and allowing 15 feet for the height of the observer's eye, will be seen at a distance of 16 miles in clear weather.

Initially the light was maintained by the Hokitika Harbour Board,[56] but in 1887 it was taken over by the Marine Department.[57]

The Hokitika light remained in operation until 1 November 1924. The following year the Marine Department decided to dismantle the light. The Public Works Department (PWD) was asked to advise on the cost of dismantling the light and sending it to Wellington. The secretary also mentioned that there were 'some lamp chimneys and triangular glasses which might be seen at the same time'. The 'triangular glasses' were probably spare panes for the lantern.[58]

The light was dismantled in September 1925 and the equipment and lantern sent to the Marine Department store in Wellington.

Gas burners were later installed at Napier Bluff (1881) and Timaru Harbour (1884) lighthouses.

Endnotes

1 Hassall C, *A Short History of the Port of Timaru,* p 30 Cadsonbury Publications, Christchurch 2001
2 AJHR 1874 H-23 Correspondence Relative to Selection of Site for Projected Coastal Lighthouses No 2
3 Acc 10706 vol 45 p 110 letter to Sec Customs Wgtn re New Zealand Lighthouses 17 December 1874
4 Acc 10706 vol 45 p 106 letter to Milne & Son re New Zealand Lighthouses 17 December 1874, also to Dove & Co on same page, same date
5 Acc 10706 vol 45 p 107 letter to Milne & Son re New Zealand Lighthouses 18 December 1874
6 Acc 10706 vol 45 p 115 letter to Dove & Co re New Zealand Lighthouses 22 December 1874
7 Acc 10706 vol 113, pp 357–360 Specification of a Lighthouse Lantern for Timaru Lighthouse New Zealand
8 Acc 10706 vol 45 p 121 letter to Barbier & Fenestre re New Zealand Lighthouses 26 December 1874
9 Acc 10706 vol 45 p 139 letter to Barbier & Fenestre re New Zealand Lighthouses 4 January 1875
10 Acc 10706 vol 45 p 381 letter to Milne & Son re New Zealand Lighthouses 18 March 1875
11 Acc 10706 vol 45 p 391 letter to Milne & Son re New Zealand Lighthouses 24 March 1875
12 Acc 10706 vol 45 p 392 letter to Dove & Co re New Zealand Lighthouses 24 March 1875
13 Acc 10706 vol 45 p 401 letter to Smiths Blair Street re New Zealand Lighthouses 30 March 1875
14 Acc 10706 vol 45 p 409 letter to Agent General London re New Zealand Lighthouses 3 April 1875
15 Acc 10706 vol 45 p 554 letter to Smith Morrison & Co re New Zealand Lighthouses
16 AJHR 1876 H-26 Eleventh Report of the Marine Department for year ended 30th June 1876. Appendix A,
17 Timaru Herald 13 Dec 1876 p 3

18 AJHR 1877 H-29 Twelfth Report of the Marine Department for year ended 30 June 1877. Enclosure No 1
19 *NZG Gazette* 1877 No 83 Tender p 999 1 October 1877
20 AJHR 1878 H-12 Annual Report of the Marine Department Enclosure 1
21 *NZG Gazette*,1878 No. 57 Notice to Mariners No 14 of 1878, p 893 17 June 1878
22 *NZG, Gazette*,1882 No.27 , , Notice to Mariners No10 of 1882 p 457 15 March 1882
23 AJHR 1888 H-19. Marine Department Annual Report for 1887–88 Report on Port of Timaru by W J Tennant Secretary to the Timaru Harbour Board
24 *NZG Gazette* 1891 No.33 Notice to Mariners 18 of 1891,, p 511, 5 May 1891
25 Acc 10706 vol 85 p 349 memo from Agent General London 28 January 1884
26 AJHR 1903, H-15 Marine Department Annual Report for 1902–1903
27 *NZG Gazette* 1924 No. 54Notice to Mariners No 45 of 1924, p 1866 1 August 1924
28 Journal of Proceedings of the Provincial Council, Canterbury Session XL-1874 p 91
29 AJHR 1903, H-15 Marine Department Annual Report for1902–1903
30 *NZG Gazette* 1904 No. 48, Notice to Mariners No 44 of 1904 p 1499 , 6 June 1904
31 Chance's Archive BS6/12/1/1/12 p 116 21 September 1903
32 AJHR 1906, H-15 Marine Department Annual Report for 1905–06
33 *NZG Gazette* 1906 No 63 Notice to Mariners No 59 of 1906 p 2046 20 July 1906
34 Chances Archives BS6/12/1/1/15 p 155 11Dec 1907
35 AJHR 1908 H-15 Marine Department Annual Report for1907–08 12 June 1908
36 AJHR 1930, H-15 Marine Department Annual Report for1929–30 13 September 1930
37 *NZG Gazette* 1930 No 32. Notice to Mariners No 16 of 1930 p 1536 30 April 1930
38 Graham E. *Clyde Built Blockade Runner Cruisers and Armoured Rams of the American Civil War*, Birlinn, Edinburgh 2006, ISBN 9781841585840 pp 121–122, 170, 212
39 AJHR 1876, H-26. Eleventh Report of the Marine Department for year ended 1876
40 AJHR 1877 H-29 Twelfth Report of the Marine Department for yr. ended 30 June 1877. Enclosure 1
41 NZG Archives M27 MD167 Specification Hokitika Lighthouse Contract, signed Blackett, Wellington, November 1878
42 Acc 10706 vol 85pp 184,185 copy dated 10 January 1878 in Vol 49
43 Acc 10706 vol 49, p 71 letter to Dove & Co re New Zealand Lighthouses 11 January 1878
44 Acc 10706 vol 49 p 72, letter Barbier & Fenestre re New Zealand Lighthouses 11 January 1878
45 Acc 10706 vol 85 p 192 letter Barbier & Fenestre to Stevenson, 26 January 1878
46 Acc 10706 vol 49 p 164, letter Barbier & Fenestre re New Zealand Lighthouses 30 January 1878
47 Acc 10706 vol 49 p 164, letter to Barbier & Fenestre *re New Zealand Lighthouses* 27 February 1878
48 Acc 10706 vol 85 p 200 letter Barbier &Fenestre to Stevenson, 1 March 1878
49 Acc 10706 vol 49 p 748 letter Stevenson to Agent General 10 September 1878
50 Acc 10706 vol 85 p 211 letter Dove & Co to Stevenson, 15 May 1878
51 Acc 10706 vol 49 p 503 letter Agent General London re New Zealand Lighthouses 13 June 1878
52 Acc 10706 vol 49 p 635 letter Stevenson to Agent General, 27 July 1878
53 Acc 10706 vol 49 p 698 letter Agent General London re New Zealand Lighthouses 22 August 1878
54 AJHR 1879 H-10. Annual Report of the Marine Department, The Marine Engineer to the Secretary Marine Department
55 *NZG Gazette*,1879 No.88 Notice to Mariners, no 42 of 1879 p1162 16 August 1879
56 AJHR 1880, H13 Annual Report of the Marine Department r, also Report on Lighthouse Works by Marine Engineer (John Blackett) 31 March 1880
57 NZG Archives Extract from M 1912/2780 memo Allport to Minister 21 August 1912
58 NZG Archives AADX W 3148 889 Box 45 63/112 letter Godfrey Sec MD to Under Sec PWD, 15 May 1925

9 The New Lightning Rods

Akaroa Head, Cape Saunders

The lighthouses for Akaroa Head and Cape Saunders were another pair of similar lighthouses, procured in the same order. A distinctive, although not very obvious, new feature of these lighthouses was the new type of lightning conductor which was being introduced in Scotland at the same time. This was used on the lighthouses ordered in this period.

Akaroa Head

Akaroa Harbour is, like Otago Harbour, a long inlet on the south side of Banks Peninsula. The small town of Akaroa part way up the harbour is best known in New Zealand as the site of a French effort to colonise the South Island of New Zealand before the British.

In their 1873[1] report Edwin and Johnson had suggested only a fixed white signal at Akaroa Head, and the site was not discussed at all in Johnson's 1874 report. However in his 1875 report, Johnson described three possible sites at Akaroa and recommended the site on Akaroa Head, at the eastern side of the entrance to Akaroa Harbour. Given the height of the headland, a very low wooden tower would suffice.[2] The land was secured by an Order in Council dated 4 July 1877.[3]

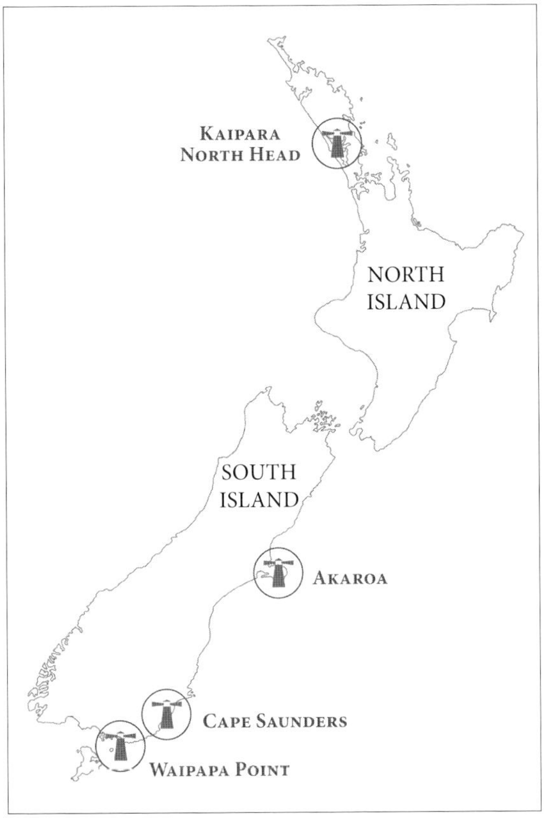

Figure 9.1: Locations of Cape Saunders and Akaroa Head, Waipapa and Kaipara North Head lighthouses © Author

The specifications for Cape Saunders and Akaroa were prepared in May 1877 and the requisition and specifications were sent on to Stevenson in July 1877.[4]

The specification for Akaroa contained the following:

Light for Akaroa Head, Banks Peninsula

- The apparatus to be second order fixed light varies by flashes – the period of flashes to be such as to prevent the light from being mistaken for the neighbouring light of Godley Head, which is a fixed light.
- The lantern to the Northern Lighthouse Board pattern of the second order.
- The light will be placed on ground elevated 247 feet above the level of the sea. The tower will be of wood 28 feet high.
- The light will show around an area of 250°, dormant panes for the balance of the 110° degrees to be supplied.
- The lamps to be of the best kind for burning paraffin oil.
- Two tinned tanks of about 50 gallons each for paraffin oil to be sent and similar quantities of stores, tools etc as sent for Cape Foulwind except that no dome riveting tools need be sent. Lightning conductors to be supplied 60 feet long.- fitted with necessary fixings for timber tower earth plate etc. A tracing of the sole plate and the bottom of the machine case showing the size of the well for weight and the distance apart required between the two wooden beams which carry the machine, to be sent as soon as possible.
- All cases to be marked and numbered. The cases for Lantern and apparatus to be marked respectively with L or A sic
  Akaroa Heads L No 1 etc
  Akaroa Heads A No 1 etc
- The position of this light can be seen on Sheet 8 Admiralty Chart of New Zealand.

Like the one for Cape Saunders this was a short (266 words) straightforward specification. It shows how obtaining new lighthouses was comparatively simple and straightforward because of the access to Scottish technology. The Marine Department could leave Stevenson to distribute detailed specifications to possible tenderers.

In March 1877 D & T Stevenson sent to the Marine Department a copy of the report made jointly with Professor Tait for the Northern Lighthouse Board on the subject of lightning protection for lighthouses. This report had been commissioned after a lightning strike on the Skerryvore light.[5] Seed expressed the Marine Department's thanks for the 'valuable report'. His final paragraph was not unexpected:

I am to add that you will no doubt be good enough to see that the two lighthouse lanterns ordered through you by the last homeward mail are properly supplied with suitable lightning rods after the plan suggested in your report.[6]

Once again New Zealand's lighthouses were being built with the latest technology because of the Scottish contribution. The same technology was being employed in Japan.[7]

D & T Stevenson streamlined the ordering process as follows:

- Milne & Son were asked to supply two lanterns as previously supplied for The Brothers and Portland island at the same price, of £509/11/- each.[8] They agreed.[9] Their tender is of interest as it includes the following schedule of prices for extra work
- Schedule of Prices

All malleable iron work complete in its place
Including fitting & all labour &(?) workmanship 1/2 per lb
All gun metal work complete in its place
Including fitting & all labour &(?) workmanship 4/6 per lb
All cast iron work complete in its place
Including fitting & all labour &(?) workmanship 45/- per cwt
All copper work complete in its place
Including fitting & all labour &(?) workmanship 2/10 per lb

Wages of an artisan	1/- per hour
Each frame of glass	20/-
Each dormant frame	9/-
Wire rope for lightning conductor	2/- per foot

- Barbier & Fenestre were asked to manufacture two second order Revolving apparatus similar to Brothers and Portland Island at their former price of £760 each.[10] They agreed.[11]
- Dove & Co were asked to supply two second order revolving machines as were supplied for the Portland Island and Brothers lighthouses (i.e. revolving once in 8 minutes) at their former contract price of £548 each and also a set of lamps for each at £135.[12] They agreed.[13]
- Smiths were asked to supply stores, tools, lamp glasses etc for two lighthouses similar to those furnished for Portland Island and Brothers lighthouses at the price of £78/6/1 each.[14] Smiths agreed.[15]

Barbier & Fenestre were later asked to supply a second order dioptric mirror of 125°,[16] which they did for £136.[17] Bryden supplied lighthouse signals for £20/10/-.[18]

Stevenson also recommended a change in the characteristic:

> With reference to the lights for Akaroa Head we propose that instead of being a fixed light varied by flashes, it should be a Flashing light attaining its greatest brilliance every 10 seconds (pencil note 10 seconds in margin of vol 85 136). The fixed light varied by flashes is a characteristic which ought not to be adopted unless under very peculiar circumstances as it detracts largely from the power of the lights and possess no advantage over a light whose flashes appear rapidly after each other. On this account it has not been adopted in Scotland since the holophotal light was introduced. We hope you will approve of the flashing light which we *now propose.*[19]

The proposal was accepted.[20]

The records indicate that the glass for the New Zealand lanterns was supplied by A D Love and Company and it seems that Stevenson were invited to inspect it before it was cut for the lantern.[21]

The Akaroa lantern was completed and ready for inspection by late November 1877. Milne & Son particularly made the familiar point that:

> [w]e would be obliged if you could get this done for us anytime today as we are much pushed for room.[22]

Barbier & Fenestre despatched the glassware from Paris on 11 December[23] and Dove & Co had machine and apparatus ready for inspection by 7 January 1878.[24]

Usually the lantern was sent first, with apparatus, lamps and stores etc. coming on a later ship. However, for the Akaroa lighthouse all the material was sent in one shipment. It was sent by General Steam Navigation Company steamer from Granton on 6 February 1878, to be shipped to Lyttelton on the *Duke of Argyle* from South West India Docks.[25] Unfortunately, on arrival it was found that the apparatus was damaged. Stevenson wrote to Dove:

> We have a note from New Zealand in which it is said 'The Akaroa Head apparatus etc must have been packed in very damp weather as it came out in very bad order – the frame of the machine too broken in two places – the machine rope was quite rotten.' We shall be glad to hear whether you have any explanation to offer as to the packing.[26]

There does not appear to be an explanation in the Stevenson papers, but Dove & Co continued to manufacture lighthouse apparatus for New Zealand.

Work had already begun on the site. Blackett visited it in March 1877 to mark out the positions of the tower and the other buildings, and stake out the line of the road and position of the landing place.[27] A work party was despatched to the site in April 1878.[28] Blackett described the construction of the road from the landing place as 'a heavy piece of work, very much of it in hard solid rock'. The project was also marred by the death

Figure 9.2: Akaroa Heads lighthouse on original site (ABPL 8848 W5221/19/1) © Archives New Zealand, the Department of Internal Affairs, Te Tari Taiwhenua

of the overseer William Black on 30 March 1879.[29] Construction finished at the end on 1879 and the light was exhibited 1 January 1880, the same day as Cape Saunders. The description in the Notice to Mariners[30] was:

> The tower is 28 feet in height, and is built of timber painted white. The light will be a 2nd order flashing white light, visible all round as far as the land admits, and will show a flash every 10 seconds. The light will be elevated 270 feet above the sea … will be seen at a distance of 228/10 (22.8) nautic miles in clear weather.

The tower was a Blackett weatherboard.

By 1903 the rollers had become badly worn and that year's annual report stated that 'New rollers are being procured from England for Akaroa Head and Puysegur Point lighthouses'.[31] However as has been shown new rollers for Puysegur were ordered in 1895. The 1905 annual report announced that

> The rings and rollers on which revolving carriage travels are showing signs of wear, and new ones have therefore been obtained from Scotland to replace them'.[32]

Installation began on 27 March 1906, and the light came back into operation on 10 April 1906.[33]

Akaroa Head was one of the eight lighthouses for which incandescent lamps were procured in 1911.[34] They appear not to have been installed and it seems Akaroa's were sent to Farewell Spit. A memo from the principal keeper dated 1914 indicates that the Doty wick burners were still in use.[35] The *Akaroa Mail and Banks Peninsula Advertiser* of 8 December 1914 announced that the newly installed incandescent burner 'replaces the old three wick burner'.[36]

The Chance burners gave considerable trouble, and D & C Stevenson suggested trying a burner made by the Diamond Company of Montreal.[37] The burner was despatched to

New Zealand on the *Remuera* in July 1914.[38] After testing in the Marine Department store, it was decided to give the lamp a full scale test in a lighthouse.[39]

This was the incandescent lamp installed at Akaroa in late 1914 and it was greeted in the local media in the same column as discussions of the weather, the state of the cocksfoot market, and the recent decisions of the Directors of the Little Akaroa Co-operative Dairy Factory Company.

The burners remained at Akaroa right up until electrification in 1952. They seem to have given satisfactory service over a long period as Principal Keepers at Akaroa in 1921 and 1944 both commented on the qualities of the burner. Voyle stated that 'Diamond Incandescent superior, more brilliant, more economical and wear better as vapour tube made of hardened steel (not Cu)',[40] and Newson gave the opinion:'I would say that the 'Diamond ' burner in my experience is the most efficient & trouble free types used in the service, provided the necessary spare parts are obtainable'.[41]

In 1938 the Stores Clerk suggested that as the new overseer had no first-hand knowledge of the Diamond burner and the lighthouse expert was about to retire, the Akaroa burner should be replaced with a Smith-Ford burner.[42] When Petersen, the chief engineer of the government steamer *Matai,* arrived to install the new burner in December, he found that the pillar and table were 6 inches too high.[43] Although he took measurements for a new one the matter was not pursued – no doubt to the satisfaction of the keepers, who continued to use the Diamond burners until 1951.[44] The Smith-Ford burner must have been left at the lighthouse, as six years later the assistant-in-charge reported that not all the spare parts could be found and the instruction book was missing.[45]

The original fixed cowl was replaced in late 1918 /early 1919 by a revolving one.[46] There was a problem with the bearings in 1951, and a telegram sent to the Marine Department showed how isolated the lighthouse still was. It read 'supply mechanic with good horse impossible to get here otherwise'.[47]

The first suggestion that Akaroa Head lighthouse be electrified was made in 1948 following a private discussion between the Chairman of the Banks Peninsula Power Board and the then Principal Keeper.[48] In 1951 it was reported that there had been some progress on preliminary work for electrification at Akaroa Head.[49]

The No 1 cottage burned down on 14 March 1952[50] and so electrification was pushed ahead as it would allow the lighthouse to be staffed with one fewer keeper. A temporary electrical system was installed by 28 November 1952.[51] The following year the annual report indicated that there was a delay in obtaining electric plant for seven lighthouses including Akaroa.[52] Electrification seems to have been completed in mid-1955, as in October of that year the lighthouse plant supervisor reported installing a Chance Brothers kerosene light as a standby.[53]

In mid-1970s the Marine Department decided to replace the lighthouse with an automatic beacon. In 1978 local residents formed the Akaroa Lighthouse Preservation Society, and in August the tower was gifted to the society. The society moved quickly to remove the deteriorating structure, but due to a fortunate misunderstanding they removed the tower and all equipment except the lens. This ensured the preservation of the winding mechanism and restoration of the original tower and not, as in the case of Manukau and Portland Island, a new version. After restoration the tower was reassembled in Akaroa township.

Cape Saunders

Although the lantern and apparatus originally purchased by the Otago Provincial Council for a first order lighthouse at Cape Saunders was eventually used at Nugget Point (Chapter3), Cape Saunders did not cease to be of interest.

By 1873 thought was again being given to Cape Saunders and it was one of the seven lighthouses for which Reynolds intended to ask the legislature to grant the necessary funds.[54] The instructions to Captains Johnson and Edwin indicated that the Marine Department were considering completing the Cape Saunders lighthouses and down grading the Taiaroa Head light to a simple harbour light.[55]

Edwin and Johnson recommended that a second or third order light, white, revolving every minute, should be placed at Cape Saunders. They also suggested that a sixth order white light would be sufficient for the entrance to Otago Harbour and the existing apparatus moved to Tory Channel.[56] They also suggested a signal at Akaroa.

Johnson examined the site during his 1874 site visits. He decided that the proposed site had a serious drawback; because of the altitude the light would be obscured by fog, so he recommended an alternative site, 50 feet lower. He repeated the recommendation of a first order dioptric holophotal apparatus with a range of 25 nautical miles.[57]

The specification for Cape Saunders consisted of only 264 words and was as follows:

Light for Cape Saunders, Otago Peninsula

- The apparatus to be second order revolving white light obtaining its greatest brilliance once every minute.
- The lantern to the Northern Lighthouse Board pattern of the second order and generally the light is to be the same as that supplied for Farewell Spit and Cape Campbell.
- The light will be placed on ground elevated 180 feet above the level of the sea. The tower will be of wood 28 feet high.
- The light will show around and area of 190°, dormant panes for the balance of the 170° degrees to be supplied.
- The lamps to be of the best kind for burning paraffin oil.
- Two tinned tanks of about 50 gallons each for paraffin oil to be sent and similar quantities of stores, tools etc as sent for Cape Foulwind except that no dome riveting tools need be sent.
- Lightning conductors to be supplied 60 feet long.- fitted with necessary fixings for timber tower, earth plate etc. A tracing of the sole plate and the bottom of the machine case showing the size of the well for weight and the distance apart required between the two wooden beams which

carry the machine, to be sent as soon as possible. (Like the Akaroa lantern the lightning conductor specification was upgraded following the Tait and Stevenson report.)

- All cases to be marked and numbered. The cases for Lantern and apparatus to be marked respectively with L or A sic
- Cape Saunders L No 1 etc
- Cape Saunders A No 1 etc
- The position of this light can be seen on Sheet 9 Admiralty Chart of New Zealand.

The French glass makers were able to announce on 31 October 1877 that the second order apparatus was completed[58] and was despatched to 'MM Dove' on 10 November.[59]

The completed lantern was sent by rail to Glasgow on 13 February 1878 and shipped to Otago on the *William Davie*.[60] The apparatus and machine were ready for inspection by 6 June and packed ready for shipment by 9 July. Stevenson particularly pointed out to the Agent General that

> if delivered in Glasgow instead of London the Contractors are willing to give a deduction of £4 in their contract price.[61]

Figure 9.3: Current Cape Saunders lighthouse (ABPL 8848 W5221/19 10) © Archives New Zealand, the Department of Internal Affairs, Te Tari Taiwhenua

The apparatus and machine, along with the lamps, signals, and stores were sent to Glasgow in the later half of July to shipment to Otago on the *Timaru* which sailed in early August.[62]

By the time of 1878 annual report, surveys had been made of the site, 'but the question of determining the best landing place has delayed the completion of the plans'.[63] By the time of the next report, Blackett reported that the necessary land has been acquired and a working party was then erecting buildings which were expected to be completed in mid-September.[64] The completed light was exhibited on 1 January 1880. The Notice to Mariners described it as follows:

> The tower is 28 feet in height, and is built of timber painted white. The light will be a second order revolving white light, visible between the bearings of S.W. by W. round by W. and N. to N.E.½N., attaining its greatest brilliancy once every minute. The light will be elevated 210 feet above the sea … visible at a distance of about 20½ nautic miles in clear weather.[65]

The 1912 annual report indicated that incandescent burners had been installed.[66]

In 1947 it was suggested that Cape Saunders should be electrified[67] although it was 1951 before the Secretary of Marine recommended the purchase of two generating sets.[68] Subsequently it was decided to establish a new steel tower adjacent to the site of the existing lighthouse for a rotating beacon.[69] This was because the original Blackett tower had become dangerous due to lack of proper maintenance. Electrification was finally completed in late 1956.[70] Eleven years later the steel tower was itself replaced by a concrete and steel structure.

The 1884 Lights: Waipapa Point and Kaipara North Head

The next two lighthouses to be erected were at Waipapa Point on the Foveaux Strait and Kaipara North Head on the west coast harbour of the same name, north of Auckland. The first was exhibited on 1 January 1884 and the latter on 1 December 1884. Both were standard second order flashing lights as supplied for Cape Saunders with standard New Zealand timber towers. They were the last two lighthouses in New Zealand to be built with wooden towers. Future lighthouse towers were either iron, cast in local foundries, or concrete.

As neither represented any particular innovation they will be discussed briefly, although the procurement of the Waipapa Lantern did have its light-hearted aspects.

Waipapa Point lighthouse

While Waipapa Point on Foveaux Strait had been recommended as a site for a lighthouse as early as 1863,[71] it took a major shipwreck to force the Marine Department to consider constructing a lighthouse on the point. At 5 a.m. on 29 April 1881, the passenger steamer *Tararua* with 39 crew and 112 passengers struck the reef off Waipapa Point; 131 people died and 64 of them are buried in a cemetery near the lighthouse. The cemetery is known as the Tararua Acre.

[Page 208] Figure 9.4a: Design for Waipapa Point lighthouse dome (Ms 5885) © National Library of Scotland

[Page 209] Figure 9.4b: Design for Waipapa Point lighthouse lantern (Ms 5885) © National Library of Scotland

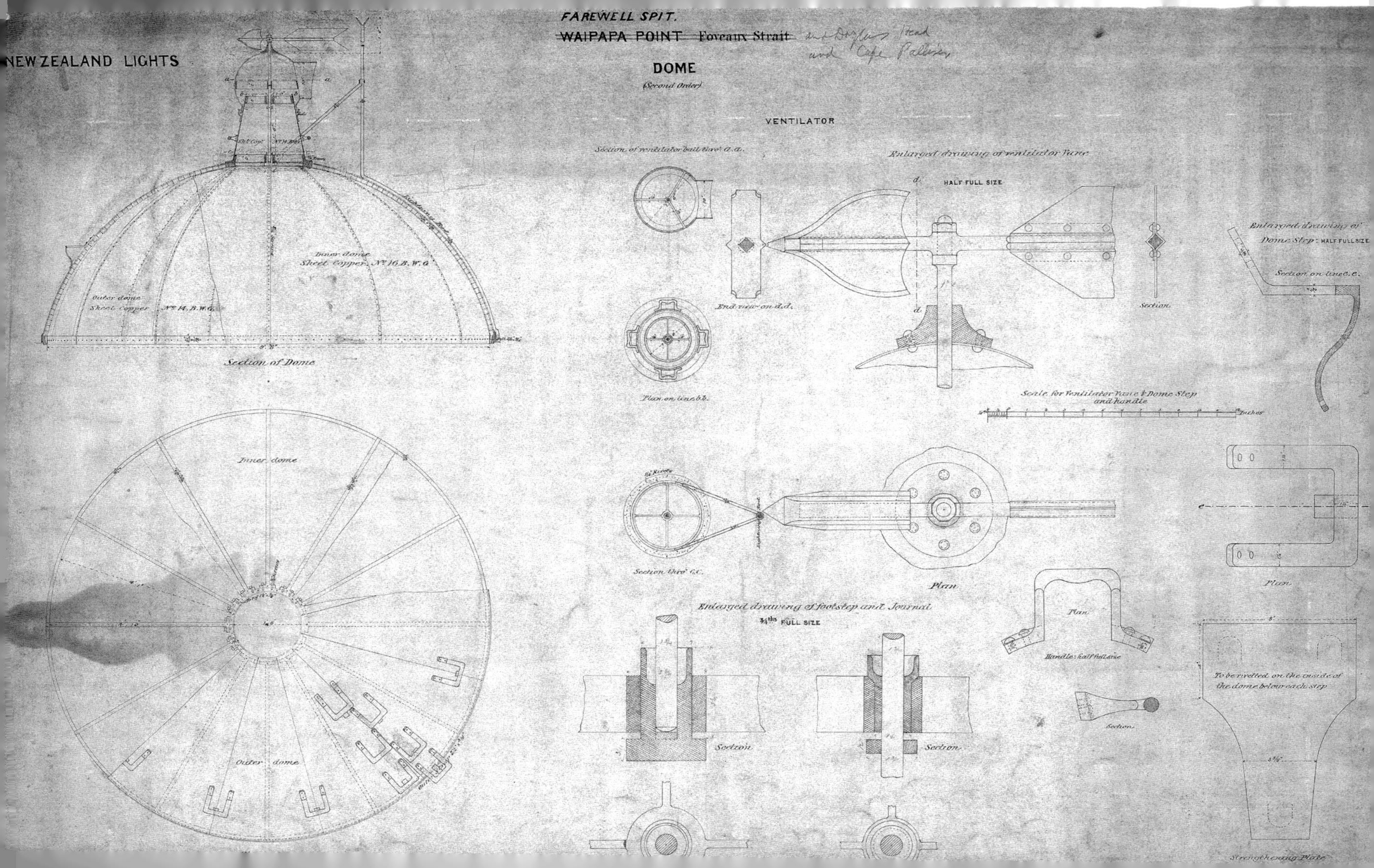
NEW ZEALAND LIGHTS
FAREWELL SPIT.
WAIPAPA POINT Foveaux Strait
and Douglas Head and Cape Palliser
DOME
(Second Order)
VENTILATOR
Section of ventilator ball thro' a.a.
Enlarged drawing of ventilator Vane
HALF FULL SIZE
End view on d.d.
Section
Plan on line b.b.
Scale for Ventilator Vane & Dome Step and handle
Inches
Enlarged drawing of Dome Step: HALF FULL SIZE
Section on line c.c.
Inner dome
Sheet Copper No 16 B.W.G.
Outer dome
Sheet Copper No 14 B.W.G.
Section of Dome
Inner dome
Outer dome
Section thro' c.c.
Plan
Plan
Enlarged drawing of footstep and Journal
3/4ths FULL SIZE
Section
Section
Plan
Handle: half full size
Section
To be rivetted on the inside of the dome below each step
Strengthening Plate

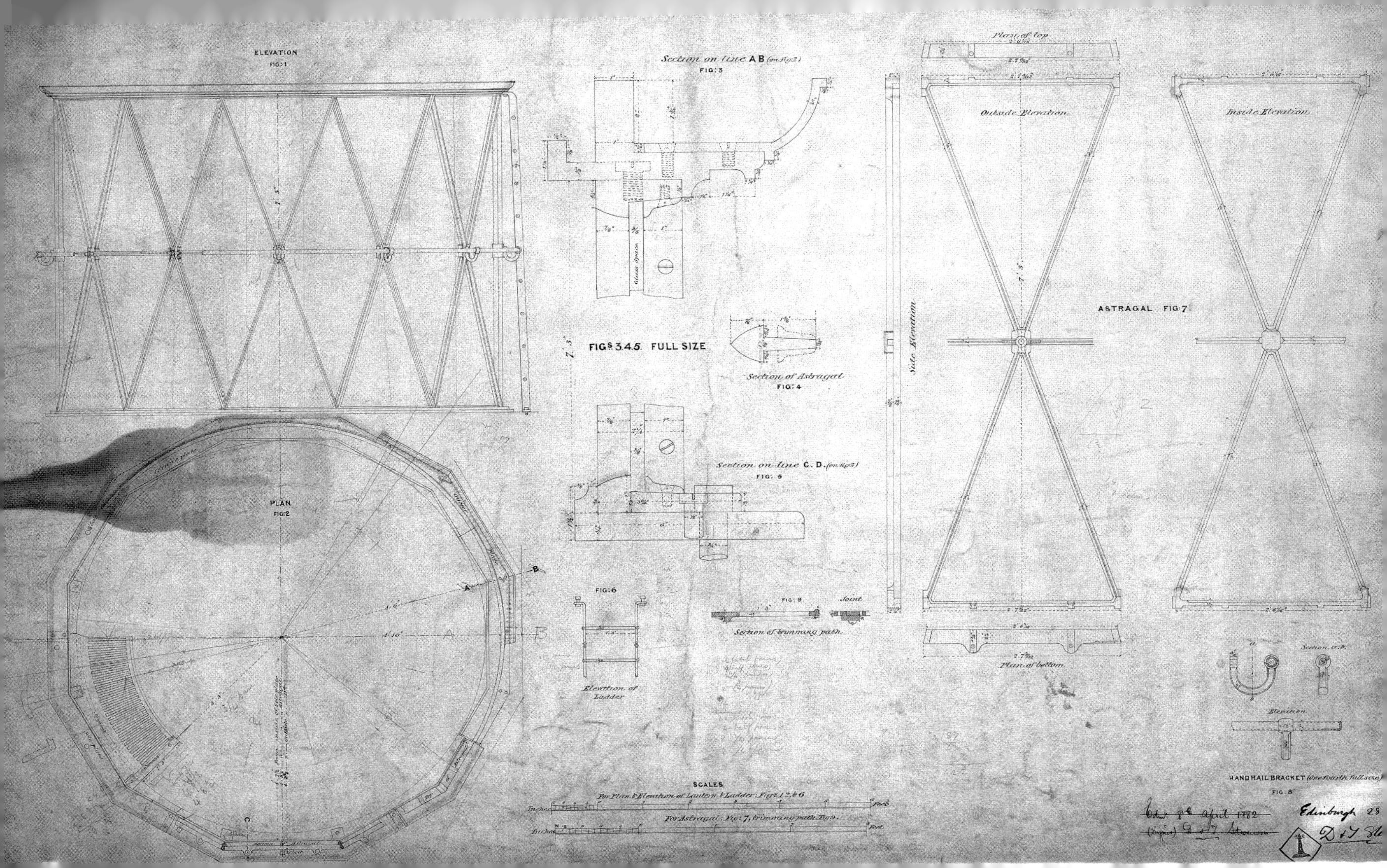
ELEVATION
FIG: 1
PLAN
FIG: 2
Section on line A B (on fig. 2)
FIG: 3
FIGS. 3.4.5. FULL SIZE
Section of Astragal
FIG: 4
Section on line C. D. (on fig. 2)
FIG: 5
FIG: 6
Elevation of Ladder
FIG: 9
Joint
Section of trimming path
Plan of top
Outside Elevation
Inside Elevation
Side Elevation
ASTRAGAL FIG: 7
Plan of bottom
Section a.b.
Elevation
HANDRAIL BRACKET (One fourth full size)
FIG: 8
SCALES
For Plan & Elevation of Lantern & Ladder: Figs 1, 2, & 6
For Astragal: Fig: 7, trimming path Fig 9.
Edinburgh 28 April 1882
D & T St

The requisition (CS 10/83) for the light was sent to D & T Stevenson by the Agent General on 18 March 1882,[72] although the requisition itself has not survived. Barbier & Fenestre were asked to provide 'a second order Revolving apparatus same as that furnished by you in 1878 for Cape Saunders Lighthouse at your former contract price of £760. A Dioptric mirror for 14 deg to be supplied for the sum of £152'.[73] They agreed.[74]

Smiths were asked to provide 'lamp glasses, tools etc in all respects the same as supplied for Cape Saunders lighthouse at your former contract price of ' £78/6/1,[75] and Bryden & Son were asked to supply 'a set of lighthouse signals & tubes same as you furnished for Cape Saunders lighthouse at your former contract price of' £20/10/-.[76]

In December 1881 William Seed wrote to D & T Stevenson asking them to include the firm Messrs W Wilkins and Co Long Acre, London amongst those invited to tender for New Zealand's lighthouse orders, although with the rider

> I do not however desire that the pattern of apparatus etc at present used should be in any way altered, nor do I wish to fetter your judgement in any way in dealing with the tenders you may receive from the firms named.[77]

Wilkins and Co were invited to tender for the lantern. Although they courteously thanked D & T Stevenson for considering them, they indicated that they could not estimate by the time stated 'our Mr Farquhar being in Paris'.[78] In contrast Milne & Son were asked

Figure 9.5: Waipapa Point Lighthouse (ABPL 8848 W5221/19 13 WP?)
© Archives New Zealand, the Deparment of Internal Affairs, Te Tari Taiwhenua

'to furnish a lantern for Waipapa Lighthouse, New Zealand for the sum of £509/11/- being your contract price for Cape Saunders Lantern'.[79]

The drawing for the Waipapa Point lantern dome is of interest for its subsequent career. It was originally drawn for the Waipapa contract and was signed D & T Stevenson and dated 8 April 1882. Subsequently the title was crossed out and replaced by 'Farewell Spit'. The initial date and signature were crossed out and replaced with the D & C Stevenson signature and new date 23 October 1891. A pencilled title 'Cape Palliser' suggests that the drawing was also used for that contract in 1895.

Dove & Co were asked 'to furnish a machine, machine case, revolving carriage etc for Waipapa Lighthouse, New Zealand for the sum of £548/-/- being your contract price for Cape Saunders machine etc'.[80] Dove agreed.[81]

In November 1882 the agent general wrote enquiring about progress as the contracts were now about three months overdue.[82] (The usual contract period was twelve weeks, with a penalty of £5 for each week over the twelve). Milne & Son had the lantern ready by early December and it was sent by steamer from Granton on 13 December 1882 for shipment by the *Waimea* from South West India Docks.[83]

The major delay was the glassware from Barbier & Fenestre which was eventually despatched from Paris on 30 December 1882.[84] The apparatus and machine were shipped on the *Timaru* from the East India Docks in late March 1883.[85]

Meanwhile McKellar from the Agent General's office apparently queried a payment of £10 to Mr Gordon for inspection services. D & T Stevenson replied

> We may explain that we invariably have inspectors in daily attendance at the workshops of the contractors whose work is being executed to see the details in the Working Plans and specifications are carried out. But the work for Waipapa Lighthouse was rather too small to occupy the entire time of an inspector as was the case in 1875 and 1876 when six lighthouses were in progress for New Zealand Government, and Mr Gordon who was at the different workshops daily inspected the construction of the lantern, machine lamps etc for Waipapa, and his charge of £10 we consider as very moderate. We hope this explanation will be satisfactory.[86]

It probably was as there are no further letters on the subject in the archive.

In December 1881 Blackett and Johnson visited Waipapa and chose a site on the Point.[87] Construction was completed by the end of 1883 and the light was exhibited 1 January 1884.[88] The Notice to Mariners[89] stated that

> The Lighthouse is situated on Waipapa Point, at the eastern entrance to Foveaux Strait, lat 46 deg 39' 40"S., long. 168 deg 52' 30" E. The tower is 44 feet in height from base to vane, and is painted white. The light will be a Second Order Flashing White Light, visible all round seaward, and will show a flash every ten seconds.

> The light is elevated 70 feet above the sea, and, allowing 15 feet for the height of the eye, will be seen at a distance of about thirteen and a half nautic miles in clear weather.

By May 1911 incandescent burners had been acquired for eight lighthouses including Waipapa Point.[90] These had been installed when the next annual report was signed in May 1912.[91]

Electrification seems to have begun in 1942, although there is no mention of Waipapa Point in the annual reports. A memo from the principal keeper dated 3 April 1944 indicated that the power had failed and the 'redundant' incandescent lamp had been fitted temporarily.[92]

Kaipara Head

In the nineteenth and early twentieth centuries, Kaipara Harbour was an important port particularly for shipping out the massive kauri trees logged from the northern part of the North Island. As early as 1855, residents petitioned the Auckland Provincial Government to provide a pilot station.[93]

In 1873 Edwin and Johnson proposed a white light revolving every two minutes at Kaipara.[94] This would be both a harbour light and coastal light. Johnson visited Kaipara on his 1874 site visits, but his only recommendation was that the pilot and signal station be moved from the southern side of the harbour entrance to the northern.[95] The station commenced operations on its new site at Poutu Point on 1 January 1875.[96]

In early 1883 Blackett and Johnson visited the proposed site at Poutu Point and the requisition for a second order white light flashing every 10 seconds was sent about the same time to the Agent General for onward transmission to Edinburgh.[97]

The specification was the usual short and straightforward document that resulted from the Scottish connection and included the following:

- Northern Lighthouse pattern 2nd order,
- 2nd order white light flashing every 10 secs w' dioptric mirror of 140 degrees,
- site 245 ft above h.w.mark
- timber tower 30 ft high.
- Light will show over 220 degrees dormant panes for 140 degrees.
- Lamp best kind for paraffin oil.
- 3 tinned tanks, 50 gals each for paraffin,
- usual supply tools, no whistles for calling keepers or dome riveting tools required.
- Lightning conductor 700 ft long with necessary fixing for timber tower.
- Position near flagstaff North Head Kaipara Harbour shown on accompanying Admiralty Chart.[98]

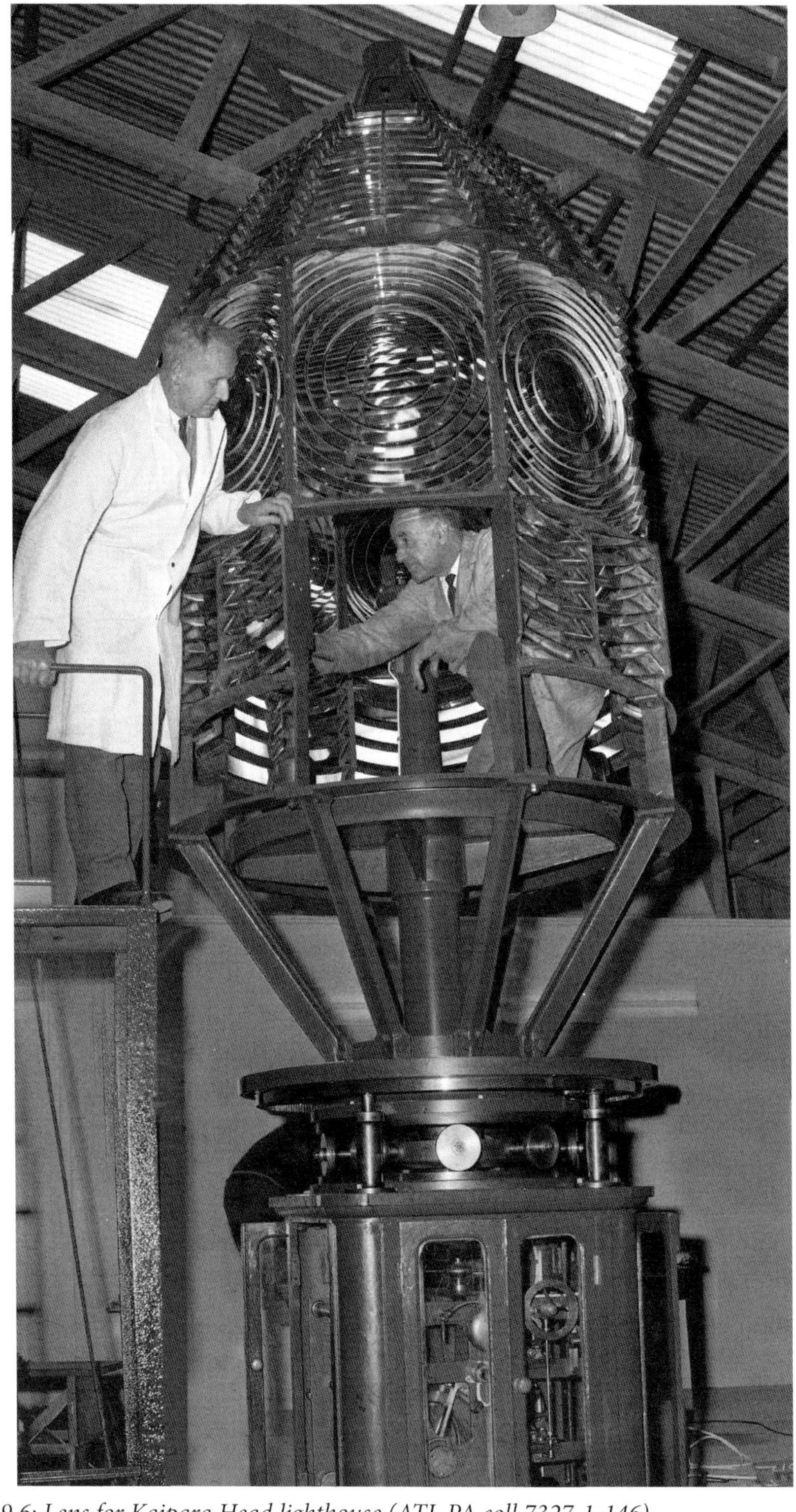

Figure 9.6: Lens for Kaipara Head lighthouse (ATL PA coll 7327-1-146)
© Evening Post Collection, Alexander Turnbull Library, Wellington, New Zealand

Figure 9.7: Typical lighthouse clock manufactured by Barrie of Edinburgh © Mike Chandler and Museum of Scottish Lighthouses, Fraserburgh, Scotland

William Seed also sent a letter to D & T Stevenson informing them that the order had been sent to the Agent General and asking D & T Stevenson to include 'a clock of same pattern & description as those used by NLB'.[99] Seed's letter enabled D & T Stevenson to move promptly when the official requisition arrived. The day following its despatch from London, Milne & Son acknowledged an order for a lantern similar to that supplied for Waipapa Point at a sum of £500/15/-. Also a 100 ft lightning conductor charged extra.[100] The following day (18 August 1883) Dove & Co acknowledged an order for the machine and lamps.[101] Barbier & Fenestre wrote on 21 August accepting the order for a '2o for Kaipara and dioptrique mirror 140°' similar to Waipapa Point. Their price was £912.[102]

The lantern was despatched to New Zealand on the *Floride* from the East India Dock in January 1884.[103] The machine and lamps, along with an order of stores, were shipped on the *Ionic* from Royal Albert Docks in June 1884.[104]

Even the best Consulting Engineering firms slip up and while the stores arrived in good order on the *Ionic* they did not include the clock[105] which had been manufactured by Barrie of Edinburgh. It took about 2 months to track the clock down, but by 25 October it was packed and ready to go.[106] It was shipped to Auckland on the *Arawa* from the Royal Albert Dock leaving a slightly confused shipping agent who had no record of to which order the clock belonged.[107]

A work party was landed on the site in November 1883, but construction was delayed due to illness of the first foreman appointed.[108] The light was first exhibited on 1 December 1884 and was the last timber lighthouse tower built in New Zealand.

The Notice described the light thus:

> A light will be exhibited from the lighthouse which has been built at the North Head Kaipara Harbour … The lighthouse is situated on the North head of Kaipara Harbour on the west coast of the North Island in lat. 36deg 21' S and long. 174deg 7' 45" E.

Figure 9.8: Original Kaipara Head lighthouse (AWNS_19020925_8_5)
© Sir George Grey Special Collections, Auckland Libraries, New Zealand

> The tower 44 feet in height from base to vane, and is painted red. The light will be a 2nd order flashing white light visible all around seaward as far as land will allows or between the bearings seaward of N.35°W. round to S. 73°E. and will show a flash every 10 seconds … light is elevated 275 feet above sea level … will be seen as a distance of 23 miles in clear weather.[109]

Presumably, the clock was installed later

However, when announcing the start of construction the 1884 annual report indicated that

> [t]he land on which the lighthouses, signal station, &c., stand has been reserved from sale.

When the site for the lighthouse was chosen the Marine Department decided that the land was valueless and so made no effort to contact its Māori owners – a not uncommon attitude at the time. Once the lighthouse had been erected however, the ignored land owners surfaced and demanded ground rent for the property. Eventually the Marine Department was able to buy the land but carefully did not mention the transaction in the annual report.

In October 1919 the Doty burner was replaced by a Chance incandescent burner and the ventilating cowl was overhauled.[110] The Chance burner was later replaced by a Smith-Ford burner in 1938.[111]

By 1939 the necessity for a major light at Kaipara had lessened considerably. The timber industry had declined considerably and few ships entered the harbour at night. Hence, it was suggested that an automatic acetone-acetylene light be installed.[112] It was decided to automate the light by removing the existing lantern and apparatus and replacing lantern with the fifth order lantern that had been removed from Cape Foulwind.[113] The new light was then installed inside the smaller lantern. The downsizing of the lantern led to a decrease in height of the tower. The new light was exhibited on 14 May 1947. The Notice to Mariners described it as

> an unwatched automatic light flashing white every 10 seconds with a flash of 1 second. Elevation altered to 276 feet. Height of tower altered to 40 ft. Harbour arc of visibility 339° to 301°.[114]

The redundant apparatus was moved to Wellington, where it was overhauled and set up to be used in the lighthouse keeper training school.

The automatic apparatus was dismantled in 1955,[115] and the tower and lantern are now cared for by Heritage New Zealand.

Endnotes

1 JHR 1873 H-19. Report on the Lighthouses on the Coasts of New Zealand, Enclosure No 2
2 AJHR 1875 H-12A Ninth Report of the Marine Department , Appendix A Enclosure No 3
3 *NZG Gazette* 1878 No. 7 order in Council p 57– 4 January 1877
4 Acc 10706 vol 85 p 135 letter from Agent-General 7 July 1877 [135]
5 Acc 10706 vol 48. letter to William Seed, secretary Customs, re New Zealand Lighthouses 22 March 1877
6 Acc 10706 vol 85. letter William Seed Sec Customs [129–131] 30 May 1877
7 Acc 10706 vol 48 p 612, Specification for Lightning Conductor
8 Acc 10706 vol 48. p 514 letter to Milne & Son re New Zealand Lighthouses 16 July 1877
9 Acc 10706 vol 48 p 529. letter Milne & Son 18 July 1877 re New Zealand Lighthouses 18 July 1877
10 Acc 10706 vol 48. p 511letter to Barbier & Fenestre re New Zealand Lighthouses 16 July 1877
11 Acc 10706 vol 48 p 538 letter B & F re New Zealand Lighthouses 20 July 1877
12 Acc 10706 vol 48. p 515 letter to Dove & Co re New Zealand Lighthouses 16 July 1877
13 Acc 10706 vol 48 p 524. letter Dove & Co re New Zealand Lighthouses 17 July 1877
14 Acc 10706 vol 48. p 516 letter to Smiths George Street New Zealand Lighthouses 16 July 1877
15 Acc 10706 vol 48 p 523. letter Smiths 17 July 1877 re New Zealand Lighthouses 17 July 1877
16 Acc 10706 vol 48 p 564 letter Barbier & Fenestre re New Zealand Lighthouses 31 July 1877
17 Acc 10706 vol 48 p 584 letter to Barbier & Fenestre re New Zealand Lighthouses 13 August 1877
18 Acc 10706 vol 48 p 539. letters William Bryden & Son 20 July 1878
19 Acc 10706 vol 48 p 540. letter to William Seed, sec Customs Wgtn. re New Zealand Lighthouses 19 July 1877
20 Ac 10706 vol 85. P 164 copy letter Wm Seed 21 September 1877,
21 Acc 10706 vol 85 p 172 letter from Milnes 7 November 1877
22 Acc 10706 vol 85 p 178 letter from Milne & Son 29 November 1877
23 Acc 10706 vol 85 p 179 letter from Barbier & Fenestre 12 Dec ember1877]
24 Acc 10706 vol 85 p 182 letter from Dove & Co 7 January 1878]
25 Acc 10706 vol 49 p 188. letter to Smith Morrison re New Zealand Lighthouses 6 February 1878
26 Acc 10706 vol 51 p 344. letter Dove & Co re New Zealand Lighthouses July 12 1880
27 AJHR 1877 H-29 Twelfth Report of the Marine Department for year ended 30 June 1877. Enclosure No1
28 JHR 1878 H-12. Marine Department Annual Report Enclosure 1,t.
29 AJHR 1879 H-10 Marine Department Annual Report Report of Marine Engineer (Blackett) d

30 *NZG Gazette* 1879No.124 Notice to Mariners, No 60 of 1879, p 1879 9 December 1879
31 AJHR 1903 H-15 Marine Department Annual Report for 1902–03
32 AJHR 1905 H-15 Marine Department Annual Report for 1904–05
33 *NZG Gazette* 1906 No.20Notice to Mariners, No 22 of 1906, p 802 14 Mar 1906
34 AJHR 1911 H-15 Marine Department Annual Report for 1910–11
35 NZG Archives ADOE 16612 M1/715 8/10/11, part 1 memo J W Arthur, PK Akaroa to Sec Marine Department November 30 1914
36 http://paperpast.natlib.govt.nz
37 NZG Archives ADOE 16612 M1/715 8/10/11 copy letter Stevenson to high commisioner dated 25 August 1913
38 NZG Archives ADOE 166112 M1/ 715 8/10/11 memo high commissioner toprime minister No 3195 17 July 1914
39 NZG Archives ADOE 16612 M1/715 8/10/11 (formerly M8/9/29) memo Fraser (lighthouse artificer) to secretary Marine Department. 14 September 1914
40 NZG Archives ADOE 16612 M1/715 8/10/11 letter Voyle, principal keeper Akaroa, to secretary Marine Department 17 July 1921
41 NZG Archives ADOE 16612 M1/715 8/10/11 letter Newson principal keeper Akaroa to secretary Marine Department May 1 1944
42 NZG Archives ADOE 16612 M1/715/8/10/11 memo Jackson, stores clerk, to secretary Marine Department 13 June 1938
43 NZG Archives ADOE 16612 M1/715 8/10/11 memo M Petersen, chief engineer (SS *Matai*) to secretary Marine Department 30 December 1938
44 NZG Archive ADOE 16612 M1/715 8/10/11 letter Smith to principal keeper 18 July 1951
45 NZG Archive ADOE 16612 M1/715 8/10/11 extract from original on M 8/6/18 6 memo Olsen, asst-in-charge Akaroa, to secretary Marine Department 6 October 1944
46 AJHR 1919 H-15 Annual Report Marine Department 1918–1919 report dated 15 June 1919
47 NZG Archives ADOE 16612 M1/715 8/10/11 telegram Akaroa Lighthouse to secretary Marine Department, 18 July 1951
48 NZG Archives ADOE 16612 M1/715 8 10/11 memo Koening, principal keeper Akaroa, to secretary Marine Department 2 June 1948
49 AJHR 1951 H-15, Annual Report of the Marine Department 1950–1951,
50 NZG Archives ADOE 16612 M1/715/ 8/10/11 memo W C Smith, principal keeper, to marine engineer, 14 March 1952
51 NZG Archives ADOE 16612 M1/715 8/10/11 2 memo principal keeper to secretary Marine Department 1 December 1952
52 AJHR 1953 H-15 Marine Department report for year ended 31 March 1953
53 NZG Archives ADOE 16612 M/1/715/ Record 8/10/11 memo W Tate, lighthouse plant supervisor, to secretary 3 October 1955
54 AJHR 1873 -19AEighth report of the Marine Department for year ended 30 June 1873
55 AJHR 1873 H19 Report on the Lighthouses on the Coasts of New Zealand. No 1
56 AJHR 1873 H-19 Report on the Lighthouses on the Coasts of New Zealand. Enclosure in No 2
57 AJHR 1874 H-23 Correspondence Relative to Selection of Site for Projected Coastal Lighthouses Enclosure No 2
58 Acc10706 vol 85. p 171letter Barbier &Fenestre 31 October 1877 (in French)
59 Acc10706 vol85. p 173letter Barbier &Fenestre 12 November 1877 (in French)
60 Acc 10706 vol 85. p 198letter from Milnes 13 Feb 1878 13 February 1878
61 Acc 10706 vol 49 p 564 letter to A G London re New Zealand Lighthouses dated 9 July 1878
62 Acc 10706 vol 85. p 221 memo Agent General 12 July 1878
63 AJHR 1878H-12 of the Marine Department Annual Report Enclosure 1
64 AJHR 1879 H-10 Marine Department Annual Report The Marine Engineer to the Secretary, Marine Department
65 *NZG Gazette* 1879 No.125 Notice to Mariners No 61 of 1879 p 1729, 9 December 1879
66 AJHR 1912, H-15 Annual Report Marine Department for1911–12
67 NZG Archives ADOE 16612 M1/728 8/20/11 letter secretary of Marine to P Sec Min of Marine 3 July 1947
68 NZG Archives ADOE 16612 M1 /728 8/20/11 memo Sec Marine to Min of Marine, 18 July 1951
69 NZG Archives ADOE 16612 M1/728 8/20/11 (extract from M8/9/339) letter C W O Turner, marine engineer, (Ministry of Works) to secretary of Marine 8 March 1956
70 AJHR 1957, H-15 Marine Department Report for yeard ended 31 March 1957 .

71 NZG Archives AAAC 708 D500 144c 839 Dog Island Correspondence Report Theoph Heales, chief surveyor, Southland Province dated June 1863
72 Acc 10706, vol 85 p 296 memo Agent General 18 Mar 1882, ref CS 10/82
73 Acc 10706, vol 53 p 28 letter to Barbier & Fenestre, Paris re New Zealand Lighthouses 5 April 1882
74 Acc 10706 vol 85 letter from Barbier & Fenestre 11 April 1882
75 Acc 10706, vol 53 p 35 letter to Smiths George St re New Zealand Lighthouses 8 April 1882
76 Acc 10706, vol 53 p 35 letter to Bryden & Son George St re New Zealand Lighthouses. 8 April 1882
77 Acc 10706 Vol 86 p 285 (bound into vol 53) letter Seed to Stevenson, 1 December 1881
78 Acc 10706 vol 85 letter from Wilkins13 April 1882
79 Acc 10706, vol 53 p 51 letter to Milne &Son re New Zealand Lighthouses 14 April 1882
80 Acc 10706, vol 53 p 50 letter to Dove & Co re New Zealand Lighthouses 14 April 1882
81 Acc 10706 vol 85 p 301 letter from Dove & Co 17 April 1882 [301]
82 Acc 10706, vol 85 p 306 letter Agent General 6 Nov 1882, ref CS /10 [306]
83 Acc 10706 vol 53 p 734 letter to Agent General 12 December 1882
84 Acc 10706 vol 85 letter from Barbier & Fenestre 3 Jan 1883
85 Acc 10706, vol 85 letter from Agent General 27 March 1883
86 Acc 10706 vol 54 p 159 letter to W J McKellar, 7 Westminster Chambers. 9 May 1883
87 AJHR 1882 H-12 Marine Department Annual Report , Annual Report on Lighthouse Works etc, by the Marine Engineer
88 AJHR 1884 H-6 Marine Department Annual Report ,
89 *NZG Gazette* 1883 No 122 . Notice to Mariners No 49 of 1883 p 1666 20 November 1883,
90 AJHR 1911, H-15 Marine Department Annual Report for 1910–11
91 AJHR 1912, H-15 Marine Department Annual Report for 1911–1912
92 NZG Archives ADOE 16612 M1/767 8/55/11 part 1 (extracted from M8/6/11) memoW Smith, principal keeper, to secretary Marine Department dated 3 April 1944,
93 Acts and Proceedings Auckland Provincial Council, Session II 1854–55. Petition presented by Mr James O'Neill, 5 January 1855
94 AJHR 1873 H-19.Report on the Lighthouses on the Coast of New Zealand. dated 3 March 1873
95 AJHR, 1874 H-23. Selection of Site for proposed Coastal Lighthouses Enclosure No 3 Captain Johnson to the Hon. the Commissioner of Customs 10 June 1874
96 *NZG Gazette* 1874 No 58, Notice to Mariners No 25 of 1874 p 741,3 November 1874
97 AJHR 1883, H-15 Marine Department Annual Report dated 6 June 1883
98 Acc 10706 vol 85 letter Agent General London re New Zealand Lighthouses 16 August 1883
99 Acc 10706 vol 85 letter from Wm Seed 12 June 1883
100 Acc 10706 vol 85 letter Milne & Son re New Zealand Lighthouses 17 August 1883
101 Acc 10706 vol 85 letter Dove & Co re New Zealand Lighthouses
102 Acc 10706 vol 85 letter Dove & Co re New Zealand Lighthouses
103 Acc 10706 vol 85 letter Agent Geheral London C/s 88 re New Zealand Lighthouses 31 December 1883 also Auckland Institute & Museum Library, MS 659 71/33
104 Acc 10706 vol 85 letter Agent General London re New Zealand Lighthouses 5 June 1884
105 Acc 10706 vol 85 letter from Marine Dept 282/41 21 August 1884
106 Acc 10706 vol 85 letter Andrew Barrie re New Zealand Lighthouses 25 October 1884
107 Acc 10706 Vol 85 letter E A Smith London re New Zealand Lighthouses 10 November 1883
108 AJHR 1884 H-6 Marine Department Annual Report . Annual Report on Lighthouse Works etc by Marine Engineer 31 March 1884
109 *NZG Gazette*, 1884 No 117. Notice to Mariners No 50 of 1884, p 1532, 29 October 1884,
110 NZG Archives ADOE 16612 M1/747 8/35/9 (detached from M8/35/7) letter Fraser to Sec MD 28 Oct 1919
111 NZG Archives ADOE 16612 M1/ 747 8/35/9 (orig 8/6/24) Extract letter Sinclair, principal keeper, to secretary Marine Department 1 Dec 1938(?)
112 NZG Archive ADOE 16612 M1/ 747 8/35/9 letter chief clerk to secretary Marine Department 12 June 1939
113 NZG Archive ADOE 16612 M1/ 747 8/35/9 memo Tate (lighthouse overseer) to secretary Marine Department 26 May 1947
114 *NZG Gazette* 1947 No 24 Notice to Mariners No 10 of 1947 p 569, 5 May 1947
115 AJHR 1956 H-15 Marine Department Report for year ended 31 March 1956

10 The Big Lights and Iron Towers

Cuvier Island, Stephen's Island, Cape Kidnappers and East Cape, Cape Palliser, and Kahurangi Point

Kaipara Harbour lighthouse (1 December 1884) was the last lighthouse to be built with a timber tower. The lighthouse exhibited just prior to this, Moko Hinau (18 June 1883) had the first concrete tower.

The next lighthouse to be lit, Cuvier Island, was the first New Zealand lighthouse with a cast iron tower since the early provincial lights. The towers for these early lights were sent out from England, but as was shown in the case of Mana Island, the towers were not necessarily of an ideal design for New Zealand.

New Zealand's industrial base had expanded in the 25 years since those early lights and it was now possible to design and cast iron lighthouse towers in New Zealand. Cuvier Island and its successors East Cape, Stephen's Island, Cape Palliser, and Kahurangi Point all had iron towers, as did Cape Brett and Castlepoint (Chapter 11). (The two replacement lighthouses at Baring Head and Cape Reinga have comparatively short towers and were constructed in reinforced concrete.)

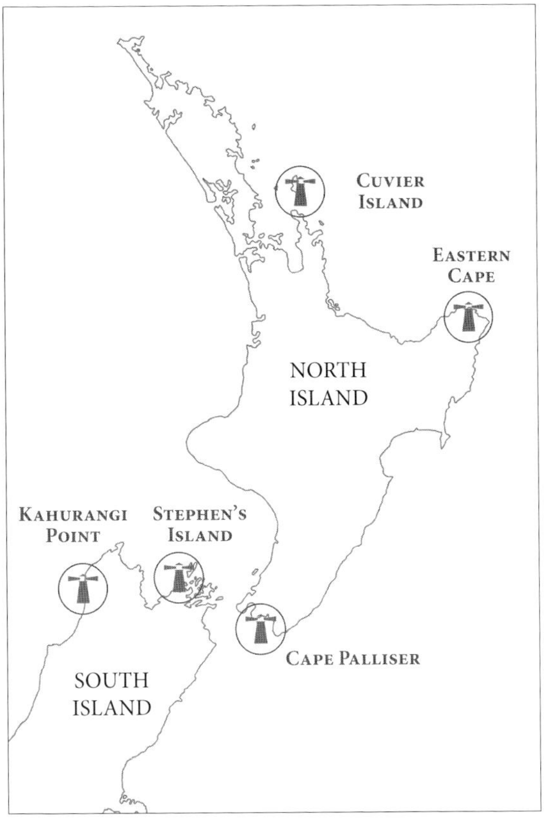

Figure 10.1: Locations of Cuvier Island, East Cape, Stephen's Island, Cape Palliser and Kahurangi Point lighthouses © Author

Cuvier Island and Stephen's Island lighthouses were originally lit with five-wick Doty burners. These were the biggest ever used in the New Zealand lighthouses.

Cuvier Island

Cuvier Island lies on the approach to the Hauraki Gulf. It was listed in Edwin and Johnson's 1873 report.[1] They recommended a second or third order white light revolving every half minute. While Johnson did not land on Cuvier Island during his 1874 site visits he did remind the Marine Department about the recommendation made the previous year.

No further action was taken until late 1883/early 1884 when according to the 1884 annual report.[2] 'A survey has been directed to be made of the best site for establishing a light in the locality'. The 'locality' was 'Eastern Entrance to Hauraki Gulf'. And the choice was between Cuvier Island and Red Mercury Island. The following year the survey was made by Blackett and Johnson who recommended Cuvier Island.[3]

Although the Annual Marine Department reports up to 1886 had noted all lighthouse building, the reports from 1887 become much more detailed. This is because Lewis Wilson had been appointed Assistant Secretary with responsibilities including preparing the report. As was shown in Chapter 1, Wilson was a nephew by marriage of Thomas Stevenson and had a particular interest in lighthouses.

As it was felt that a particularly strong light was necessary, the Marine Department decided to order a five-wick Doty burner for Cuvier, the largest to date being four-wick.[4] The lamps that were supplied were later described by D & C Stevenson as being of 'the "new pattern" being the best yet made'.[5] They also indicated to Wilson that the power of a five-wick lamp should equal 166,098 standard candles.[6]

In their letter acknowledging the order for Cuvier Island, D & C Stevenson did suggest

> The next time you wish a very(?) powerful light we would recommend you to adopt the large hyper-radiant size of apparatus for although it is a little more expensive we think it is worth the money and the Trinity House as well as ourselves intend to adopt it in the future.[7]

The Marine Department never did go to the extent of the hyper-radiant (which was bigger than first order). The future orders were for first order for Stephen's Island and Cape Brett lighthouses and second order for the remainder.

D & C Stevenson also agreed that a first order white light revolving every half minute was a good design.

The original specification would have included a comment on the height of the proposed tower so D & C Stevenson also commented

> The fall for the weights is rather short (30 Feet) but we presume that you can make a well at the base of the tower so as to allow the weights to descend a few feet further. We presume the weights can go down the centre of the tower and we shall arrange accordingly.

Invitations to tender were sent out of 23 and 24 March 1886. Milne & Son[8] and Dove & Co[9] were sent specifications for the lanterns and the machine etc., Barbier & Fenestre[10] and Chance Brothers[11] were invited to tender for the first order revolving apparatus, Smiths for stores[12] and Brydens for lighthouse signals.[13]

The contract for the lantern went to Dove & Co for £950/-/-[14] that for the machine to Milne & Son for £714/--/- who also successfully tendered for the lamps and burners with a price of £146/10/-.[15] Smiths offer for 'stores, tools, lamps glasses, paraffin cisterns etc' was for £106/13/6[16] and Bryden & Sons for the signals was £14/10/-.[17] Bryden and Sons' price was considerably lower than that charged for the six lighthouses. This is because at the time of tendering the length of tube required to connect lightroom and keepers' houses was not known. This would form extra work.

Milne & Son later suggested a change in the design of the table which supported the lamps and in the shape of the rollers. These were approved by Stevenson provided

> The heights from floor to top of ring under legs to be 9 feet 1½ inchs -The other dimensions all as usual.[18]

The packages containing the lantern were despatched from Granton by steamer on 16 October 1886 for transhipment to the *Invercargill* in London.[19] The apparatus and machine along with the stores followed on 26 February 1887 and were shipped to New Zealand on the *Aorangi* from Royal Albert Docks.[20]

Unfortunately, there was a problem. The lighthouse apparatus was left in store and when finally unpacked was found to have been damaged. In an 1891 letter to D. Alan Stevenson, Wilson described the damage as follows

> When the Cuvier Light was supplied some years ago - a new departure was made as far as packing was concerned in varnishing the bright work – this turned out a very great success – but to show the care that must be taken in having the iron perfectly dry I may mention that the only parts that were rusty were where the workman had touched the metal with his fingers& in several places the marks could be distinctly seen "in rust". Please note particularly that the keys are taken off the lenses before being packed this was – this was not done in the case of Cuvier the result being that some of the burners had every tube carried away & all required considerable repairs.[21]

Drawings and specifications for the iron tower were completed by the end of March 1888[22] and tenders called for soon after. The contract to manufacture the tower was let to Mr A Beany of Auckland for the sum of £704/3/-.[23]

The surveying of the proposed site had to be delayed as the Oamaru breakwater (on the South Island east coast south of Timaru) had first call on the Marine Engineer's time,[24] but construction began in August 1888 when a work party under the supervision of David

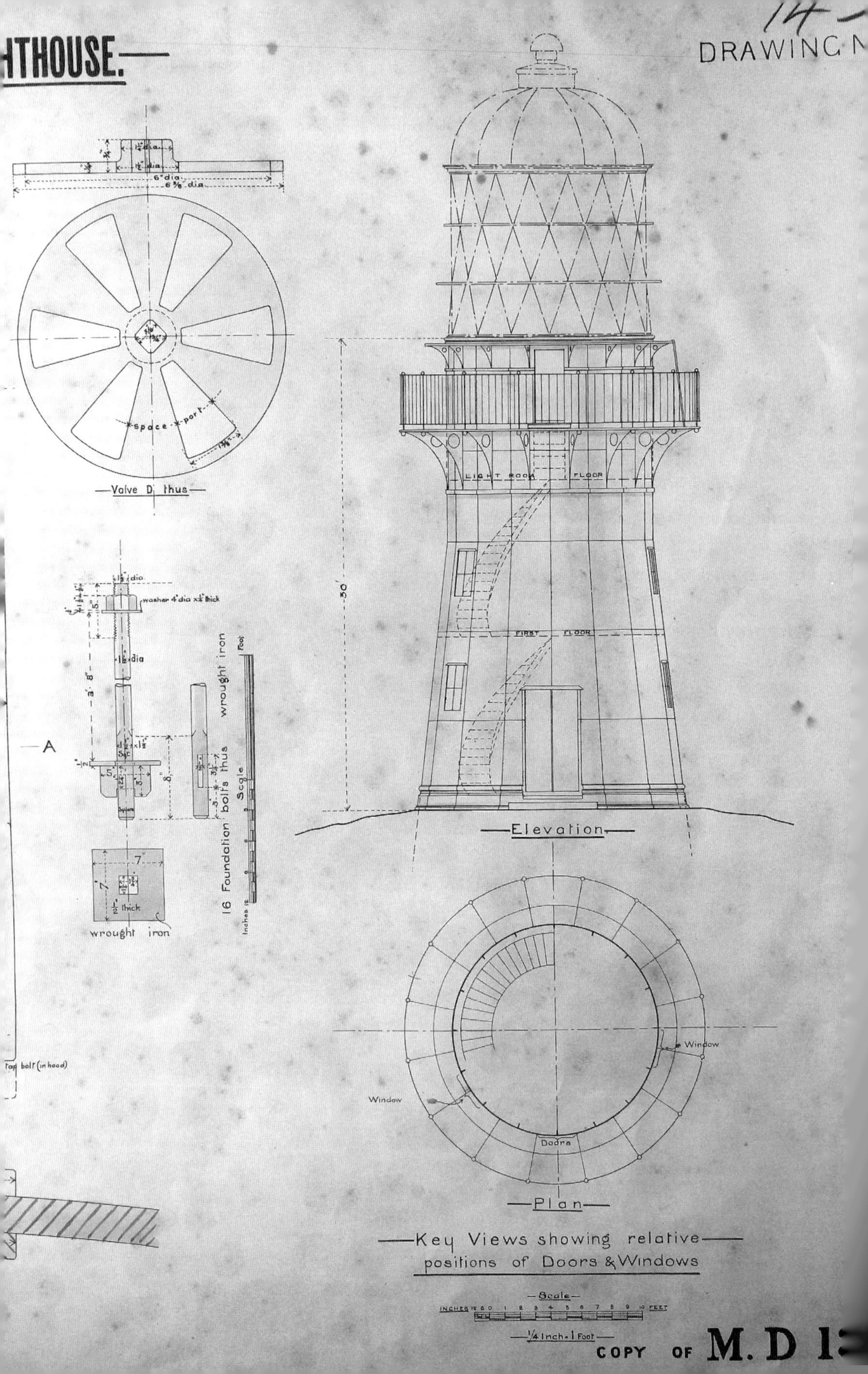

14
ITHOUSE.
DRAWING N
6" dia
6 5/8" dia
space port
Valve D thus
1 1/2" dia
washer 4" dia x 1/4" thick
1 1/2" dia
3' 8"
A
8"
wrought iron
16 Foundation bolts thus
Scale
Foot
Inches 12
7"
1/2" thick
wrought iron
Tap bolt (in hood)
LIGHT ROOM FLOOR
FIRST FLOOR
30'
Elevation
Window
Window
Doors
Plan
Key Views showing relative positions of Doors & Windows
Scale
INCHES 12 6 0 1 2 3 4 5 6 7 8 9 10 FEET
1/4 Inch = 1 Foot
COPY OF M.D 1

Scott, the lighthouse artificer arrived. Soon after this the tower was landed and the lantern in February 1889.[25]

The light was finally exhibited on 22 September 1889. The Notice to Mariners described the light as follows:

> On or after 22 Sept 1889 light exhibited from lighthouse on Cuvier Island, tower 50ft base to top of lantern, built of iron and painted white. First order revolving white light visible all round horizon as far as land allows, attain greatest brilliancy every 30 secs, light elevated 390 ft asl, eye 15 ft, vis 26 nm clear weather.[26]

The 1888 annual report stated that 'Steps have been taken to take land necessary for the erection of the Cuvier Island Lighthouse'. It was later discovered that a large number of the Māori owners had sold their share of the island to a Mr Tizard of Thames on the mistaken understanding that he was purchasing for the Crown.[27] The Land Court apparently made some ruling but this was not acted upon. It was later claimed that '[t]he award of the court has however never been communicated to this Department'.[28] This was in response to a complaint from Tizard, who wrote

> The lighthouse and keepers houses are situated on the island and the Government has had the sole use of it for many years; but I have never received any remuneration for it, nor yet payment under the agreement entered into for transferring the necessary land; although I have been compelled to pay the Land Tax for the island.[29]

The matter was not resolved until 1930.

In 1894 lamp glasses were ordered for Cuvier Island along with those for Stephen's Island and Dog Island.[30] The invitation to tender was sent to Smiths but also to an English firm, Messers Silber & Co of 57 Woad Street, Cheapside, London. The letter to Silber was more detailed than that to Smiths and then, after the description of the high standard of glass required, Silber were reminded 'in short the same as you supply to the Northern Lighthouse Board'.[31]

The 1909 annual report included Cuvier Island as being among the lighthouses where incandescent burners had been installed in the previous year.

These seem to have caused the familiar problem of condensation on the lantern panes as in 1924 it was suggested that a revolving cowl be fitted.[32] Some thought was given in 1938 to replacing the Chance burners with Smith-Ford burners.[33] Instead it was decided to electrify the light.[34] The incandescent lamp had been replaced by an electric light and a lens motor fitted by mid-March 1939.[35] Electrification was completed with the installation of radio beacons by July 1941.[36]

[Opposite] Figure 10.2: Design tower for Cuvier Island (ADOE 16630 M27/65 ?)
© Archives New Zealand, the Department of Internal affairs, Te Tari Taiwhenua

Stephen's Island

Stephen's Island was one of the lighthouse sites recommended by the Beacons and Lighthouses Committee of the House of Representatives in their 1854 report,[37] but it was over forty years before action was taken. While the West Coast gold rush had petered out, coal mining was becoming important and with this further concern for the safe passage of cargo through the Cook Strait.

On 2 June 1879 the marine engineer visited Stephen's Island (this was a second attempt) but was unable to land and examine the site due to the 'rough and precipitous' nature of the island.[38] By August 1881 a suitable site had been identified and a track had been cut from the beach through the bush to it. No further work could be done until funds became available.[39] The Marine Department was forced to wait and in the 1883 report William Seed recommended that 'provision be made in the estimates of the current year for the erection of this light'.[40] He repeated his plea in the 1884 report and then appears to have given up.[41] Wilson raised the question again in the 1888 report where he wrote:

> The necessity for the erection of a light at Stephens Island, Cook Strait, had [*sic*] already been pointed out on more than one occasion. It would be a most useful light to all vessels bound through Cook Strait from the westward, the number of which is increasing rapidly owing to the development of the coal trade at Greymouth and Westport. I would urge upon the Government the desirability of proceeding with the erection of this light as soon as possible.[42]

In the 1890 report Wilson again raised the matter.[43] These Fabian tactics seemed to have eventually worked, and December 1890 found Captain Johnson back on Stephen's Island but unable to examine the proposed site thoroughly due to the vegetation that had grown in the meantime.[44] However, Wilson remained optimistic, and in his letter to D Alan Stevenson commented:

> The money for Stephens Island is not voted yet, but I hope & expect that it will be this session of Parliament and then as soon as some necessary particulars are obtained I will have the order for the apparatus & flashing light with a 30" period'.

He then cheerily wrote 'What do you think of this?'[45]

The appropriation was finally voted and the lighthouse components ordered by August 1892.[46] It may be significant that the minister responsible for the Marine Department was Richard Seddon, who was member of parliament for Westland (as the constituency was named) and began his career in Hokitika, one of the coal mining areas of the West Coast.

The eventual design was for a first order revolving with a five-wick burner with group flashing showing a double flash every half minute[47] supported by a cast iron tower. This

Figure 10.3: Interior of lens, Stephen's Island lighthouse. (AABPL 8848 W5221 19 28) © Archives New Zealand, the Department of Internal Affairs, Te Tari Taiwhenua

was the first New Zealand light to have a group flashing light. While this was not absolutely cutting edge technology, the advice was Scots. As early as 1878, Lewis Wilson wrote to Thomas Stevenson

> I see by the papers that Dr Hopkinson's group flashing lights are all the rage at home now. I should be glad if you would be kind enough to send me copies of any papers etc about new lights.[48]

However the apparatus had obviously had some improvements since then as D & C Stevenson later wrote to Wilson

> The apparatus you will observe when you see it is novel in design – the double flashes are very sharp and distinct.[49]

The chosen pattern was flash 1¾ second, dark 3 seconds, flash 1¾ seconds dark 23½ seconds.[50]

Like Cuvier Island lighthouse, that for Stephen's Island was ordered individually. On 15 January 1892 D & C Stevenson acknowledged receiving the order from the Agent General.[51]

Plans and specification for the apparatus were initially sent to Barbier & Cie (successors to Barbier & Fenestre) and Chance Brothers for their information. Chances apparently suggested a change in the design, and D & C Stevenson replied that they had 'no objection to the use of 14 instead of 9 elements'.[52] In the end tenders were requested for the altered design from Barbier & Cie,[53] Chance Brothers[54] and another French firm, Sautter, Haile & Cie.[55] The contract went to Barbier & Cie for a tender price of £1470/-/-.

Requests for tenders for the lantern, machine, and the five-wick lamps were sent out on 25 February 1892 to Milne & Son[56] and Dove & Co.[57] The contract for the lanterns went to Dove & Co[58] for £1046/-/-, and that for the machine and the lamps to Milne & Son[59] for £956/10/-. Barrie were asked to supply a lightroom clock 'similar to that you furnish for the Commissioner of Northern Lighthouses',[60] and their tender was accepted on 28 April.[61]

The tender for the lightroom signals went not to Bryden & Son as previously but to the firm of W M B Mountain, also of Glasgow.[62] For the first time electric bells were to be used. Those for Cuvier Island lighthouse were described as 'whistles'. Mountain were asked to provide

> a bell and a push board for each of the three keepers houses, and in the tower a bell will also be required and a three indicator push board The system would be powered by a six cell battery in box and a spare battery and the necessary spare apparatus would be required.

As the actual site layout was unknown, Mountains were asked to make an allowance for this in the length of wire sent.[63]

The day after tenders for the lanterns, machine, lamps, stores, and signals closed, the firms involved were asked to resubmit their tenders on duplicate forms supplied – something that had never been done before. Previously both D & T Stevenson and D & C Stevenson had accepted tenders on behalf of the Agent General. Now they were required to transmit acceptances by the Agent General, which had been sent from London.

This change had its origin in an exchange of letters with the Agent General the year previously. On 10 December 1890, Smiths had successfully tendered to supply enamel paint for the New Zealand lighthouses. On 13 February 1891, the Agent General wrote querying the cost of packing and despatch to London. D & C Stevenson quoted Smiths as follows

> We do not consider the charge in our invoice for 1 Box zinc lined addressing and carriage to London excessive, and as this was the price quoted in our offer of the 10th Dec 1890 we cannot see our way to make any alteration.[64]

This does not seem to have satisfied London, as further enquiries were made and in a letter of 1 April 1891 D & C Stevenson quoted Smiths as follows:

> In answer to your letter regarding the cost of the package and carriage to London of the enamel paint lately ordered from us. The cost of the package lined and packing is 12/9, and the actual cost of carriage to London 3/-.
>
> When making our offer we had to estimate what the package would cost and we estimated it at 19/- and not knowing the exact size of the package might be and we estimated the carriage to London at 8/-.[65]

Smiths also took the opportunity to add a slight dig:

> We may state on the whole transaction we are loser by 10/8 as the paint cost us this sum more than our offer of the 10th Dec 1890.

London pressed the matter requesting information on shipments by Dove & Co as well as Smiths. At this point D & C Stevenson decided to be firm. They replied as follows:

> We send you inclosed the information which you wished from Messers Dove and Co and Messers Smiths relative to the sums entered in their Tender for packing, carriage and insurance etc.
>
> We think it right to call your attention to the fact that these are Tenders or offers for which the contractors undertake to supply and deliver the supplies taking the risk of the ultimate cost being greater or less.[66]

They then proposed a solution:

> In future we shall ask the contractors to exclude the carriage to London from their offers and supply vouchers for the carriage. We shall further send the tenders to you before accepting them should you desire it.

Faced with the extra work of sorting out payments for packing and shipping, the Agent General agreed to leave tenders in the existing form, although the offer to examine the individual tenders was obviously accepted.

However, the question of delivery charges was still not completely clear, as in dealing with the tenders for Stephen's Island D & C Stevenson were forced to point out:

> With reference to Messer Barbier's tender only including delivery to Edinburgh this is quite correct and their contract ends when we inspect and try the apparatus here. The cost of delivering the apparatus supplied by Messer Barbier free on board export ship in London is included in Messer Milne's offer. We therefore return Messer Barbier's offer for an acceptance.[67]

For reasons unstated, the major contracts were delayed. On 10 September D & C Stevenson wrote to Dove & Co, pointing out that the completion date had been 12 August and the lantern was still not finished. More ominously, they informed them:

> As we know of no reason which has prevented you completing the work in the specific time we fear it will be necessary to exact the penalty under which you are bound and which being at the rate of 2 per cent per week now amounts to over £100.[68]

The lantern was finished and packed by 12 October, and despatched from Granton by steamer on 15 October for shipment to New Zealand on the *Duke of Devonshire* from the Royal Albert Docks.

Unfortunately, when the lantern was unpacked it was found that the trimming path had been damaged. Stevenson quoted the secretary of the Marine Department as writing:

> two of the trimming path gratings were damaged one being cracked half way through and the other broken in two … Five of the cleaning path gratings were damaged and will have to be repaired, the case was too large and they were not blocked in any way.[69]

Stevenson indicated that they had asked the Marine Department for an estimate of the cost of repairs which Dove & Co 'will require to remit to the New Zealand Government'.

In their letter transmitting Dove & Co's reply, D & C Stevenson stated that 'this is the first complaint we have had' and pointed to 'their reply in which you will see they *offer* to refund any expense you may have been at in repairing the broken castings' [author's italics]. Here we see Stevenson safeguarding their client's interests while still maintaining the reputation of Dove & Company. Stevenson even offered to add to future specifications a clause making contractors responsible for any breakages till the packages were delivered in Wellington. This was not done.

The apparatus was also behind schedule and in a letter of 12 October Barbier & Cie were told:

> We are very much disappointed that the apparatus for Stephens Island has not been sent forward. You are aware that you have exceeded the contract period by two months and have already received a heavy penalty. We hope to hear by return of post that the packages are now on their way to Edinburgh.[70]

This letter crossed in the post a letter from Barbier & Cie indicating that the apparatus in 25 packages was on its way from Dunkirk to Leith.[71]

When the apparatus was unpacked it was found that

> you have introduced two settings into the lens panels one of which passes through the center of the bullseye. They detract from the appearance of the apparatus and to some extent impair their efficiency as an optical instrument. These settings were not shown in our drawing and appear to us to be quite unnecessary. In the future before making any such alterations in our designs we shall be obliged if you will communicate with us.

Given the time element, D & C Stevenson were prepared to accept the 'modified' version, but the whole episode suggested that Barbier & Cie were poor successors to Barbier & Fenestre. By 2 December D & C Stevenson could write that the apparatus had been tested and was ready for packing.[72] However, conscious of the recent complaints, D & C Stevenson reminded Milne & Son:

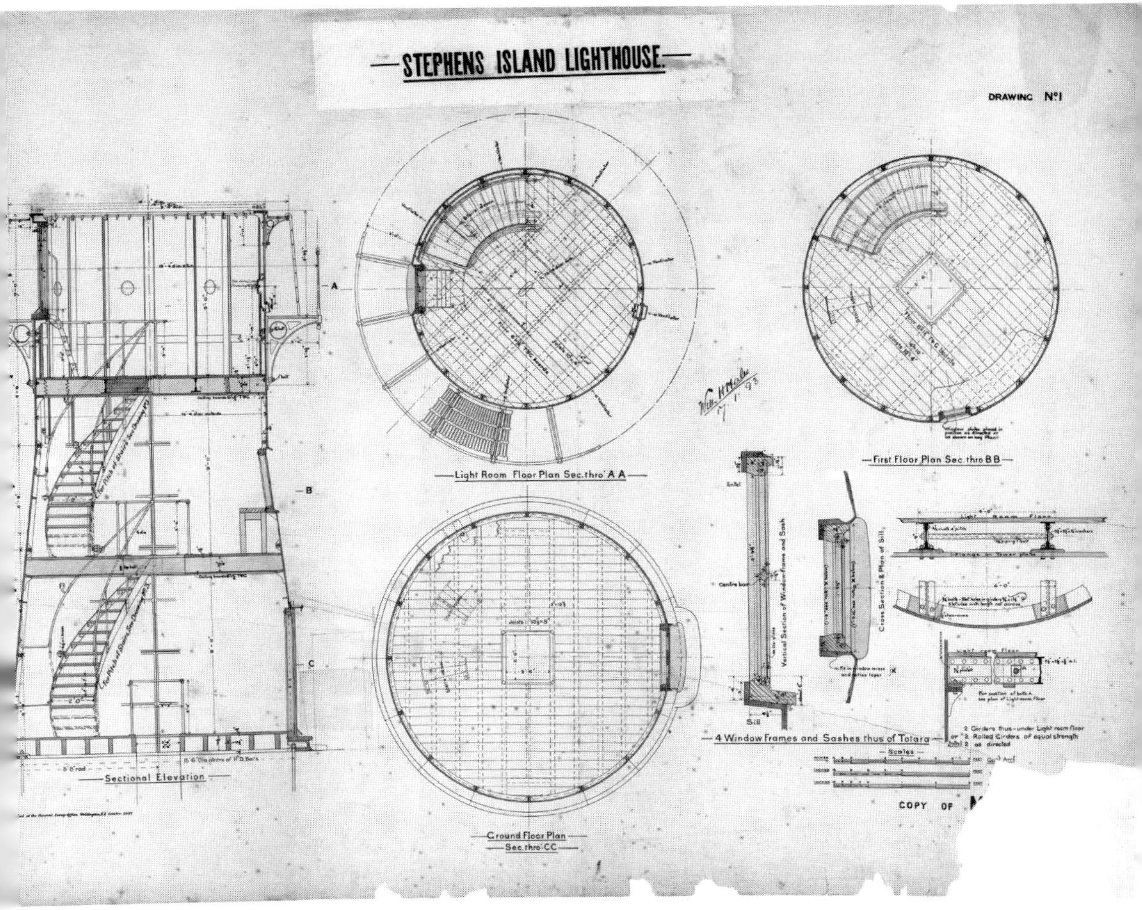

Figure 10.4: Design of Tower for Stephen's Island lighthouse (ABPL 8607 W5220/38 MD1804) © Archives New Zealand, the Department of Internal Affairs, Te Tari Taiwhenua

> Please to keep in view in varnishing the bright work of the machine that the iron is perfectly dry and that any marks of the workman's fingers are also carefully wiped off before the varnish is applied. The keys of the racks of each burner should also be taken off before the burners are packed.[73]

The electric bells and the clock were apparently both ready within the contract period, but Smiths experienced difficulties with the lamp glasses:

> Messers Smiths glasses etc will also be ready for shipment next week The English manufacturers have failed to supply glasses which we can pass and Messers Smith have been under the necessity of getting them made abroad and the package containing them has been delayed in transit in consequence of some quarantine regulations.[74]

Figure 10.5: Stephen's Island lighthouse (ABPL 8848 W5221/19/31)

Finally apparatus, machine and case, bells, clock, and store were despatched by steamer from Leith on the Saturday after 29 December 1892, to be transhipped to the *Indramayo* at the Royal Albert Dock.

The tower for Stephen's Island used the same design as Cuvier Island lighthouse and was also manufactured by Beany and Son of Auckland, in this case for the sum of £689.[75,76] One of the unsuccessful tenderers was Luke & Co of Wellington, later responsible for Castlepoint lighthouse.

Construction began between 1 April and 1 August 1892,[77] and the light was eventually exhibited on 29 January 1894.

The Notice to Mariners described the lighthouse as a

> tower 50 ft, built iron painted white, first order group flashing white light, two distinct flashes in quick succession every half minute, and will shoe over an arc of 293 deg a; around seaward as far as land will allow, elevated 600 ft above sea, allowing 15ft eye, vis 32 n m.[78]

In 1907 it was decided to install a Matthews incandescent light on Stephen's Island, and the high commissioner was instructed to procure one through D & C Stevenson.[79]

Stephen's Island was the third lighthouse to have incandescent lights installed, the first being Jack's Point Timaru (see Chapter 8) and Pencarrow. On the basis of the experience with these three lights, the Marine Department recommended in the 1908 annual report 'that provisions be made in the estimates for installations for further lights'.[80]

By 1934 this had been replaced by a Smith-Ford incandescent, which proved unsatisfactory. The principal keeper complained

> at times through out the night the burner seems to lose heat & spout kerosene through the nipple. Using old Chance burner.[81]

In December 1936 the engineer in chief indicated that[82]

> [i]t is proposed to electrify the Stephen's Island Lighthouse very shortly.

and the following year's annual report stated that the Stephen's Island light was to be converted to electricity, similarly to what was planned for Cape Campbell.[83]

Four days after the 1937 annual report was signed off, Cory-Wright and Salmon Ltd, the New Zealand agents for Chance Brothers, contacted the Public Works Department for details of the lighthouse apparatus which Chance Brothers had been unable to trace.[84] The information requested was:

- focal length optical apparatus,
- true to scale copy showing sectional plan of optical apparatus at the focal plane,
- true to scale copy showing vertical section of the optic, and
- the height of the focal plane from lens carriage.

The reply from the Public Works Department[85] indicated that no plans were available but that it was:

> understood that plans could be obtained from D&C Stevenson, Engineers, Edinburgh, through whom the makers Barbier & Fenestre, Paris, supplied apparatus in 1892.

It is not clear whether D & C Stevenson were consulted, but the 1938 annual report stated that while the delivery of plant for the electrification programme was slow, most of that required for Stephen's Island had arrived.[86] The new light was first exhibited on 29 November 1938.[87]

Cape Kidnappers and East Cape

In their 1873 report Edwin and Johnson had recommended that a third order white light revolving every half-minute be placed on Cape Kidnappers and a white light revolving every minute on East Cape. They did not specify a size for East Cape.[88]

The following year Johnson visited East Cape Island as he felt that it was evident that the island would be the best position even though it is 'not the intention of the Government immediately to erect a lighthouse' there.

Eventually it was decided to go ahead with a lighthouse for Cape Kidnappers. However, by the time the lighthouse material arrived the bureaucratic mind had changed and the lighthouse was erected on East Cape Island.

The lighthouses for Cape Kidnappers and Cape Palliser were a joint order. The Marine Department wrote on 26 April 1895 to inform D & C Stevenson of the prospective order, although the specifications were sent separately.[89] Both lights were second order, but while that for Cape Kidnappers had a flash every 10 seconds, that for Cape Palliser was group flashing. Two new firms were invited to tender.

In this case there had been recent modifications to the design for the lantern. When D & C Stevenson sent a tracing showing the design of the lantern, they commented:

> The lantern is the same as for earlier supplied for 2nd order Lights for New Zealand with the exception that the cleaning path which was formerly on same level as sole plate at bottom of glass panes, will require in the present cases to be carried by brackets fixed to the framing of the tower at a height of 7 feet from the floor of the lightroom. The brackets are shewn 12 in number and two malleable iron rings bent to the circle the mean radius of the inner ring being 3 feet 4inches and the outer 4 feet 5 inches. The least diameter at inside of cleaning path will require to be 6 feet 6 inches and the rings will require a space cut out of them for the lightroom ladder as shewn on the drawing. These brackets and rings we think may be made and fitted on the inside of the tower by yourselves and we shall send out the grating so that it can be fixed down to the rings with screws.[90]

The oil fountain was designed to be carried on the dormant panes, and the pipes led through the machine case thus giving more room inside the apparatus.

There were some improvements in the machine. The rollers were designed with conical surfaces which were thought to lead to less cutting of the roller path. In reply to a letter after the machine had been in use for about a month, Stevenson pointed out that:

> With regard to the first point of having to file the bracket which carried the fly wheel to allow a second turn of rope to go on the winding barrel, we may say this machine revolves more quickly than was in former times usual and necessitates a longer rope, and we regret we did not notice in making the trials that the bracket would come in the way if two turns had been put on. As a rule we find that if the winding be commenced on the side of the barrel away from the winding wheel sufficient number of double turns are got on.[91]

The order for the lantern and apparatus for Cape Kidnappers was sent from London on 17 July 1895.[92] Requests for tenders for the lantern, machine, and lamps were made to Dove & Co,[93] Milne & Son,[94] and for the first time Steven and Struthers of Elliot Street Glasgow,[95] all on 22 August 1895.

Requests for tenders for the apparatus were sent to Chance Brothers,[96] Barbier & Bernard[97] (as it now was), and Henri Lepaute[98] on 19 August 1895. (This was the first occasion when Lepaute were invited to tender for any of the New Zealand lighthouses). The tenders were to be for two second order apparatus, one six-sided and the other double flashing.

The contracts for the lantern and machine were given to Milne & Son at a price of £500 for each contract.[99] Although the original specification had stated contract periods of four months for both lanterns and apparatus, the final agreed period was five months with Milne & Son endeavouring to have the lanterns ready in four.[100] Sometime after the contract was let, D & C Stevenson decided to alter the design for the machine to add ball bearings to the rollers. Milne & Son were approached for a quote to supply the ball bearing on 24 May.[101] In the letter to the Agent General submitting the invoices for the machine etc., D & C Stevenson explained:

> As there have been complaints from New Zealand in reference to the rollers of machines we thought it well to give Messers Milne a Certificate of Extra Work under the Specification to introduce ball bearing in the machine for Cape Palliser and also for Cape Kidnappers Lighthouses as these ball bearings cause less friction in working.[102]

The question of payment for delivery of the optical apparatus still seemed to be exercising the Agent General mind, and D & C Stevenson wrote on 13 September 1895:

> Cape Kidnappers Lighthouse We have received your letter of the 12th Instant and have forwarded to Messers Milne your acceptance of their tender for the machine and lantern for Cape Kidnappers Lighthouse. The specification of the revolving machine states distinctly that Messers Milne are to deliver the optical apparatus f.o.b. London. The following is the clause in the specification referring to this matter and it applies to Cape Kidnappers and Cape Palliser Lighthouses. "the offer is to include the cost of unpacking the glass work, putting the whole up for trial in an experimental lighthouse and afterwards packing the whole in suitable double cases, the inner ones where necessary being tin lined and also delivery of the whole free on board export ship in London." Messers Barbier & Bernard's tender is in terms of the specification to be delivered free in Edinburgh and as explained above we think it unnecessary to add any clause to your acceptance of Messer Milne's tender for the machine.

The contract for the apparatus was awarded to Barbier & Bernard, whose price was £810. The following month they suggested an alteration – 15 prisms above the lens – to which D & C Stevenson agreed.[103]

While lanterns and apparatus were under construction, tenders were called for the smaller contracts. These were for the supply of lightroom clocks, electric bells and stores. Barrie were invited to tender for the clocks for both lighthouses 'similar to those you supply to the Northern Lighthouse Board'.[104] Tenders for the electric bells were requested from Bryden and Son[105] and also from Fleming and Co of Edinburgh.[106] D & C Stevenson had intended putting off ordering stores until nearer shipping date,[107] but they reported ordering the stores at the same time as the bells and clocks.[108] Barrie's price was £9/13/6 for each clock; Smiths' price for the stores for each lighthouse £75/ 7/ 7; and the contract for the electric bells was awarded to Bryden & Son for a price of £5/16/9 for each lighthouse.[109]

In the same letter as the summary of the tenders for stores etc for Cape Kidnappers, D & C Stevenson informed the Agent General that the lanterns for Cape Kidnappers had been passed and were in the process of being packed. By 27 February the packages containing the lantern and electric bells had been sent to London for shipment to New Zealand on the *Kaikoua*.[110]

Barbier & Bernard overran their contract period, and it was 9 March before D & C Stevenson could report that the apparatus was on its way from Paris. They however pointed out:

> But there is no great hurry in the matter as the lantern requires to be fitted up at the lighthouse and the lightroom finished before the fitting up of the machine and apparatus can be proceeded with.[111]

On 11 June D & C Stevenson reported that the apparatus was undergoing final testing and would be ready for shipment in 10 to 14 days.[112] The complete shipment of machines, apparatus, stores and lamps, and clock were sent from Leith on 4 July for shipment to Wellington by *Ionic* from Royal Albert Dock.

However, in the interim it was decided to build the new lighthouse at East Cape rather than Cape Kidnappers.

East Cape Island lighthouse

In his report in the 1896 Marine Department annual report, George Allman, the inspector of lighthouses and nautical adviser, recommended a new light at East Cape. The cape is the turning point for vessels bound for Auckland from southern New Zealand, as it is approximately midway between the Cuvier Island and Portland Island lighthouses.[113]

Some time in 1897, 'a strong opinion having been expressed by master mariners that a lighthouse at East Cape was a much more pressing necessity', it was decided to erect a lighthouse on East Cape Island and to use the lantern and apparatus obtained for Cape Kidnappers.[114] The 1898 Marine Department report indicated that the land had been acquired and the nautical advisor and the marine engineer had visited the island to select suitable sites.[115]

The island is very rocky with no natural water supply and thin topsoil, meaning that keepers were unable to grow vegetables or keep livestock. The island is also subject to erosion and earthquake-induced slips. The Māori name for the island is Whangaokeno or

Motu o Kaiawa, and although Māori land, local Māori chose not to live on it. The island was and is regarded as *tapu* or set apart. During the construction four men were drowned when a boat coming from the government steamer *Hinemoa* capsized in the rough seas. They were buried on the island in a small cemetery and were later joined by three of the lighthouse children.

The iron tower and associated buildings were designed by the Public Works department, and tenders for the tower were advertised in July 1898.[116] The contract was given to Charles Judd and Co of Thames in August 1898 for a contract sum of £988 19s 6d.[117]

By May 1899 construction was well under way.[118] The light was exhibited on 9 August 1900.

The Notice to Mariners[119] described it as follows:

> The site of the light is in latitude 37°4' south, and longitude 178°36' east, as measured from the Admiralty Chart.
>
> The tower is constructed of cast iron. The total height from the bottom of the base to the top of the lantern is 69 ft., and is painted white.
>
> The light will be 362 ft above the level of the sea at high water. It is a dioptric revolving white light of the second order, showing a white flash at intervals of tens seconds. It will be visible seaward over an arc of 280 degrees of the horizon or as far round as the land will allow, and in ordinary clear weather it will be seen twenty-two nautical miles from the deck of a vessel 15ft. above sea level.

Incandescent lamps were procured for East Cape Island in 1911[120] and a replacement was installed in 1918 along with a revolving cowl.[121]

East Cape Island was prone to erosion, which, exacerbated by earthquakes, eventually threatened the lighthouse station. It was therefore decided in 1921 to move the lighthouse to Matakawa Point on the mainland.[122]

Figure 10.6 East Cape lighthouse (ABPL 8848 W5221/19 15), © Archives New Zealand, the Department of Internal Affairs, Te Tari Taiwhenua

East Cape lighthouse

The land at Matakawa Point was all in Māori ownership ('is owned by natives') and the Marine Department showed a similar attitude to that shown at Kaipara. The Department anticipated taking 126 acres 2 roods 16 perches [51.2 ha] and having compensation fixed by the Native Land Court at £3 per acre.[123] This prompted protests from the families concerned, who were losing a substantial part of their livelihood. The protest pointed out that the land was being taken from several families to provide farms for two men, that Edwin and Johnson had recommended an area of five acres as being sufficient for a lighthouse station, and that the Marine Department had had no compunction in placing three families on the unwelcoming East Cape Island.[124]

Eventually the Department had to be contented with two parcels of land: one of 44 ac 0 rood 16 perch [17.8 ha] for the lighthouse station and one for a road of 7 ac 2 rood 1 perch [3 ha][125] at a compensation of £4.10s per acre.[126] *

In June 1922 it was reported that removal of the lighthouses and tower was under way and was expected to take about two months.[127] The completed transfer was announced in the following year's annual report.[128]

Some of the prisms in the lens were smashed when the lens was unloaded at Matakawa Point.[129] As the Marine Department did not have diagrams of the lens easily to hand, D & C Stevenson

> forwarded to the Marine Engineer with their letter of the 19th December 1922, a sketch plan of each segment with each prism numbered thereon.[130]

The contract to supply the new prisms was let to Barbier Bernard and Turenne[131] for £359.[132] ** The delivery and installation of the prisms was noted in the 1925 annual report.[133]

Cabinet approval for electrification was given in January 1949,[134] and that year's annual report stated that estimates had been approved, and specifications were being prepared for complete electrification and installation of standard radio beacons at East Cape, along with Tiri Tiri Matangi, Portland Island, Dog Island, Godley Head and Taiaroa Head.[135] The 1954 report indicated that electrification of the light and installation of radio beacons was proceeding at seven lighthouses, including East Cape lighthouse.

Cape Palliser

The second report of the Marine Department had recommended a light at Cape Palliser to guide ships into Cook Strait. The report particularly mentioned ships coming from Panama as 'being near the terminus of a long ocean run, when the fuel on board the vessel will be short, and consequently every hour saved is of moment'.[136]

Edwin and Johnson suggested a second or third order white light revolving every 2½ minutes.[137] However, nothing further was done until 1893, when the annual report for that

* It should be noted that today such blatant land grabbing by a government department would be roundly condemned, and attempts are being made to undo the damage done by this kind of economic pillaging.

** The clerk making out the application for payment unfortunately referred to the contractors as Messers Barber, Benard and Turene London.

year indicated that 'Mariners urge the erection of lights at Cape Palliser, eastern entrance to Cook Strait, and at Rocks or Kiourangi Point, [see below] on the west coast of the South Island'.[138]

The lantern and apparatus were ordered, along with those for Cape Kidnappers. However, soon after receiving the requisition D & C Stevenson raised the question of the characteristic of the Cape Palliser light; the original design had been for 10 seconds light, 5 seconds dark, 10 seconds light and 35 seconds dark. D & C Stevenson pointed out that flashes of 10 seconds length 'would necessitate a very expensive form of apparatus, and is moreover, a waste of the power of the light, and one which we could not recommend'. They suggested instead either the same pattern as Stephen's Island which was sufficiently far away not to cause confusion i.e. '2 seconds light, 3 seconds dark, 2 seconds light and 23 seconds dark' (this was the recommended option), or '3½ seconds light, 5 seconds dark, 3½ seconds light and 48 seconds dark'.[139]

They also pointed out in a later letter[140] that

> a light to a ten second flash of the same size as the two second flash we recommended would give only one third of the power.

Most of the order for Cape Palliser lighthouse was manufactured by the same firms that provided the lighthouse for Cape Kidnappers. The drawings originally done for the Waipapa lighthouse and recycled for Farewell Spit's second lighthouse may have also been used for Cape Palliser, as the drawing has Cape Palliser added to it in pencil (see Figure 9.11a). The one major difference was the group flashing apparatus which was provided by Chance Brothers, whose price was £926.[141] Even though only one side of the correspondence is available, it demonstrates a steady succession of problems which compare unfavourably with the contract with Barbier & Bernard. Chance Brothers even managed to overrun the contract period longer than the French glassmakers did theirs.

Chances recorded the order on 14 September 1895. The order was described as

> Second Order Double flashing Light
> as per specn & drawing of D&C.S
> comprising 8 side in 4 groups
> or 8 panels of 16 upper prisms, 7
> panels of 6 lower prisms, 8 central
> lenses composed of Fresnel bullseye,
> and 6 equiangular Stevenson prisms
> the whole framed and mounted in gunmetal.[142]

The gunmetal framing and mountings appear to have been Chance's 'improvement', as four days later D & C Stevenson wrote:

> We have received your letter of the 17th Instant. We do not approve of the gun-metal & settings as you propose. We would prefer if you would leave out the setting AB which was put on the drawing by mistake as it is only a temporary one for transit. The specification of Cape Kidnappers applies to Palliser. We have marked the dimensions you require on the tracing which we return herewith.[143]

Further clarification was needed, and on 16 October D & C Stevenson explained:

> The bottom ring of the apparatus should be 10 mm in thickness with good lap joints fixed by four steel screws. The lamp joints should be 4" long. The holes should be drilled in this ring, for fixing on carriage as they will be fitted here.[144]

This seems to have prompted Chances to come up with another suggestion, causing D & C Stevenson to reply on the 25 October 1895:

> We have received your letter of the 22nd Instant. The data for setting the Cape Palliser apparatus is given in the Specification, and we fear a Doty burner could be of no use to you in this operation.[145]

They were forced to repeat themselves three days later:

> We have received your letter of the 25th Instant. If you work to the information given in the Specification we think there will be no difficulty in setting the prisms etc of the Cape Palliser Light.[146]

In early January D & C Stevenson obtained details of the apparatus as it then was. (They may have despatched an inspector to Smethwick.) This prompted on 11 January 1896:

> It is evident that you have ground the inside faces to a much shorter radius of curvature than that given. It is not a matter of anything like the difference between 3726 or 3740 m/m which would be quite immaterial; but what you have done is a much more serious alteration of our design by shortening the radius of curvature necessitating thereby a departure from the co-ordinates of the outer faces to correct this departure. It is by no means an improvement and we would like to hear your explanation and reason for so doing.[147]

Chance's reply on the 13th prompted D & C Stevenson to write on the 14th:

> We have received your letter of the 13th Instant. We regret that you should have altered the co-ordinates of the Cape Palliser apparatus without first

communicating with us, and in future we hope you will adhere to the co-ordinates we give on our plans. Meantime, please forward the coordinates you have used.[148]

Chance seem not to have read the letter properly, as two days later D & C Stevenson wrote:

We have received your letter of the 15th Instant. Please forward the co-ordinates to which you have ground all the prisms, as requested in our letter of the 14th Instant.[149]

Some information came but it was not clear, and so 20 January saw the following:

We have received your letter with the co-ordinates but you do not say where the angle C is, and as it is not shewn on any of your drawings. Will you send us a drawing with this angle indicated.[150]

There seem to have been further suggestions as to design modifications, and 22 January saw D & C Stevenson's sharpest comment:

We have received your letter of the 21st Instant. We think that you should arrange the lenses to produce the best result and alter your metal work to suit if necessary. We notice what you say that you have attempted to work to our co-ordinates. M Lepaute and Messer Barbier & Bernard have both executed these equiangular lenses, to our designs, which have worked in to a nicety, when set together, with most accurate results, and it is a pity you placed your centre for the equiangular curve 200 m/m below the focal plane without communicating with us.[151]

Since the apparatus was now beginning to differ from the original drawings, D & C Stevenson checked on the following day

Will you please let us know the exact height from the bottom of the bottom ring to the focal plane of the Cape Palliser apparatus. We wish this information to enable us to get the stools made for the apparatus to stand on. Please let us also have the diameter of the bottom ring.[152]

This was followed the following day by a brisk reminder

We have received your letter of the 23rd Instant. We have to refer you to our letter of the 22nd instant. The dioptric apparatus you have made must be so arranged as to produce the best result. If it is necessary to alter the metal work this must be done.[153]

Finally Chance Brothers decided to send a Mr H I Stobart, one of their engineers, up to Edinburgh to discuss the matter with Stevenson. After the visit D & C Stevenson wrote:

> We had submitted to us yesterday by your Mr Stobart a proposal for the setting of the lens of Palliser. We think that if the deviation of 44 mm is the error with this way of setting that it will do, but what we have to impress on you is what we said in our last letter, namely, to set the prisms where they are best for focussing. We think there will be a want of security by the glass not being properly in the setting. In any case *it is a makeshift* [italics author]. Let us know exactly how you set the prisms and when you have got the apparatus ready so that we can inspect them should we see it necessary.[154]

By this time D & C Stevenson must have been regretting the convention which required them, as consulting engineers, to recommend to their clients the acceptance of the lowest tender.

About the same time that Barbier & Bernard were despatching the Cape Kidnappers apparatus, D & C Stevenson were writing to the Agent General:

> Messers Chance do not expect to be ready with the optical apparatus until about a month hence. We have been corresponding with them and have also had an interview with them on the subject. We have again written to them urging completion of the work.[155]

The apparatus had arrived in Edinburgh by 17 April,[156] but there were still problems. A month later Stevenson were pointing out:

> We find on examination of the optical apparatus for Cape Palliser that the rods for setting the burners are not fixed in the focal plane of the panels, one is fixed about 1/32 of an inch lower than the centre of the bulls eye, and the others vary to ¼ inch below. Will you please let us know why this is so. The eyelets are specified to be put in the focal plane. We regret that the glass is so badly striated.[157]

Chances also neglected to send spare glass sections.[158]

The glassware was finally despatched from Leith on 4 July along with the shipment for Cape Kidnappers, to be shipped to New Zealand on the SS *Ionic*.

Both lighthouses included auxiliary lights to mark offshore reefs. The original request was for a fifth order lamp which D & C Stevenson suggested would not be powerful enough to be of any use at five miles. Instead they proposed:

> The apparatus is part of a refracting belt of a fixed light 60 deg in azimuth the lamp being so placed in the focus as to dip the rays of light to the five mile limit. The lantern will require to be fixed to the side of the tower resting on brackets.

This would mean alterations to the cast iron towers in each case, i.e.:

> The cast iron parapet will require to have a door cut through the plate to get admittance to the lamp. In the case of the wrought iron open tower beams will require to be secured to the uprights to carry the brackets and the inside of the tower where the lantern is placed lined to protect the lamp. We estimate for each place the lantern and lens and lamp will cost £320.[159]

D & C Stevenson had already approached Barbier & Bernard for a quote to supply the two lamps.[160]

In early November the Agent General passed on a request for an alteration to the auxiliary light for Cape Palliser. This seems to have been for a more complicated light, as when D & C Stevenson presented the revised design they quoted an approximate cost of £800.[161] As the Cape Kidnappers lighthouse was erected at a different site, the only auxiliary light installed was that at Cape Palliser where it was used to mark the Black Rocks.

After the site had been selected by P S Hay, a civil engineer acting on behalf of the marine engineer, and Captain Allman, the nautical adviser, the Public Works department was instructed to commence construction of the tower, dwellings etc.[162]

The contract for the cast iron tower was let to S Luke and Co Ltd of Wellington. Their tender price was £709 0 0.[163]

The tower and buildings were completed in July 1897 and the artificer completed installation of the lantern and apparatus in October.[164] It was first exhibited on 27 October 1897, and the Notice to Mariners described it as a

Figure 10.7: Cape Palliser lighthouse (ATL1/2-092140-F) © Randle Collection, Alexander Turnbull Library, Wellington, New Zealand

> Tower constructed cast-iron, concrete base. Total height bottom of base to top lantern 58 ft. Situated 640 ft dist from high water mark at an elevation of 210 ft a.s.l. painted white, light will be 258 ft asl at high water, dioptric revolving white light of second order, flashing twice every half minute, with interval of three secs between flashes. vis to seaward over an arc on horizon 214 deg, vis 25 statute miles, warning light shows over Black Rocks, not approach within four or five miles of light until the dangers are passed.[165]

The 1912 annual report indicated that incandescent lamps had been installed in Cape Palliser lighthouse.[166]

In 1951 the Ministry of Works recommended that the light be electrified by the installation of two diesel generator sets at a cost of £3,250. It was anticipated that this would lead to a saving in the wages of one keeper, and of maintenance cost.[167] Although the minister of marine approved the expenditure within a week,[168] electrification was not completed until June 1954.[169] The lighthouse was connected to the mains supply in 1966.[170]

Kahurangi Point

The Kahurangi Shoal (also known as Stewart's Breakers) lies off the north-west coast of the South Island between Cape Farewell and Westport.

In 1889, 23 masters of ships engaged in the coasting trade petitioned the Marine Department for a lighthouse at 'Kiourangi Point' to mark the shoal. Wilson replied that

> there are no funds available for this purpose, and that moreover there are many places in the colony where it would be more desirable in the interests of navigation to erect a light.[171]

Despite this, the west coast shipping community began a campaign to have a lighthouse placed at either Kahurangi Point or Rocks Point, to the south. In 1893 the member for Buller raised the question of the lighthouse in the House and was apparently promised one.[172] Beginning in 1895, R McKenzie, the member for Motueka, annually asked (until 1900) the minister of marine if he would

> make provision on this year's estimates for the erection of a lighthouse on Kiourangi Point or Rocks Point, on the west coast of the Middle Island?[173,174,175,176,177]

The Westport Harbour Board passed a resolution in 1894

> to call attention to the necessity of having lighthouse erected at Rocks Point on the Coast of this island for the guidance of mariners who are engaged in the very extensive coastal shipping trade.[178]

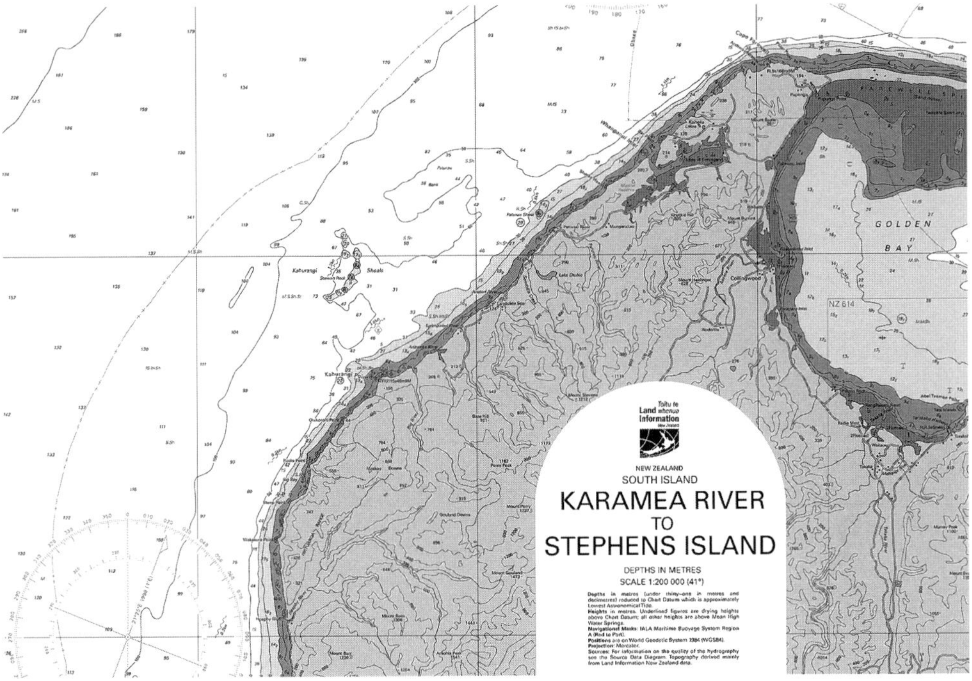

Figure 10.8: Location Kahurangi Point and Shoal (Maritime NZ chart nz61)
© Creative Commons 3.0 New Zealand licence

Seed was able to brush this aside with the comment that as the vote had been taken '[t]his letter has therefore come too late'.

Three years later the Board again wrote, to announce that it had unanimously passed a resolution that the

> Department's consideration should be given to the necessity of erecting a lighthouse at Kahurangi Point. The trade at this Port has increased to a very large extent, during the past year or two and the urgency of the matter in question, affecting as it does, the shipping industry connected with Westport, warrants the Board in bringing it before you in the hope that you will have the question duely attended to.

This campaign gradually bore fruit.

The 1893 annual report acknowledged that Mariners were urging the erection of a lighthouse at 'Kiorangi Point'on the west coast of the South Island to aid the coal trade from Westport.[179]

In the 1896 annual report, the inspector of lighthouses again urged the establishment of a light at the point, as the 'Kiourangi Shoal' was

> a source of danger to all classes of shipping, but more particularly to larger vessels with heavy draft of water making for the coast of New Zealand.[180]

Eventually, in 1900 the Shipmasters' Association was consulted as to the position of a new lighthouse between Farewell Spit and Westport.[181] The association recommended Kiourangi Point, and McKenzie asked his last question in the House, i.e.:

> Mr R McKenzie to ask of Minister of Marine 'If he will make provision on this year's estimates for the erection of a lighthouse, as recommended by the Shipmasters' Association on Kurangi or Rocks Point.?'[182]

When consulted, the Shipmasters' Association recommended siting the new lighthouse at Kiourangi Point.[183] Initially the Marine Department had considered using the redundant lighthouse from Matiu/Somes Island on the grounds that coastal shipping would use the channel between the shoal and the coast, the nautical adviser going so far as to say

> I do not think that any vessels bound for the Straits from the Westward would make for this light if it were erected on Kiourangi Point.[184]

When consulted the Shipmasters' Association rejected the Matiu/Somes option as too small and resolved:

> That the Light to be placed on 'Kiourangi Point' should not be of less power than a second order.[185]

Hales, the marine engineer, and Blackburne, the nautical adviser, visited the site in October 1900 and recommended a site on a small rounded knob near the limestone cliff at the southern side of the point.[186] They proposed

> a cylindrical cast iron tower 43 feet high on a concrete base, the total height of 110 feet from high water sea level to the focal plane of the light, which would be visible from the bridge of a vessel at sea for a distance of 16 miles (nautical) assuming the eye of the observer to be at a height of 15 feet above the surface of the water.

For the light itself they recommended a

> fixed 2nd dioptric fitted with reflecting dioptric mirror, lamp to have four wick burner; ruby coloured glass shade be provided for fixing the inside of the lantern frame and adjusted to shew a sector of 20 degrees of red light over the Stewart Breaker, which is a distance 4½ miles from the lighthouse site.
>
> The lantern to be second order constructed to shew the light the light over 212 degrees of the horizon to seaward; the remaining 148 degrees to be filled with dormant panes.

At the end of 1900 the Agent General was requested to order through D & C Stevenson a second order lantern and apparatus for Kahurangi lighthouse. (From 1901 onward the Marine Department spelled the name of the site 'Kahurangi'. Such changes are not unusual: Balfour's Tehamaki Strait is now Tamaki Strait, reflecting a better understanding of and appreciation for the Māori language.)

The specification closely followed the suggestions of Hale and Blackburne, and began with expected reference to the Northern Lighthouse Board design for the lantern, i.e.:

> The lantern is to be of the Northern Lighthouse pattern of the second order constructed to show the light over 212 degrees seaward: dormant panes for the remaining 148 degrees to be supplied. The lantern is to be designed for attachment to a cylindrical cast iron tower, of which a plan is enclosed.
>
> The lamp is to have a four wick burner. A ruby coloured glass shade is to be supplied for fixing to the inside of the lantern frame and adjusted to show a sector of 20 degrees of red light over the Stewart Breaker, distant 4½ miles from the lighthouse site.
>
> The apparatus is to be for a second order fixed white light, and a reflecting dioptric mirror is to be fitted to the back section of the apparatus.
>
> The lighthouse will be constructed on land about 65 feet above the level of the sea. The tower will be about 45 feet in height, making the lantern sole plate 110 feet above sea level.
>
> A full supply of the ordinary tools and stores is to be supplied, including a lighthouse clock, but electric bells are not required as telephones which can be procured in the Colony will be used.
>
> The lantern and apparatus are to be shipped to Wellington, addressed to the Secretary, Marine Department.
>
> I forward under separate cover, a chart showing the locality.[187]

Two days later the Secretary Marine Department wrote to D & C Stevenson to inform them of the impending order and to indicate that:

> The red shade should not be sent out fixed to the frame as that will be done here. You will observe from the plan which will be sent to you by the Agent General that the lightroom is seven feet high, but the pillar should be made two inches lower than that, and if necessary it can be raised a little off the floor when being fixed in position in the lighthouse.

And also to ensure that

> Messers Chance Bros & Co Ltd of Birmingham should have an opportunity of tendering for the light.

The requisition for a second order light was acknowledged by D & C Stevenson on 23 January 1901.[188] However, they took the opportunity to suggest improvements:

> We think it our duty, however to point out that in our opinion it is, unless there is some reason we are not cognizant of, undesirable to place a fixed light at such an important place as Kahurangi Point, both because it of weak power, & being fixed is easily mistaken for vessels & steamers lights. You will understand this when we say that the power of the proposed light will only be about 4000 candles, while if it were made a flashing light and apparatus of the same size or order would have a power of 48,000 candles & we never now use a fixed light if it can possible by avoided, & indeed are gradually converting our old fixed lights to flashing

They also suggested that the arc over Stewart's Breakers could be shown ('shewn') equally well by a small light in the tower. The letter finished with suggested wording for a telegram to New Zealand:

> Stevensons deprecate fixed lights as out of date for important station like Kahurangi, suggest group flashing light shewing 3 flashes in quick succession every ½ minute power would be 4,800 in place of only 4,000, & a separated small fixed red light shown from window in lowest flat of tower to cover Stewarts breaker, probable additional cost £900. Answer flashing or fixed.[189]

A rather garbled telegram was received in Wellington. When consulted Hales agreed that a flashing light was better but fixed would be adequate, as

> there is little prospect of any settlement on or works being established on this part of the coast, the lights at which might be mistaken for that in the lighthouse.[190]

On 4 February 1901 a telegram was sent to the Agent General confirming the original order.[191]

D & C Stevenson had already been working on the design but found that they needed further instructions. On 1 February they wrote:

> Kindly say if a cleaning path grating should be supplied for resting on brackets shown on accompanying sketch (tracing attached in file). We

[Opposite] Figure 10.9a: Lantern Kahurangi Point lighthouse © Archives New Zealand, the Department of Internal Affairs, Te Tari Taiwhenua

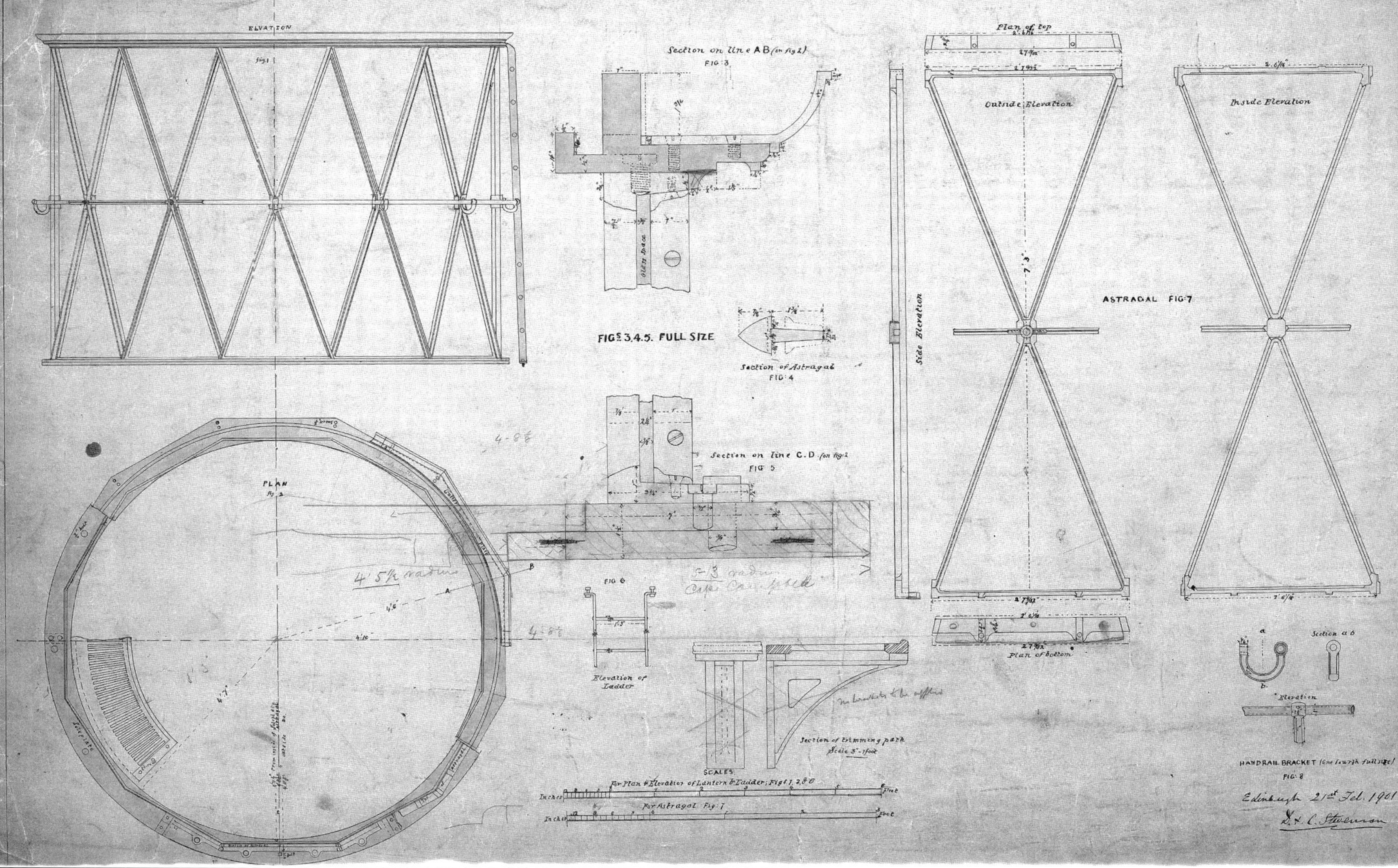
ELEVATION
Fig 1
Section on line AB (on fig 2)
FIG 3
FIGS 3.4.5. FULL SIZE
Section of Astragal
FIG 4
Section on line C.D (on fig 2)
FIG 5
PLAN
Fig 2
4' 5½ radius
FIG 6
Elevation of Ladder
Section of trimming path
Scale 3" - 1 foot
SCALES
For Plan & Elevation of Lantern & Ladder; Figs 1, 2 & 6
For Astragal Fig 7
Side Elevation
Plan of top
Outside Elevation
Inside Elevation
ASTRAGAL FIG 7
Plan of bottom
Section a b
Elevation
HANDRAIL BRACKET (one fourth full size)
FIG 8
Edinburgh 21st Feb. 1901
D. & C. Stevenson

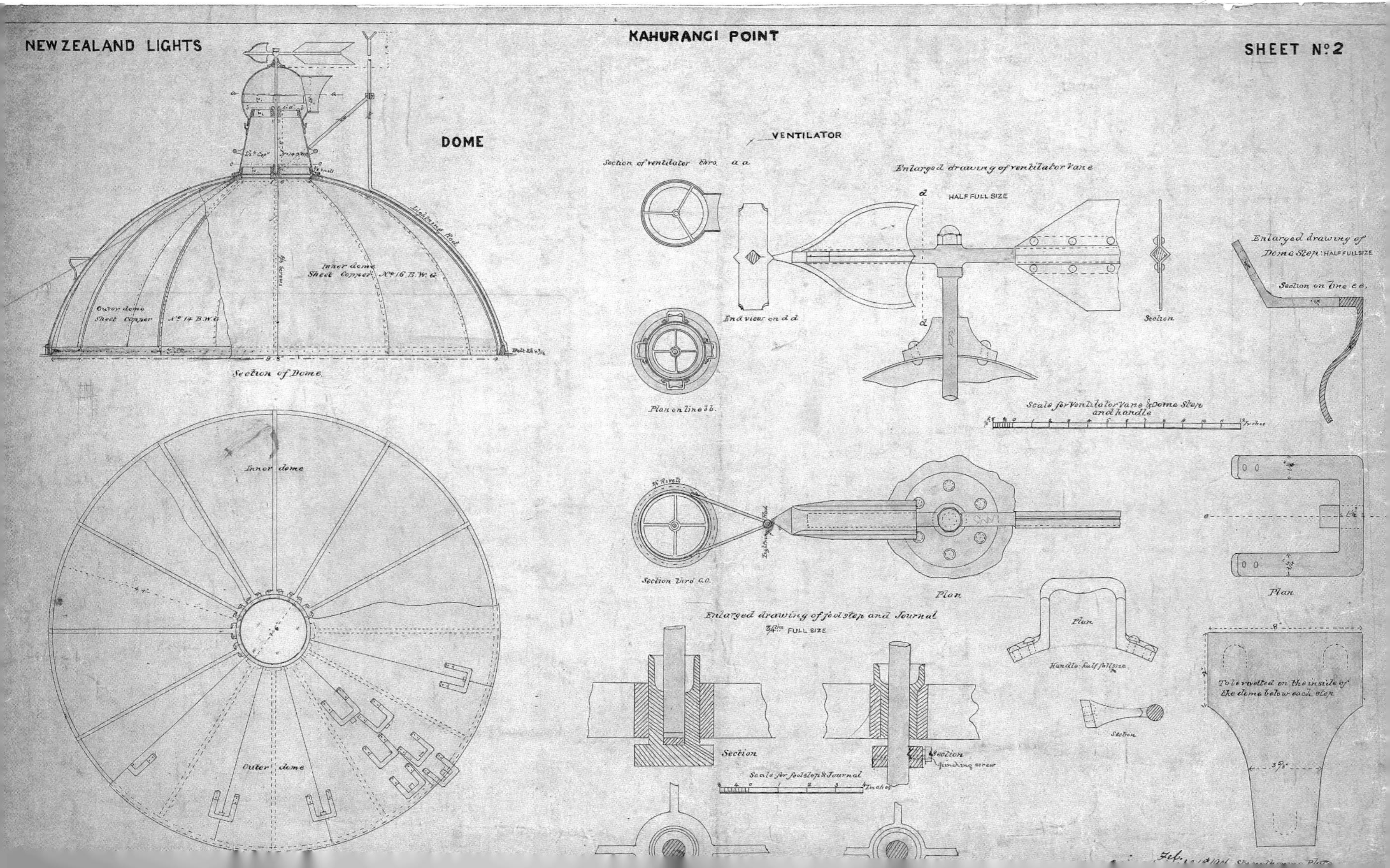

NEW ZEALAND LIGHTS
KAHURANGI POINT
SHEET No 2
DOME
Inner dome
Sheet Copper No 16 B.W.G.
Outer dome
Sheet Copper No 14 B.W.G.
Lightning Rod
Section of Dome
Inner dome
Outer dome
VENTILATOR
Section of ventilator thro a a
Enlarged drawing of ventilator vane
HALF FULL SIZE
End view on d d
Section
Plan on line b b.
Scale for Ventilator Vane & Dome Step and handle
Enlarged drawing of Dome Step: HALF FULL SIZE
Section on line e e.
Section thro' c.c.
Plan
Plan
Enlarged drawing of foot step and Journal
3/4ths FULL SIZE
Section
Section
Binding screw
Scale for foot step & Journal
Inches
Plan
Handle: half full size
Section
To be rivetted on the inside of the dome below each step

presume you are fixing brackets on the cast iron tower plates to carry the grating. We usually send the gratings but in this case we are in doubt whether you wish them sent or not. Will you also say what length of lightning conductor will be required from the foot of the tower.[192]

In his reply, the secretary confirmed that the grating would be needed, brackets and rings for supporting it would be provided with the tower and the length of lightning conductor required from the foot of the tower was 100 feet.[193]

Drawings for the lantern were signed on 21 February. Sheet No 1 of these was recycled for the new Cape Campbell lantern by simply being retitled.

The tender for lantern, table, burners etc was awarded to Milne & Son (£688.10/-), that for the apparatus to Henri Lepaute (Société des Establissements Henry-Lepaute) for £400; and Chance Brothers provided stores, lamps, glasses for £67.07.2. A Barrie & Son supplied a Sprin timepiece for £10.3.6.[194]

The lantern was inspected and packed in mid-October[195] and shipped to New Zealand on the *Aotea*.[196] The stores etc supplied by Chance Brothers probably went at this time, as they were in general stock items. The clock was added to the Chance Brothers shipment.[197]

Manufacture of the apparatus was delayed. An accident in their works in August meant that part of the glass for the prisms had to be recast,[198] and a breakage two months later delayed finishing.[199] The optical apparatus was finally shipped to New Zealand on *Gothic* in January 1901.[200]

By May 1903 the tower and ancillary buildings were approaching completion and the lantern and apparatus had arrived.[201] The tower was cast iron, manufactured by Judd Engineering at Thames (Figure 10.10). Judd had earlier supplied the East Cape lighthouse tower and later manufactured the tower for the Cape Brett lighthouse.

The light was exhibited on 20 November 1903[202] and described thus:

> The light, which is a dioptric fixed white light of the second order, is about 110 ft above the level of the sea at high water, and is visible to seaward over an arc of 212 degrees of the horizon for a distance of about 16½ miles … with a red sector of 20 degrees, so adjusted as to show over the Stewart Breaker, distant about 4½ miles N.,28° W (magnetic) from the lighthouse site. The tower, which is cast iron, and painted white is 59 ft in height from the bottom of the base to the top of the lantern.

By July 1912, incandescent burners had been obtained for Kahurangi Point,[203] but they were not installed until December 1915, being first lit on 27 December.[204]

The 1924 annual report indicated that plans had been drawn up and estimates made to convert the light to an automatic acetone-acetylene light.[205] The procurement of this equipment showed the imperial attitude which in New Zealand is now denominated

[Opposite] Figure 10.9b: Dome Kahurangi Point lighthouse © Archives New Zealand, the Department of Internal Affairs, Te Tari Taiwhenua

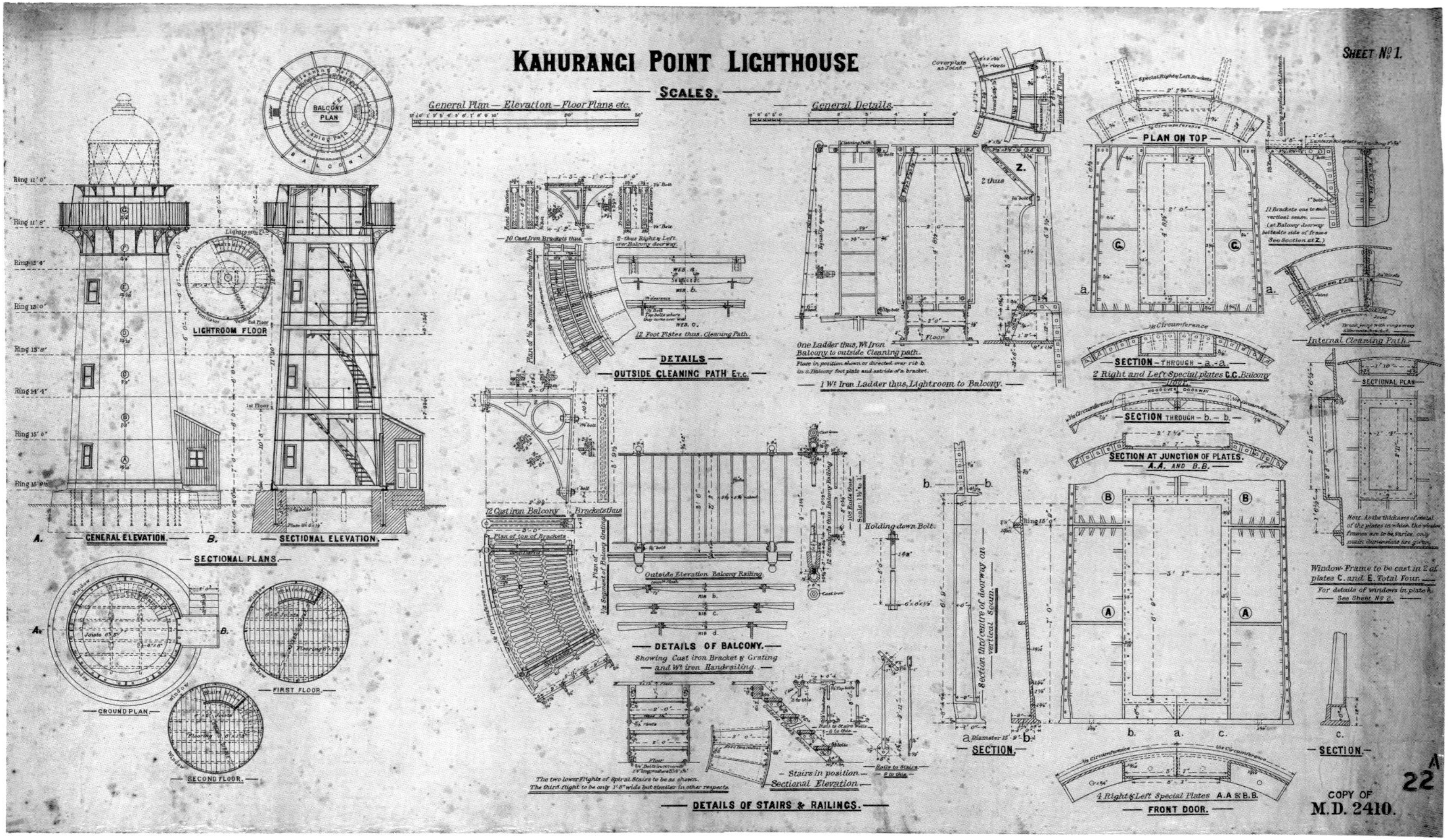

Figure 10.10: Design for Kahurangi steel tower © Archives New Zealand, the Department of Internal Affairs, Te Tari Taiwhenua

'Colonial Cringe'. On 15 November 1923 the Marine Engineer wrote to Acetone Welding Company of Napier, agent for the Swedish company Aga, who manufactured the Dalen burner, and to Cory, Wright & Salmon, agents for Chance Brothers, requesting a detailed price for an automatic fl asher with sun valve.[206] The letter indicated the desirability of 'British Manufacture'. This caveat prompted two reactions. All communication with Cory Wright and Salmon and their principals emphasise that the burners were 'ENTIRELY BRITISH' [capitals theirs]. Acetone Welding saw this as blatant preference for Chance Brothers and complained. While the Secretary for the Marine Department preferred the Aga equipment, a memo to the engineer in chief PWD (who was also the marine engineer) pointed out that:

> There is no doubt that the Aga Company have always given us the very best of service, but at the same time I cannot see that the Department would have a leg to stand on if the issue was raised that we were refusing to utilise a British article and accepting a foreign made article without very thorough and complete tests as to the efficiency of it.[207]

(This may have been a question of preferential tariffs.)

The matter was resolved the following year when the Aga Company established an English branch and the Acetone Company were able to advise that they could comply with the requirement that 'all material is to be of British manufacture'.[208] If the lightkeepers, struggling with their Chance incandescent burners, had been approached, the reply would probably have been a chorus of 'Buy Swedish!' if it had not meant redundancies.

The actual conversion was finished in the latter half of 1926 and the new light was first exhibited on 3 September 1926. The light was changed from fixed to group flashing: two flashes every 15 seconds made up of flash 0.75 sec, eclipse 4.5 sec, flash 0.75 sec, eclipse 9sec.[209] Although this was nominally an automatic light, one keeper and his family were retained due to the lighthouse's remoteness.

The new system was well bedded in when the lighthouse was extensively damaged and the light extinguished by the Murchison earthquake of 17 June 1929.[210] (The Murchison earthquake was a 7.8 Richter magnitude earthquake centred at to the east of but close to Kahurangi Point.)

The shaking destroyed the apparatus and the steadying gear, but most of the damage was due to a landslide. This demolished the keeper's house, oil store, workshop, and dry store, covering the area in debris. Fortunately the keeper, and his wife and son were not in the house at the time and escaped with their lives although losing everything else. The demolished house struck the base of the tower, stoving in the porch and damaging the plates at the base of the tower where the porch was attached.[211] The tower base remained in its original position; it was later found to have developed a slight list of 2.9" to the west and 1.25" to the north. This was dismissed as 'a mere trifle'.[212]

Examination of the site indicated that the landslide was partly of human origin. When the station was established, the hillside behind had been stripped of bush. The roots left

Figure 10.11: Kahurangi lighthouse © Archives New Zealand, the Department of Internal Affairs, Te Tari Taiwhenua

behind had rotted leaving an unstable slope. The earthquake was sufficient to send the whole hill moving.[213]

A temporary light was installed using the lens from the Hokitika lighthouse which had been decommissioned in 1925 and a 35 mm flasher which could be connected to the existing cylinder, which had survived the earthquake.[214] This temporary light was first planned to come into operation on 31 July with the same height of light and red sector and to have the flashing pattern of 1 flash every 8 secs, thus: flash ¾ sec, eclipse 7¼ second, although the arc was expected to be less than previously.[215] The temporary light was exhibited to schedule but with a pattern of flash 1 sec and eclipse 7 seconds, and an arc of only180° (048° through east to 228°).[216] It was later found that the arc of visibility of the red sector was 25°, between 151° and 176°. At a distance of eight miles south-westward the light was obscured at 045°.[217] This change in arc was ascribed to the list on the tower and it was anticipated that the new equipment could be adjusted to allow for this.[218,219]

Acetone Welding Company and Cory, Wright & Salmon were asked for quotes from their principals for a one catadioptric lens 800 mm in diameter with a light sector of 210° with remaining 150° being filled up with a reflector.[220] The tender was given to Chance Brothers for a catadioptric lens with a chromium plated reflector, for £416.0.0.[221] It arrived in New Zealand in mid-May 1930 on the *Huon*.[222]

Six firms, including Judd who had cast the tower originally, were asked for quotes for replacement plates. The tender was given to the local firm, Despatch Foundry of Greymouth, who quoted £88.[223] The plates were completed by early April 1930, but could not be shipped to Wellington immediately as the Greymouth bar was unworkable by large vessels at the time.[224]

Repairs began on 2 February 1931 and the new light was first exhibited on 7 March 1931. The Notice to Mariners[225] described the light as:

> group flashing every 15 secs thus: flash 0.75 sec; eclipse 4.5 sec; flash 0.75 sec; eclipse 9 sec. Light now shows white from 038 deg through east to 151 deg, red thence to 176 deg; white thence to 248 deg, obscured elsewhere.

Less than a month after the exhibition of the new light, another earthquake (4 April 1931) caused further damage to the lower prisms of the new lens.[226] In August the damaged portion was sent to England by *Matakana* to be repaired by Chance Brothers.[227] It was returned and reinstated the following year.[228]

Despite the trauma of the earthquake the lightkeeper and his family were prepared to return to Kahurangi Point, and a new house was erected for them in 1934.[229] However, the erosion that was so dramatic in 1929 continued; in 1936 the house had to be shifted[230] and a further shift was contemplated in 1940.[231]

In 1943 the engineer in chief suggested using the equipment that had been obtained to electrify Godley Head at Kahurangi Point.[232] The 1945 annual report indicated that electrification was in progress at Kahurangi.[233]

Endnotes

1 AJHR 1873 H-19 Report on the Lighthouses on the Coasts of New Zealand
2 AJHR 1884 H-6 Marine Department Annual Report
3 AJHR 1885, H-13 Marine Department Annual Report
4 AJHR 1887 H-4 Marine Department Annual Report for 1886-87d
5 Acc 10706 vol 59 p 14 letter to LHB Wilson 26 May 1890
6 Acc 10706 vol 59 p 62 letter to Wilson 17 June 1890
7 Acc 10706 vol 56 p 558 letter to William Seed 11 March 1886
8 Acc 10706 vol 56 p 576 letter to Milne 23 March 1886
9 Acc 10706 vol 56 p 577 letter to Dove 23 March 1886
10 Acc10706 vol 56 p 576 letter to Barbier & Fenestre 23 March 1886
11 Acc 10706 vol 56 p 578 letter to Chance Brothers 23 March 1886
12 Acc 10706 vol 56 p 578 letter to Smiths dated 23 1886
13 Acc 10706 vol 56 p 579 letter to Messers Bryden 24 March 1886
14 Acc 10706 vol 56 p 591 letter to Dove Greenside Lane. 2 April 1886
15 Acc 10706 vol 56 p 592 letter to Milne Milton House 2 April 1886
16 Acc 10706 vol 56 p 595 letter to Smiths & Co 2 April 886
17 Acc 10706 vol 56 p 596 letter to Bryden & Sons, George Street 2 April 1886
18 Acc 10706 vol 56 p 623 letter to Milne 15 April 1886
19 Acc 10706 vol 56 p 878 letter to Agent General 16 Oct 1886
20 Acc 10706 vol 57 p 81 letter to EA Smith 26 February 1887
21 Acc 10706 Vol 88a. letter Wilson to D A Stevenson 15 July 1891
22 AJHR 1888 H-4 Annual Report Marine Department report . Annual Report on Lighthouse Works etc by marine engineer 31 March 1888
23 AJHR 1888, H-4 Annual Report Marine Department Annual Report for 1887-88
24 AJHR 1887 H-4 Marine Department Annual Report for 1886-87. Annual Report on Lighthouse Works etc by marine engineer 31 March 1886
25 AJHR 1889 H-31 Marine Department Annual Report for 1888–89 also Report of Acting Marine Engineer
26 *NZG Gazette* 1889 No.52 Notice to Mariners, No 23 of 1889 p956 2 Sept 1889

27 AADX W3148 889 Box 42 63/219 part Letter Thomas Cheeseman to C Y O'Connor Public Works Department 10 October 1888

28 AADX W3148 889 Box 42 63/219 part 1 Letter Blow, Under-Secretary to registrar Native Land Court Auckland PW1900/4859 128/856 29 Dec 1900

29 in AADX W3148 889 Box 42 63/219 part 1 Letter Tizard to Hall-Jones 1900/4859 26 November 1900

30 Acc 10706 vol 63 p 772 letter to Smiths 18 March 1894

31 Acc 10706 vol 63 p 785 letter to Messer Silber & Co 57 Woad Street, Cheapside 22 March 1894

32 NZG Archives ADOE 16612 M1/ 733 8/25/2 letter W A Fraser to secretary Marine Department 24 Feb 1924

33 NZG ArchivesADOE 16612 M1/ 7338/25/2 letter W A Fraser (at Cape Brett) to secretary Marine Department 6 Jan 1938

34 AJHR 1938, H-15 Marine Department Annual Report for year ended 31st March 1938r

35 NZG Archives ADOE 16612 M1/733 8/25/2 letter E Bowley, principal keeper, 27 Feb 1939 (copy extract originally M 8/6/33)

36 AJHR 1941, H-15 Marine Department Annual Report for year ended 31st March 1941

37 Votes and Proceedings of the House of Representatives Session 1 1854

38 AJHR 1879 H-10 of the Marine Department Annual Report, report of The Marine Engineer to the Secretary, Marine Department

39 AJHR 1881, H27 Marine Department Annual Report

40 AJHR 1883, H-15 Marine Department Annual Report

41 AJHR 1884 H-6 Marine Department Annual Report

42 AJHR 1888 H-19 Marine Department Annual report for 1887-88

43 AJHR 1890, H-18 Marine DepartmentAnnual Report for 1889–90

44 AJHR 1891 H-30. Marine Department Annual Report for 1890–91,

45 Acc 10706 Vol 88a letter Wilson to D A Stevenson 15 July 1871

46 AJHR 1892 H-29 Marine Department Annual Report for 1891-92

47 AJHR 1892, H-29 Annual Report Marine Department Annual Report for 1891-92

48 Acc10706 vol 88a letter Wilson to Thomas Stevenson 4 Jan 1878

49 Acc 10707 vol 61 letter Stevenson to Wilson 7 January 1893

50 Acc 10706 p 302 letter Stevenson to Wilson 10 May 1892

51 Acc 10706 vol 61 p 104 letter to Agent General 15 January 1892

52 Acc 1006 vol 61 p 140 letter to Chance & Co 15 February 1892

53 Acc 10706 vol 61 p 149 letter to Barbier & Cie 19 February 1892

54 Acc 10706 vol 61 p 150 letter to Chance & Co 19 February 1892

55 Acc 10706 vol 61 p 150 letter to Sautter, Haile et Cie 19 February 1892

56 Acc 10706 vol 61 p 163 letter to Milne 25 February 1892

57 Acc 10706 vol 61 p 160 letter to Dove 25 February 1892

58 Acc 10706 vol 61 p 215 letter to Dove 11 March 1892

59 Acc 10706 vol 61 p 216 letter to Milne 11 March 1892

60 Acc 10706 vol 61 p 157 letter to Barrie George Street, Glasgow, 24 February 1892

61 Acc 10706 vol 61 p 354 letter to Agent General 28 April 1892

62 Acc 10706 vol 61 p 211 letter to Mountain & Co, Glasgow, 11 March 1892

63 Acc 10706 vol 61 p 156 letter to W M B Mountain, 7 Bothwell St, Glasgow, 24 February 1892

64 Acc 10706 vol 59 p 519 letter to Agent General 19 February 1891

65 Acc 10706 vol 59 p 597 letter to Agent General 1 April 1891

66 Acc 10706 vol59 p 523 letter Stevenson to Agent General 10 April 1891

67 Acc 10706 vol 61 p 212 letter to Agent General 11 March 1892

68 Acc 10706 vol 61 p 694 letter to Dove. 10 September 1892

69 Acc 10706 vol 62 letter to Dove 13 October 1892

70 Acc 10706 vol 61 p 752 letter to Barbier & Cie 12 October 1892

71 Acc 10706 vol 61 p 760 letter to Milne 13 October 1892

72 Acc 10706 vol 61 p 891 letter to Agent General 2 December 1892

73 Acc 10706 vol 61 p 898 letter to Milne & Son 7 December 1892

74 Acc 10706 vol 61 p 891 letter to Agent General 2 December 1892

75 AJHR 1893, H-31 Annual Report Marine Department 1892–1893

76 *NZG Gazette* 1893,No.10 , Tenders Received for Ironwork Contract, Stephen's Island Lighthouse p 183, 7 February 1893
77 AJHR 1892 H-29 of Marine Department Annual Report 1891–92
78 *NZG Gazette*Notice to Mariners No 49 of 1893, 15 December 1893, Marine Department Wellington,
79 NZG Archives ADOE 16612 M1/749 8/40/11 pt1 letter Hall-Jones, acting pm to high commissioner, London, 26 January 1907
80 AJHR 1908 H-15 t Marine Department Annual Report for1907–08
81 NZG Archives ADOE 16612 M1/759 8/51/2 (orig 8/9/28) letter Schofield, principal keeper, to secretary Marine Department dated 6 October 1934
82 NZG Archives AADX W3148 889 / 29 63/74 pa letter engineer in chief to district engineer Public Works Department Wellington labelled PW63/74 dated 10 December 1936
83 AJHR 1937, H-15 Marine Department Annual Report for year ended 31st March 1937
84 NZG Archives ADOE 16612 M1/759 8/51/2 (orig 8/9/28) letter Cory-Wright & Salmon Ltd to engineer in chief, PWD 24 August 1937
85 NZG Archives, ADOE 16612M1/759 8/51/2 (orig 8/9/28) copy letter J Wood, engineer in chief Public Works Department, to Cory-Wright & Salmon Ltd dated 8 Oct 1937
86 AJHR 1938, H-15 Marine Department Annual Report for year ended 31st March 1938
87 NZG Archives ADOE 16612 M1/759 8/51/2 (orig 8/9/28) Telegram L Whiteman, principal keeper, to secretary Marine Department 30 November 1938
88 AJHR 1873 H-19 t Report on the Lighthouses on the Coasts of New Zealand
89 Acc 10706 vol 63 p 933 letter to sec MD 1895 12 June 1895
90 Acc 10706 vol 64 p 146 letter to secretary Marine Department 29 August 1895
91 Acc 10706 vol 68 p 842 letter to secretary Marine Department 3 November 1900
92 Acc 10706 vol 64 p letter to Agent General 18 July 1895
93 Acc 10706 vol 64 p 124 letter to Dove 22 August 1895
94 Acc 10706 vol 64 p 124 letter to Milne 22 August 1895
95 Acc 10706 vol 64 p 125 letter to Messers Steven & Struthers Elliot Street Glasgow 22 August 1895
96 Acc 10706 vol 64 p 109 letter to Chance Brothers 19 August 1895
97 Acc 10706 vol 64 p 108 letter to Barbier & Bernard 19 August 1895
98 Acc 10706 vol 64 p 107 letter to Lapaute 19 August 1895
99 Acc 10706 vol 64 p 168 letter to Agent General 3 September 1895
100 Acc 10706 vol 64 p 196 letter to Agent General 11 September 1895
101 Acc 10706 vol 64 p 921 letter to Milne 24 May 1895
102 Acc 10706 vol 65 p 84 letter to Agent General 22 July 1896
103 Acc 10706 vol 64 p 252 letter to Barbier & Bernard 22 October 1895
104 Acc 10706 vol 64 p 529 letter to Barrie 9 January 1896
105 Acc 10706 vol 64 p 520 letter to Bryden 9 January 1896
106 Acc 10706 vol 64 p 521 letter to Fleming & Co 9 January 1896
107 Acc 10706 vol 64 p 456 letter to Agent General 9 December 1895
108 Acc 10706 vol 64 p 526 letter to Agent General 10 January 1896
109 Acc 10706 vol 64 p 578 letter to Agent General 29 January 1896
110 Acc 10706 vol 64 p 676 letter to Agent General 17 February 1896
111 Acc 10706 vol 64 p 718 letter to Agent General 9 March 1896
112 Acc 10706 vol 64 p 973 letter to Agent General 11 June 1896
113 AJHR 1896, H-15t Marine Department Annual Report for 1895–96 . Report George Allman, Inspector of Lighthouses and Nautical Adviser
114 AJHR 1898 H-15 Marine Department Annual Report for 1897–98
115 AJHR 1898 H-15 Marine Department Annual report for 1897–98 Report of Nautical Adviser and Inspector of Lighthouses 30 April 1898
116 New Zealand Herald Saturday 9 July 1898 p 5
117 *NZG Gazette* 1898 No 61 Tenders p 1309 9 Aug 1898
118 AJHR 1899 H-15 Marine Department Annual Report fot 1898–99
119 *NZG Gazette* 1900 no 64 Notice to Mariners No 36 of 1900 p 1386, 11 July 1900
120 AJHR 1911 H-15 t Marine Department Report for 1910–11

121 AJHR 1918 H-15 Marine Department Report for 1917–1918

122 AJHR 1921/22 H-15 Marine & Inspection of Machinery DepartmentReport for year ended 31st March 1921 1920 . Report by Marine Engineer F W Furkert 23 Aug 1921

123 NZG Archives AADX W3148 889/ 28 63/60 note to undersecretary Public Works Department 4 November 1921

124 NZG Archives AADX W3148 889/ 28 63/60 memo Poauau and 11 others to Coates, minister public works, 10 December 1921

125 *NZ Gazette* 1923 No 16 Notice of Intention to Take Land p531 12 February 1923

126 NZG Archives AADX W3148 889/ 28 63/60 part 1 handwritten note on copy memo/telegram to W Graham PW 6/62 11 December 1924

127 AJHR 1922 H-15 Marine and Inspection of MachineryDepartment Report for year ended 31st March 1922

128 AJHR 1923, H-15 Marine Department Annual Report 1922-23

129 NZG Archives AADX W3148 889 28 8/60/4 letter secretary Marine Department to minister dated 6 November 1923

130 Acc 10706 vol 87 memo to high commissioner from prime minister's office re Castle point prisms 6 April 1934

131 NZG Archives AADX W3148 889 / 28 63/60 memo secretary Marine Department to marine engineer 16 July 1924

132 NZG Archives AADX W3148 889 / 28 63/60 Application PW 63/60 23 March 1925

133 AJHR 1925 H-15 Marine Department Annual Report for 1924–25

134 NZG Archives AADX W3148 889 / 29 63/60 part 2 copy memo Chief Accountant Public Works Department to Resident Engineer Gisborne Pw 63/11/1 Authority No J.166 in

135 AJHR 1949 H-15 Marine Department Report for year ended 31st March 1949

136 AJHR 1866 E - No 3 Second Report of the Marine Board Department 1865–6 Enclosure No 1

137 AJHR 1873 H-19.Lighthouses on the Coasts of New Zealand

138 AJHR 1893, H-31 Marine Department Annual report for 1892–93

139 Acc 10706 vol 63 p 978 letter to Agent General 1 July 1895

140 Acc 10706 vol 63 p 991 letter to Agent General 3 July 1895

141 Acc 10706 vol 64 p 168 letter to Agent General 3 September 1895

142 Chance Archives BS6/12/1/1/10 p 45 *Second Order Double Flashing Lights*

143 Acc 10706 vol 64 p 219 letter to Chance 18 September 1896

144 Acc 10706 vol 64 p 290 letter to Chance 16 October 1895

145 Acc 10706 vol 64 p 314 letter to Chance 25 October 1895

146 Acc 10706 vol 64 p 327 letter to Chance 28 October 1895

147 Acc 10706 vol 64 p 530(?) letter to Chance 11 January 1896

148 Acc 10706 vol 64 p 532 letter to Chance 14 January 1896

149 Acc 10706 vol 64 p 538 letter to Chance 16 January 1896

150 Acc 10706 vol 64 p 540(?) letter to Chance d 20 January 1896

151 Acc 10706 vol 64 p 552 letter to Chance 22 January 1896

152 Acc 10706 vol 64 p 555 letter to Chance 23 January 1896

153 Acc 10706 vol 64 p 558 letter to Chance 24 January 1896

154 Acc 10706 vol 64 p 583 letter to Chance 31 January 1896

155 Acc 10706 vol 64 p 717 letter to Agent General 9 March 1896

156 Acc 10706 vol 64 p 837 letter to Agent General 17 April 1896

157 Acc 10706 vol 64 p 903 letter to Chance 18 May 1896

158 Acc 10706 vol 64 p 912 letter to Chance 22 May 1896

159 Acc 10706 vol 64 p 144 letter to secretary Marine Department 29 August 1895

160 Acc 10706 vol 64 p 138 letter to Barbier & Bernard. 26 August 1896

161 Acc 10706 vol 64 p 346 letter to Agent General 6 November 1876

162 AJHR 1896, H-15 Marine DepartmentAnnual Report for 1895–96 30 May 1896

163 *NZG Gazette* 1896 No 18 Tenders p 447 7 March

164 AJHR 1898, H-15 Marine DepartmentAnnual report for 1897–98

165 *NZG Gazette* 1897 No.87 Notice to Mariners No 17 of 1897,p1778 12 Oct 1897,

166 AJHR 1912, H-15 Marine Department Annual Report for 1911–12

167 NZG Archives ARDJ 18994 /TR1/ 579 45/1/7 part 1 memo secretary for Marine to Minister of Marine labelled M8/19/6 2 March 1951

168 NZG Archives ARDJ 18994/ TR1/579 45/1/7 part 1 memo secretary Marine Department to marine engineer, M.8/19/6 in 8 March 1952

169 AJHR 1954, H-15 t Marine Department Report for the year ended 31 March 1954.

170 NZG Archives ARDJ 18994/TR1/579 45/1/7 part 1 telegram nautical adviser to secretary Marine Department 1 December 1966

171 NZG Archives ADOE 16612 M1/744 8/34/5 part 1 letter Wilson to Todd & others originally M 401/61 4 October 1889

172 NZG Archives ADOE 16612 M1/744 8/34/5 Quoted in clipping either Hansard or a Wellington newspaper d 10 August 1899 Marine 1900/679

173 NZG Archives ADOE 16612 M1/744 8/34/5 order paper and memo to Minister M95/1550 dated 26 July 1895

174 NZG Archives ADOE 16612 M1/744 8/34/5 copy order paper House of Representatives for 10 July 1896 M96/1400

175 NZG Archives ADOE 16612 M1/744 8/34/5 copy order paper House of Representatives 2 November 1897 M7/1510

176 NZG Archives ADOE 16612 M1/744 8/34/5 copy order paper House of Representatives 5 July 1898 M98/1392

177 NZG Archives ADOE 16612 M1/744 8/34/5 copy order paper House of Representatives 9 August 1899 M1900/679 No 584/106

178 NZG Archives ADOE 16612 M1/744 8/34/5 part 1 memo secretary of Westport Harbour Board to Minister of Marine M99/N18 October 19 1894

179 AJHR 1893, H-31. Marine Department Annual report for 1892–93

180 AJHR 1896, H-15 Marine Department Annual report for 1895–96 report. The Inspector of Lighthouses to the Secretary, Marine Department 13 May 1896

181 AJHR 1900, H-15 Marine DepartmentAnnual report for 1899–1900

182 NZG Archives ADOE 16612 M1/744 8/34/5 copy order paper House of Representatives M1900/1283 26 June 1900

183 NZG Archives ADOE 16612 M1/744 8/34/5 letter Kennedy, secretary Shipmasters' Association, to secretary Marine Department M1900/679 2 April 1900

184 NZG Archives ADOE 16612 M1/744 8/34/5 memo Minister Hall-Jones to secretary of Marine (Glasgow) with handwritten additions by nautical adviser. Marine 1900/679 15 May 1900

185 NZG Archives ADOE 16612 M1/744 8/34/5 letter Kennedy, secretary Shipmasters' Association of New Zealand, to secretary Marine Department M1900/1283

186 NZG Archives ADOE 16612 M1/744 8/34/5 Report Hales and Blackburne to Minister for PW. M1900/2264 orig *N0128/144* October 26 1900

187 NZG Archives ADOE 16612 M1/744 8/34/5 letter Premier to Agent-General 1900/2264 no 63/109 5 December 1900

188 Acc 10706 vol 68 p 998 letter to secretary Marine Department 23 January 1901

189 NZG Archives ADOE 16612 M1/744 8/34/5 letter D & C Stevenson to Agent General M1900/2664 23 January 1900

190 NZG Archives ADOE 16612 M1/744 8/34/5 telegrams Marine 1900/2264 24 January 1901

191 NZG Archives ADOE 16612 M1/744 8/34/5 letter Premier's office to Agent General-M1900/2264 No 463/110 4 February 1901

192 NZG Archives ADOE 16612 M1/744 68/34/5 letter D & C Stevenson to secretary Marine Department M1900/616 1 February 1901

193 NZG Archives ADOE 16612 M1/744 8/34/5 letter secretary Marine Department to D & C Stevenson NO 347/111 28 March 1901

194 NZG Archives ADOE 16612 M1/744 8/34/5 memo Agent General to Premier 10 August 1901 along with tender for Requisition No R.B. 107/31 M1901/2285

195 NZG Archives ADOE 16612 M1/774 8/34/5 letter D & C Stevenson to secretary Marine Department M1901/2452 12 October 1901

196 NZG Archives ADOE 16612 M1/744 8/34/5 memo Agent General to Premier via Suez No 3459 M1902/90 29 November 1901

197 NZG Archives ADOE 16612 M1/744 8/34/5 invoice M1902/546 30 December 1901

198 NZG Archives ADOE 16612 M1/744 8/34/5 copy letter Stevenson to Agent General M1901/2452 22 August 1901

199 NZG Archives ADOE 16612 M1/774 8/34/5 letter D & C Stevenson to secretary Marine Department M1901/2452 12 October 1901

200 NZG Archives ADOE 16612 M1/744 8/34/5 memo Agent General to Premier via Frisc No 156, M1902/546 25 January 1901

201 AJHR 1903, H-15 Marine Department Report for the year 1902–03

202 *NZG Gazette*,1903 No 86, Notice to Mariners No 79 of 1903, pp 2386 , 5 November 1903

203 AJHR 1912, H-15 Annual Report Marine Department Report for 1911–12

204 NZG Archives ADOE 16612 M1/744 88/34/3 letter Fraser to secretary Marine Department Marine 8/34/3 5 January 1916

205 AJHR 1924, H-15 Marine Department Annual Report 1923–24

206 NZG Archives ABKK W4357 889 /462 63/116 letters marine engineer to Acetone Welding Napier (agents for Aga) and Cory Wright & Salmon (agents for Chance) original 8/34/5 15 November 1923

207 NZG Archives ABKK 889 W4357 /462 63/116 letter Sharp, staff engineer (later DE) to engineer in chief. PW63/116 12 March 1924

208 NZG Archives letter ABKK 889 W4357 / 462 63/116 Furkert, marine engineer, to secretary Marine Department, PW63/116 25 February 1925

209 *NZG Gazette* 1926 No 56, Notice to Mariners No46 of 1926, pp 2604 25 August 1926

210 *NZG Gazette* 1929, No.46 , Notice to Mariners No 30 of 1929, p 1737 ,19 June 1929,

211 NZG Archives ADOE 16612 M1/745 8/34/6 21 copy report Page, principal keeper, to secretary Marine Department, labelled M 8/34/7 21 June 1929

212 NZG Archives ADOE 16612 M1/744 M8/34/7 Report Fraser to secretary Marine Department labelled M8/34/7 dated 23 September 1929

213 NZG Archive ADOE 16612 M1/745 8/34/5 memo Sharp, inspecting engineer, to secretary Marine Department M8/34/5 21 February 1931

214 NZG Archives ABKK 889 W4357/ 462 63/116 letter Sharp to marine engineer PW 63/116 (also M1 745 8/34/7) 16 July 1929

215 *NZG Gazette* 1929, No 52,Notice to Mariners No 37 of 1929, p 1912 23 July 1929,

216 *NZG Gazette* 1929, No 58, Notice to Mariners No 42 of 1929, p2154, 19 Aug 1929

217 *NZG Gazette* 1929 No.63 Notice to Mariners No 47 of 1929, p2503, 17 Sept 1929,

218 NZG Archives ADOE 16612 M1/744 8/34/7 letter Fraser to secretary Marine Department labelled M8/34/7 1 October 1929

219 NZG Archives ADOE 16612 M1/744 8/34/7 letter engineer in chief to Cory-Wright & Salmon, labelled M8/34/7 22 November 1929

220 NZG ArchivesADOE 16612 M1/744 8/34/7 letter engineer in chief to Acetone, labelled M8/34/7 22 November 1929

221 NZG Archives ABKK W4357 889 / 462 63/119 memo Kahurangi Lighthouse- Replacement of Lens damaged by Earthquake from engineer in chief to secretary Marine Department. 63/116 20 December 1929

222 NZG Archives ADOE 16612 M1/744 8/34/7 letter Cory, Wright &Salmon to store manager Public Works Department. Labelled M8/34/7 22 May 1930

223 NZG Archives ADOE 16612 M1/ 744 8/34/7 copy official order to Despatch Foundry labelled M8/34/7 23 December 1929

224 NZG Archives ADOE 16612 M1/744 8/34/7 telegram Public Works Department to secretary Marine Department 7 April 1930

225 *NZG Gazette* 1931, No 21, Notice to Mariners No 10 of 1931, p648, 18 March 1931

226 NZG Archives ABKK W4357 889 / 492 63/116 copy memo principal keeper to secretary Marine Department M8/34/7 4 April 1931 (original in M1 744 8/34/7)

227 NZG Archives ADOE 16612 M1/744 8/34/7 letter sending Bill of Lading, secretary Marine Department to Cory, Wright & Salmon. labelled M8/34/7 25 August 1931

228 AJHR 1932/3 H-15 Marine Department Annual report for year ended 31st March 1932

229 AJHR 1934/35, H-15 Marine Department 1 Annual report for year ended 31st March1934

230 NZG Archives AATE W3403/ 37 8/86/4 letter district engineer Public Works Department Nelson to permanent head, Public Works Department Wellington 8/86 17 December 1936

231 NZG Archives AATE W3403 Box 37 8/86/4 letter district engineer Nelson to permanent head Public Works Department, Wellington 3 October 1940

232 NZG Archives ABKK W4357 889 Box 462 63/119 memo engineer in chief Public Works Department to district engineer Christchurch. Public Works Department 63/116

233 AJHR 1945 H-15 Marine Department Annual report for year ended 31st March 1945

11 Mercury Baths: Cape Brett and Castlepoint

The twentieth century saw four new major lighthouses: Cape Brett, Castlepoint and the 'replacement' lighthouses at Cape Reinga and Baring Head which have already been discussed.

Cape Brett and mercury baths

The next technical innovation to be introduced to New Zealand was to place the lighthouse apparatus in a bath of mercury rather than on rollers. Cape Brett lighthouse was the first light to have this technology.

A white light revolving every two minutes had been recommended for Cape Brett by Edwin and Johnson in their 1873 report,[1] but no further comment was made until 1896. In that year's annual report, the inspector of lighthouses recommended the construction of the Cape Brett lighthouse.

> Placed almost mid-way between Moko Hinau Light and Cape Maria van Diemen, it would be a guide to the Bay of Islands as well as being a good coastal light.[2]

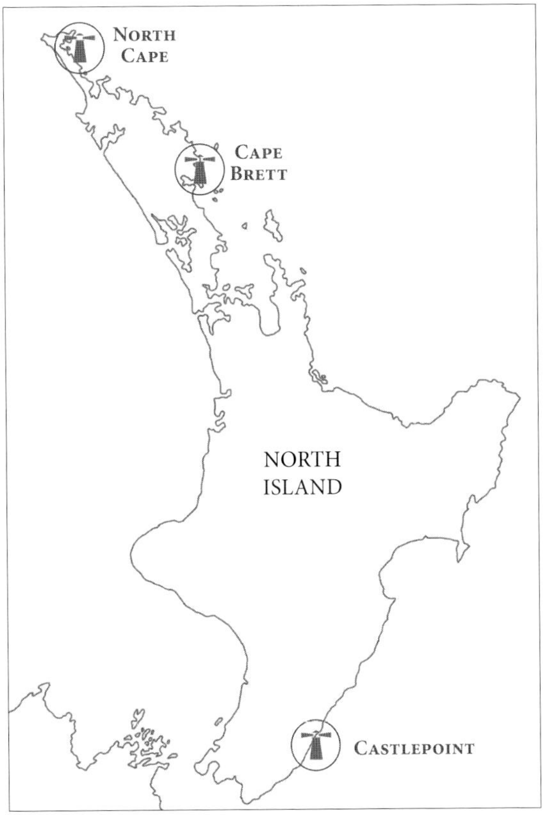

Figure 11.1: Location of Cape Brett and Castlepoint lighthouses © Author

Another ten years passed before further action was taken. The 1906 annual report recommended that financial provision be made for the construction of the Cape Brett light.[3] The necessary sum was voted in the parliamentary session following the 1906 report.

However, there continued to be delays. Shipmasters were again consulted, and in late 1907 a party consisting of J A Wilson, district engineer for the Public Works Department, Auckland, Captain Bollons, master of the government steamer, and G Allport, the secretary of the Marine Department, made site visits to Cape Brett and two other sites recommended by shipmasters. The party recommended a first order group flashing incandescent light at Cape Brett and the recommendation was subsequently approved.[4] This was ordered in early 1908.

The contract for the apparatus and lamps was awarded to Chance Brothers for a tender price of £10936/5/0. The order was entered in their records on 25 March 1908 and was for a First Order 920m/m apparatus.[5] Their description was:

> To be a two flash light 50° deg between the flashes two panels extending over 180 degrees. The other 180 degrees to be taken up by a dioptric mirror 790 m/m focal distance & having a vertical angle of 60 degrees. The mirror to consist of three frames of 60 deg one of which is to be made into a double door
>
> The prisms to be adjusted to a focus of six feet above the level of the sea.

A marginal note dated 2 April 1908 added 'Top gm steadying frame to be supplied per Mr Stobart'. Stobart had previously been sent to Edinburgh to discuss the Cape Palliser light with D & C Stevenson.

The lamp was to be

> 55 m/m Chance incandescent installation comprising 3 burners, air and oil containers, and accessories as per printed list. Vapourisers to be of the type recently approved by the N.L.B.

A 1948 memorandum from the Marine Department's plant overseer describes the apparatus as 'made by Chance Bros and Jas. Milne & Son'.[6] A later 1937 report, by a mechanical engineer, wrote of 'Clockwork equipment by James Milne of 'Edinborough'. The lantern was probably made by Dove. As no specifications have survived it is not possible to state when and by whom the decision was made to use a mercury bath instead of rollers to support the apparatus. Experience suggests that the Marine Department might have specified rollers and made the change on Stevenson's suggestion. By August 1909 the lantern, apparatus and lamps had arrived in New Zealand.[7]

On 8 January 1908 the contract for the cast iron tower was let to Judd of Thames, who had also cast the towers for Palliser and Kahurangi lighthouses.

The November 1909 *Gazette*[8] gave notice that the lighthouse was in the course of construction, and the light was first exhibited on 21 February 1910.

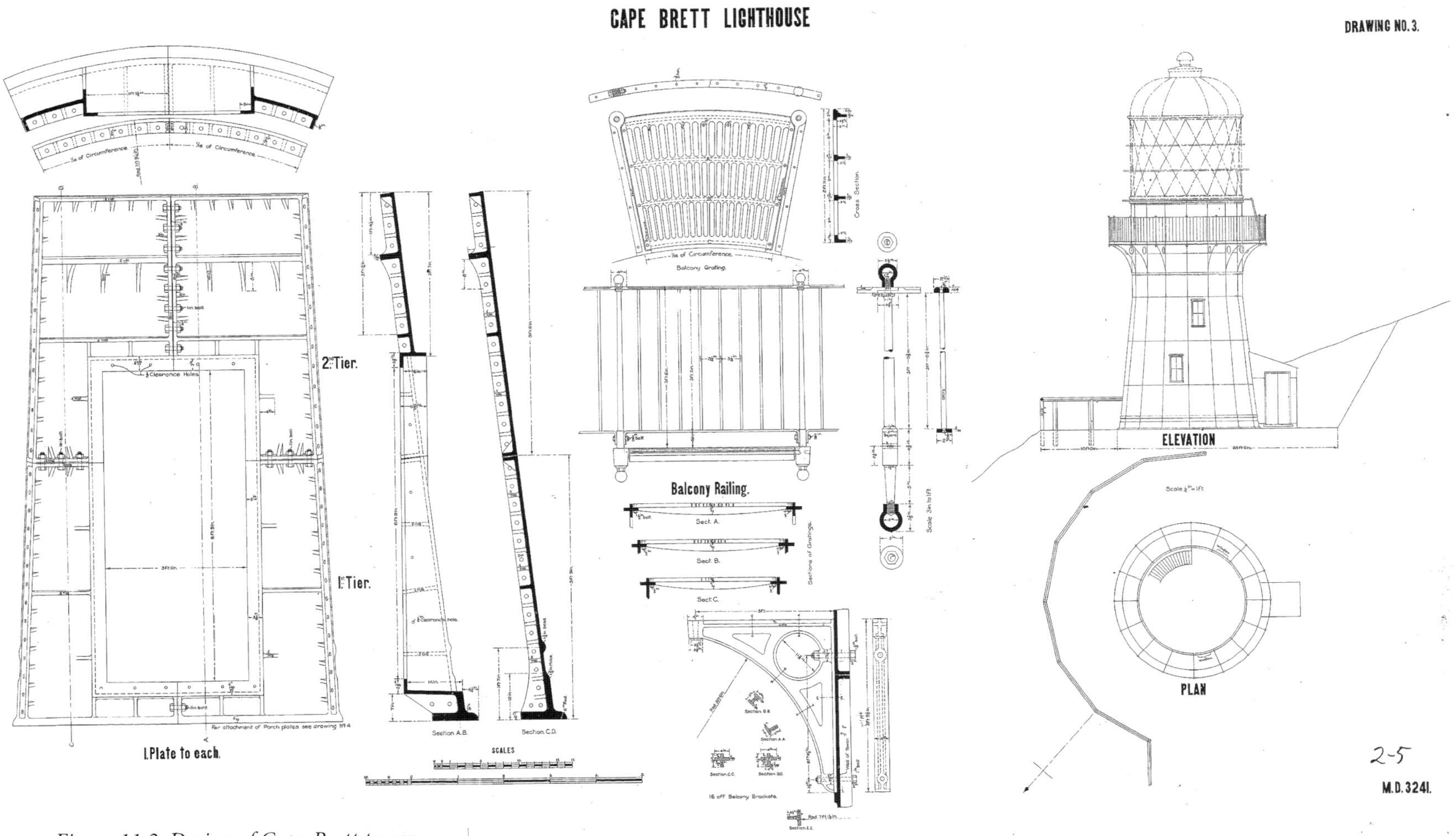

Figure 11.2: Design of Cape Brett tower

Figure 11.3: Cape Brett lighthouse (ATL WA-10357) © Whites Aviation Collection, Alexander Turnbull Library, Wellington, New Zealand

It was described in the Notice to Mariners as

> group flashing, white light, first order with 2 flashes in quick succession every half minute. The focal plane of the light will be 510 ft above high water, and it should be visible, except when obscured by land over an arc of 240° for a distance of thirty miles and a half.[9]

The keepers at Cape Brett, along with those at Castlepoint, were to have problems with the mercury baths. The apparatus was sensitive to mercury levels[10] and showed a tendency to slow down or gain according to the temperature, which was also ascribed to insufficient mercury.[11] Vibration from the undersupported apparatus was also thought to damage the mantles on the incandescent lamps.[12]

By 1931 a major problem had developed. The light continued to turn erratically, and it was necessary to place weights on the apparatus. Sinclair, the principal keeper at the time, suggested that the problem was probably the mercury bath. When his review of old journals showed that similar problems had been described but no cause clearly ascertained, he asked whether atmospheric conditions would affect the mercury.[13] This suggestion was dismissed and the keeper was recommended to check the governor. He did so but reported that the problem continued.[14] Peterson, the chief engineer from the *Matai*, the government steamer, was ordered to check and he found that clumps of mercury had formed against the apparatus. He put this down to grease, vaseline or other lubricant working down the side of the bath and gathering up the mercury.[15] The principal keeper designed a tool to scrape the sides of the bath, and when two were made they were duly entered in the departmental property register as 'Spare Tools for Cleaning Mercury Bath'.[16]

Four years later there was a recurrence of the problem. The principal keeper described the mercury as 'pasty', and the keepers were forced to turn the light by hand. Skimming the surface followed by washing the mercury with kerosene improved the condition slightly.[17] In 1935 the lighthouse artificer visited and suggested that the pastiness could be due to oxidation of the mercury.[18] When a sample was eventually analysed the government analyst put it down to the adsorption of kerosene vapour.[19]

Finally the New Zealand Government consulted D & C Stevenson.[20] They noted that

> [t]he formation of the so-called paste is understandable to us, and it has been found to arise from lubricating oil from above finding its way into the mercury trough [as had been suggested by Peterson five years earlier]. It is due to over-oiling by the lightkeepers. From our experience it is certainly not due to petroleum vapour, but actual oil.

They pointed out that it was sometimes difficult to detect when the keepers were using too much lubricating oil. And that it might have occurred several years previously. Their recommended solution was

> to jack up the apparatus and take out all the mercury completely. The trough should then be cleaned by some method so as not to introduce additional foreign substances such as the hairs of cotton waste.

They also suggested that surface oil and dirt needed only to be skimmed off each year or six months, but if 'an exceptionally large amount of dirt' entered the lightroom, the space over the mercury could be covered by a cover plate.

The memo to the high commissioner also raised the question of steadying rollers on the top of the apparatus. Although these were standard with the older style machine on bevelled rollers, they had not been provided with the mercury bath design. Consequently there was a risk of damage in an earthquake.

D & C Stevenson commented that steadying rollers were supplied with some mercury installations where they appeared to be called for, and

> there should be no difficulty in fitting such rollers in any new installation if desired. The additional expense would be small.

In 1934 an earthquake damaged the lens at Castlepoint lighthouses (another with a mercury bath), and the question of steadying rollers at Cape Brett was raised again.[21] Although the likelihood of an earthquake at Cape Brett was considered remote, the cost of steadying rollers and spider to top of the lantern house was estimated as not more than £75, and thus the expenditure was warranted as an insurance against possible damage.[22] However, it seems that the steadying rollers were not installed.

By 1951, thought was being given to electrification, and it was suggested that Cape Brett could be electrified with two small diesel generating sets. This would allow the station to be reduced from three keepers to two.[23] Expenditure for electrification was authorised in July 1951,[24] and electrification finally completed by April 1956.[25]

North Cape

Edwin and Johnson had recommended that a lighthouse be placed at North Cape in their 1873 report.[26] However, it was not until 1911 that funds were voted for the lantern and apparatus. This enabled an order for a second order light to be placed the following year.[27] The apparatus was manufactured by Chance Brothers, probably to a Stevenson design.

Although the complete lantern and apparatus arrived in 1914 (after the Castlepoint apparatus) it was not used at North Cape. After remaining in storage for about 10 years, it was sent south to be used to upgrade the light on Dog Island, (see Chapter 3), thus enabling the Dog Island light to make the jump from antique to cutting edge without intermediate steps.

North Cape was eventually lit by an acetone-acetylene automatic light, in July 1928.[28]

Castlepoint

In their 1873 report[29] Edwin and Johnson suggested that a fifth order white light flashing every ten seconds should be placed at Castlepoint.

However, it was not until 1910 that the decision was made to build the light. It was to be second order group flashing, flashing every 45 seconds with 8 seconds of darkness between each flash, with a range of 21 miles.[30]

The 1911 annual report indicated that 'lantern and apparatus have been ordered from England' apparently in Memorandum No 3766 to the high commissioner, dated 18 August 1911.[31] The optical apparatus was supplied by Barbier, Bernard & Terenne, the machine by Milne & Son,[32] and the lantern probably by Dove & Co. The incandescent burners were supplied by Chance Brothers. By May 1912, lantern and apparatus had arrived in New Zealand.[33]

As at Cape Brett, the tower was to be cast iron. Eight firms were able to offer tenders, showing how New Zealand was becoming more technologically self-sufficient. The successful tenderer was S Luke and Co (Ltd) of Wellington. C Judd of Thames (see above) was one of the unsuccessful tenderers However, the marine engineer's 1912 report indicated that the contract was 'very much behind'.[34]

The light was first exhibited on 12 January 1913.[35]

The Notice to Mariners described it as

> group flashing light with three white flashes every 45 seconds ... The height of the light is 170 ft., and should be visible for a distance of 19½ miles.

In 1922 the then principal keeper was killed in a fall from the lighthouse while fixing the telephone wire.[36] It took 11 years for one of the contributing factors to be officially recognised; in his January 1933 memo, the acting secretary of the Marine Department pointed out that when the tower was erected

> it was found that the usual cleaning path had not been provided round the outside of the tower near the base of the lantern.

Grating and brackets were never sent, and

> [t]he usual iron ladders for reaching the path and for access from the path to the top of the dome are in the oil store at the station, awaiting erection

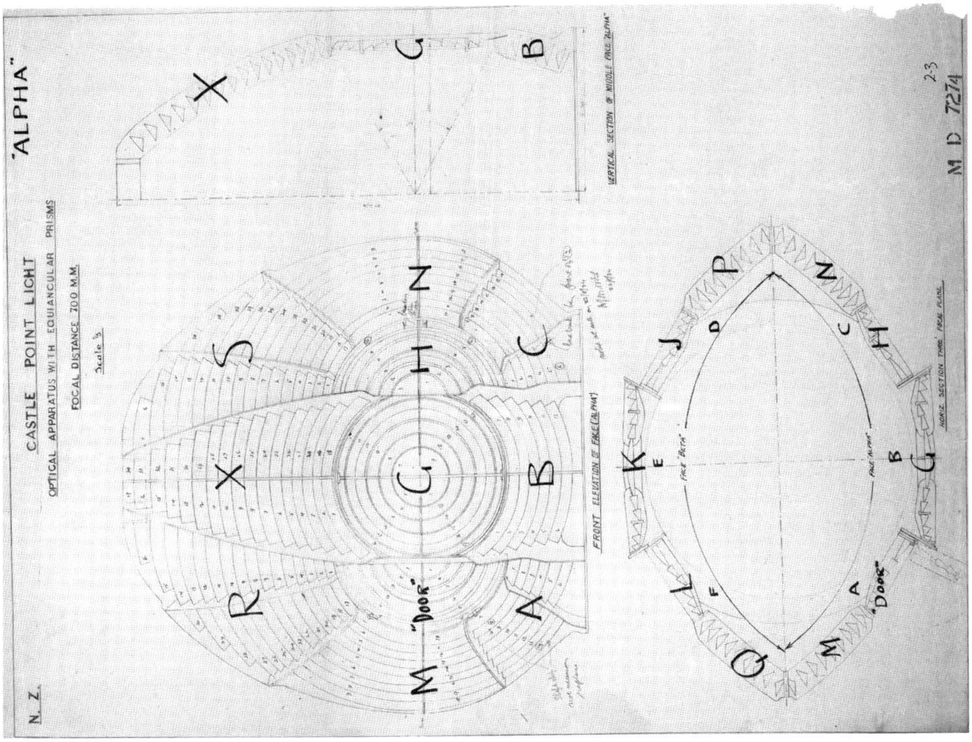

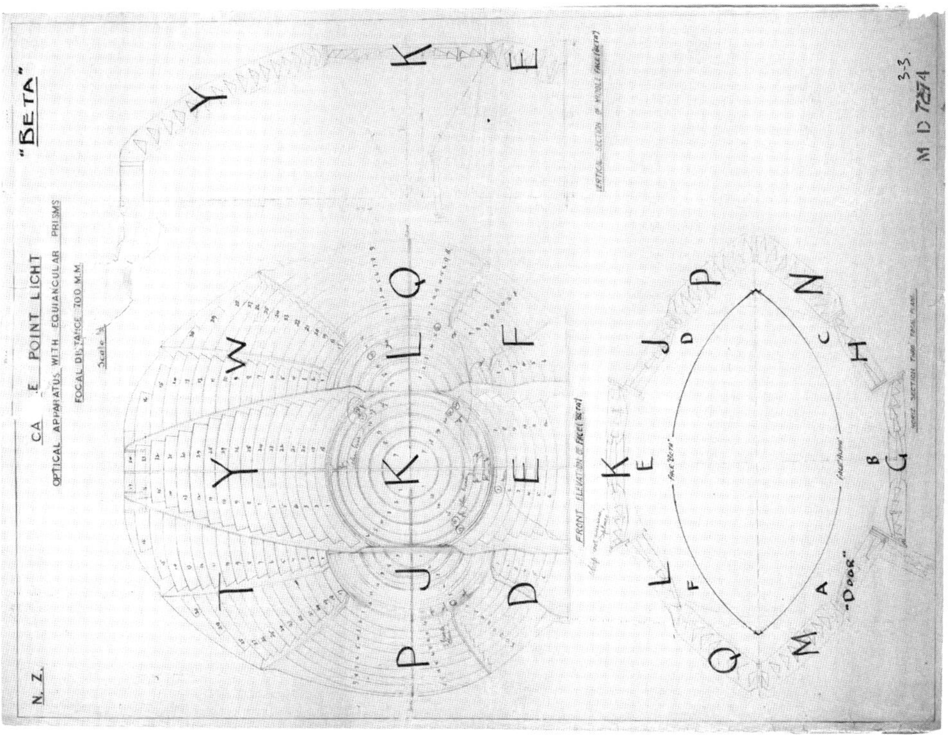

Figure 11.4 a, b: Lens for Castlepoint lighthouse
© Archives New Zealand, the Department of Internal Affairs, Te Tari Taiwhenua

[Left] Figure 11.5: Castlepoint tower in Luke's Yard, Wellington © Archives New Zealand, the Department of Internal Affairs, Te Tari Taiwhenua

[Below] Figure 11.6: Castlepoint lighthouse (ATL1/4-021305-F) © S C Smith Collection, Alexander Turnbull Library, Wellington, New Zealand

of the cleaning path. The base of the lantern is 60 feet from the ground and with only the provision of the usual brass hand-rail round the outside of the lantern it is obvious that considerable risk is run by any keeper endeavouring to clean the outside of the lantern. This matter should now be attended to, and if the gratings and brackets cannot now be located, it will be necessary to have new ones made and supplied.

When the Kahurangi lantern was being constructed, D & C Stevenson raised the question as to whether a cleaning path grating was required,[37] and were told that the grating was required and that brackets and rings for supporting it would be supplied with the tower.[38] When the order was sent for Castlepoint lighthouse, D & C Stevenson apparently did not raise the matter and the tower was built without supporting brackets.

A search was made at the lighthouse station, but no gratings or brackets were found.[39] The contract for new casting was let to Murphy & Son in April 1933[40] and the gratings and brackets had been landed from the government steamer *Matai* by 19 July.[41] The gratings and brackets remained in store for 12 months. In March 1924 the District Engineer Public Works Department offered 'to send up a suitable mechanic to carry out the work and fit the cleaning path'.[42] The offer was not taken up, and four months later the principal keeper wrote to the Secretary Marine Department complaining that the lack of cleaning path made keeping the lantern clean difficult, and

> [s]hould an older & probably less active keeper relieve me at this station & the gratings not attended to, some unforeseen fatality may occur.

This seems to have worked, as there is a handwritten note on back of the letter 'Urgent. ME [Mechanical Engineer] to arrange to have grating fitted'.[43]

Tradesmen were sent but were not able to complete the job as two castings were still required: a shorter one for the ladder access and a longer one for over the door.[44] The extra castings arrived by 3 September and were probably erected by the lightkeepers themselves.[45] The completion of the cleaning path was acknowledged in the 1935 annual report.[46]

The lighthouses suffered an earthquake in 1917, which led to the machine being dislodged and several cracks in the lens. However the tower was not damaged. The machine was jacked back into place and was easily levelled.[47]

Very late (11.45 p.m.) on 6 March 1934 the Castlepoint lighthouse was shaken by a major earthquake. Mercury was spilled from the bath and further minor cracks occurred in the lens. The keepers were able to get the light re-established by 3 a.m.

The inspection report from the district engineer of the Public Works Department commented on the cracks and requested (with geographical confusion):

> If you could let me have the maker's lens diagram which should be in your files, I will arrange to mark off on it the damaged lenses and new ones can then be obtained from Messrs Stevenson Bros in England [*sic*].[48]

The high commissioner was instructed to 'obtain a copy of the drawing and specification from Messers Stevenson'.[49] D & C Stevenson no longer had any original tracings, but were able to provide new drawings as '[t]his light was designed by us and we remember it well'.[50] D & C Stevenson assumed that Marine Department staff would be able to insert the replacement prisms into the existing frames, and that it would not be necessary for panels to be sent to Scotland for repair. They also suggested that if damage was more extensive than was able to be dealt with locally they could supply the prisms set into new smaller panels which could be fitted into the existing framework as had been supplied for the East Cape lighthouse. Figures 11.4 a and b show the drawings which were sent, along with the pencil markings showing the damage to the lens.

However, it was subsequently decided that as the light was still working satisfactorily the cracked prisms would not be replaced.[51]

Of greater concern was the general stability of the apparatus, and in mid-April 1934 the marine engineer was asked to consider the feasibility of installing steadying rollers, possibly the set of rollers taken out of Cape Foulwind lighthouse.[52] Although by August drawings were being prepared to recycle the Cape Foulwind rollers[53] the whole project seems to have slipped into abeyance until a report of 20 April 1936 urged that the

> fitting of these rollers should not be delayed as the tower being fairly lofty and on the top of a spur the structure is subject to a great amount of sway in a 'shake'.[54]

A week later the acting engineer in chief indicated that he

> considered that it will be satisfactory to fasten the steadying roller spider to the upper frame of the lantern house, which is cylindrical and well braced diagonally and therefore sufficiently rigid to take thrust from the steadying rollers.[55]

When Fraser, the lighthouse artificer, came to install the steadying rollers it was found that the stay rods had to be altered as the older Cape Foulwind lantern differed from Castlepoint. A different design of extension bracket to carry the top roller path was also required, and this was cast in Wellington as well.[56]

Despite the steadying rollers, the Castlepoint lighthouse still continued to be affected by subsequent earthquakes. The problem was usually spillage of the mercury. After two strong earthquakes in quick succession in June 1942[57] and another in August 1942,[58] it was decided that retaining the mercury bath was impractical and plans were made to replace it with roller bearings and tracks.[59]

By this time the New Zealand High Commission had moved to offices in the Strand, and procuring lighthouse materials was the responsibility of the stores indent officer. Despite being three years into World War II, obtaining the necessary rollers was comparatively straightforward due to the long-established relationship with the Stevenson firm and the Scottish manufacturers.

On 26 October 1942, the store indent officer sent the minister's memo to the high commissioner on to the firm, which was now known as A & C Stevenson. (David A Stevenson having died in 1938, his place in the partnership was now taken by Charles Stevenson's son, D Alan Stevenson.) In their reply of 24 November 1942, A & C Stevenson estimated that the cost would be £273 to £300 and the weight 12 cwt in a box 4ft × 4ft × 1ft. They justified the estimate by referring to the replacement rollers and paths supplied by them for The Brothers in 1921 at a tender cost of £211, and pointed out that the Castlepoint order had additional circular rings of 'complicated section' and that there had been an 'increased rise in wages and materials'. The eventual tender was for £264, and A & C Stevenson's fee was about £25.

While A & C Stevenson still had their original 1912 drawings, they felt it best to have certain measurements checked, and prepared a simple diagram showing the six measurements required.

The reality of wartime production only intruded in the last paragraph:

> We take it that you would have a high priority and would obtain a Board of Trade certificate authorising steel and cast iron supplies.

By 5 December A & C Stevenson had obtained a firm quote of £264 from Milne's Meters, as Milne & Son was now known. While the high commissioner's memo to the minister of marine was dated 9 December 1942,[60] it was not until 13 July 1943 that the required six measurements were cabled to the high commissioner.[61] The completed order was shipped from the Clyde in April 1944,[62] and A & C Stevenson sent detailed installation instructions.[63]

The new equipment was installed in the first week of February 1945.[64] It was found that heavier weights were required to ensure the correct rotation speed, but this was to be expected.[65] A month later the lighthouse maintenance foreman was called in, as the new rollers had become very difficult to turn. This was found to be due to the 'main rollers being taken up too tight after removal for greasing'.[66] In contrast to rollers that the keepers were used to at other lights, these rollers were tapered and fitted with tapered roller bearings.

This was the last order sent to Stevensons for the New Zealand lighthouses. It was appropriate that here, at the end, the technology supplied was the most up-to-date available.

Castlepoint light was electrified by June 1954.[67]

Endnotes

1 AJHR 1873 H-19.Report on the Lighthouses on the Coast of New Zealand. 3 March 1873

2 AJHR 1896 H-15 Marine Department,Annual Report for 1895-96 The Inspector of Lighthouses to the Secretary, Marine Department

3 AJHR 1906 H-15 Marine Department Annual Report for1905–06

4 AJHR 1908, H-15 Marine Department Annual Report for 1907–08

5 Chance Archives BS6/12/1/1/15 p 309 Government of New Zealand (Cape Brett) Marine 267/176 P.B.181/50

6 NZG Archives (memo plant overseer (Tate?) to secretary Marine Department ADOE 16612 M1/ 720 8/13/16 22 September 1948)

7 AJHR 1909 H-15 Marine Department Report for 1908–9

8 *NZG Gazette* 1909 , Preliminary Notice to Mariners, No 95 of 1909 p 2967, 11 November 1909

9 *NZG Gazette* 1910 27 January, Notice to Mariners No 7 of 191 p 293, 0, 24 January 1910

10 NZG Archives ADOE 16612 M1/720 8/13/16 memo Wilson, principal keeper Cape Brett, to secretary Marine department15 June 1927

11 NZG Archives ADOE 16612 M1/720 8/13/10 memo Wilson principal keeper to Secretary Marine Department 1 July 1927 extract from M 8/6/29

12 NZG Archives ADOE 16612 M1/720 8/13/16 memo Sharp, inspection engineer, to Secretary Marine Department Cape Brett Lighthouse 5 September 1930

13 NZG Archives ADOE 16612 M1/720 8/13/16 memo G Sinclair, principal keeper Cape Brett, to Secretary Marine Department 6 February 1931
14 NZG Archives ADOE 16612 M1/720 8/13/16 telegram principal keeper to secretary Marine Department 20 February 1931
15 NZG Archives ADOE 16612 M1/720 8/13/16 memo Peterson, chief engineer Matai ,to Burgess lighthouse engineer, 7 March 1931
16 NZG Archives ADOE 16612 M1/720 8/13/16 memoprincipal keeper to secretary Marine Department 1 May 1931
17 NZG Archives ADOE 16612 M1/720 8/13/16 memo principal keeper to secretary Marine Department 30 Sept 1935
18 NZG Archives ADOE 16612 M1/720 8/13/16 memo Fraser to secretary Marine Department Oct 11 1935
19 NZG Archives ADOE 16612 M1/720 8/13/16 copy extract report Mr Tanner's inspection report (original on 8/9/213) 7 March 1936
20 Acc 10706 vol 87 memo to high commissioner from prime minister's office re MARINE DEPARTMENT Cape Brett Lighthouse; Apparatus 11 May 1936., also NZG Archives, ADOE 16612 M1/720 8/13/16 Cape Brett Lighthouse
21 NZG Archives ADOE 16612 M1/720 8/13/16 memo Fraser to secretary Marine Department Sept 16 1937
22 NZG Archives ADOE 16612 M1/720 8/13/16 copy report of mechanical engineer to J Wood, marine engineer Public Works Department, 19 November 1937 (Original on 8/13/22) also ADOE 16612 M1 760 8/13/16 memo Fraser to secretary Marine Department 16 September 1937
23 (NZG Archives ADOE 16612 M1/760 8/13/16 memo secretary Marine Department to Minister of Marine 18 July 1951
24 NZG ArchivesADOE 16612 M1/720 8/13/16 memo secretary Marine Department to marine engineer Ministry of Works 23/07/1951
25 AJHR 1956, \H-15 Marine Department Report for year ended 31 March 1956
26 AJHR 1873 H-19. Report on the Lighthouses on the Coast of New Zealand.
27 AJHR 1912, H-15 Marine Department Report for 1911–12
28 *NZG Gazette*,1928 April 12,, p 1121 Notice to Mariners, 23 of 1928 10 April 1928
29 AJHR 1873 H-19. Report on the Lighthouses on the Coast of New Zealand.
30 AJHR 1911 H-15 Annual Report Marine Department1910–1911
31 NZG Archives ADOE 16612 M1/729 8/21/6 pt 1 memo to high commissioner (M.8/21/5) 6 April 1934
32 NZG Archives AADX 889 W3148 /40 63/193 report mechanical engineer (McLeod) Public Works Department also copy in ADOE 16612 M1/729/ 8/21/6 pt 1
33 AJHR 1912, H-15 Marine Department Annual report for 1911–12
34 AJHR 1912 H-15 t Marine Department Annual report for 1911–12 r. Report of Marine Engineer
35 *NZG Gazette* No 88 p 3414 Notice to Mariners No 142 of 1912, 4th Dec 1912
36 *Lighthouses of New Zealand, Castlepoint* Maritime New Zealand
37 NZG Archives ADOE 16612 M1/744 8/34/5 letter D & C Stevenson to secretary Marine Department M1900/616 1 February 1901
38 NZG Archives ADOE 16612 M1/744 8/34/5 letter secretary Marine Department to D & C Stevenson NO 347/111 28 March 1901
39 NZG Archives ABPL 7457 W5932/ 147 45/11/10 telegram principal keeper to secretary Marine Department 31 January 1933
40 NZG Archives ABPL 7457 W4932/147 45/11/10 memo district engineer to acting secretary Public Works. 8/47 (labelled M8/21/7) 27 April 1933
41 NZG Archives ABPL 7457 W 4932 /147 45/11/10 memo acting secretary to Fraser (originally M8/21/17) 19 July 1933
42 NZG Archives in ABPL 7457 W4932 /147 45/11/10 extract memo district engineer to secretary Marine Department (original on file M.8/21/6) 29 March 1934
43 NZG Archives ABPL 7457 W4932/147 45/11/10 letter principal keeper (Smith) to secretary Marine Department (originally M8/21/17) 6 July 1934
44 NZG Archives ABPL7457 W 4932/147 45/11/10 letter principal keeper to secretary Marine Department (originally M 8/21/17) 3 August 1934
45 NZG Archives ABPL 7457 W4932/147 45/11/10 extract from report principal keeper to Marine Department 8/6/23 (orig M 8/21/17) 3 September 1934

46 AJHR 1934/35 H-15 Annual Report Marine Department Annual Report for year ended 31st March 1934

47 NZG Archives ADOE 16612 M1/729 8/21/5 letter Fraser to secretary Marine Department 14 August 1917

48 NZG Archives ADOE 16612 M1/729 8/21/6 pt 1 memo district engineer Public Works Department to secretary Marine Department 29 March 1934

49 NZG Archives ADOE 16612 M1/729 8/21/6 pt 1 memo M. 8/21/5 prime minister to high commissioner 6 April 1934, also Acc 10706 vol 87

50 NZG Archives ADOE 16612 M1/729 8/21/6 pt 1 copy letter Stevenson to high commissioner, (M 8/21/6) 7 June 1934

51 Ac 10706 vol 87 letter office NZG officers, signed assistant store indents officer W R Castlepoint Lighthouse dated 17 June 1935

52 NZG Archives ADOE 16612 M1/729 8/21/5 1 memo secretary Marine Department to marine engineer, Public Works Department 16 April 1934

53 NZ G Archives ADOE 16612 M1/729 8/21/6 pt1 memo marine engineer to secretary Marine Department 28 August 1934

54 NZG Archives ADOE 16612 M1/729 8/21/6 copy extract Mr Tanner's report on lighthouse original in 8/9/213 20 April 1936

55 NZG Archives ADOE 16612 M1/720 8/13/16 letter Wood, acting engineer in chief, to district engineer Public Works Department 27 April 1936

56 NZG Archives ADOE 16612 M1/720 8/13/16 letter W A Fraser to secretary Marine Department 13 February 1937

57 NZG Archives ADOE 16612 M1/729 8/21/6 memo McFarlane, principal keeper, to secretary Marine Department 26 June 1942

58 NZG Archives ADOE 16612 M1/729 8/21/6 memo principal keeper to secretary Marine Department 2 August 1942

59 NZG Archives ADOE 16612 M1/720 8/13/16 memo minister of marine (Bob Semple) to high commissioner (T/B M 8/21/6) 10 August 1942

60 NZG Archives ADOE 16612 M1/720 8/13/16 memo high commissioner to minister of marine 9 December 1942

61 NZG Archives ADOE 16612 M1/720 8/13/16 telegram to high commissioner London 13 July 1943

62 NZG Archives ADOE 16612 M1/720 8/13/16 specification from Milne's Meters of three packages for Castlepoint

63 NZG Archives ADOE 16612 M1/729 8/21/6 part 1 memo A & C Stevenson 21 March 1944

64 *NZ Gazette*, No 4, Notice to Mariners No 1 of 1945 p 62 24 January 1945,

65 NZG Archives ADOE 16612 M1/720 8/21/16 memo Egger, lighthouse expert, to chief mechanical engineer dated 1 April 1945

66 NZG Archives ADOE 16612 M1/720 8/21/6 memo Tate Foreman, lighthouse maintenance, to district engineer Public Works Department Wellington, (PW 63/193)

67 AJHR 1954, H-15 Marine Department Report for the year ended 31 March 1954.

INDEX